Experimental Methods in Inorganic Chemistry

John Tanaka
University of Connecticut

Steven L. Suib
University of Connecticut

Prentice Hall, Upper Saddle River, New Jersey 07458

To Jumbo Wells, friend, colleague, and mentor.

ACQUISITION EDITOR: ***Matthew Hart***
PRODUCTION EDITOR: ***Alison Aquino***
MANUFACTURING MANAGER: ***Trudy Pisciotti***
COVER DESIGNER: ***Kiwi Design***
ART DIRECTOR: ***Jayne Conte***
ARTISTS: ***Charles Pelletreau and Patty Gutierrez***
COVER ART: ***Charles Pelletreau***

Illustrations reprinted from* Structural Inorganic Chemistry *by A. F. Wells (1984) by permission of Oxford University Press; illustrations reproduced by permission of The McGraw Hill Companies; illustrations courtesy Richard P. Muldoon/Lucent Technologies/Bell Labs reprinted from* Chemical and Engineering News; *illustrations from Lewis,* Modern Coordination Chemistry, *reprinted by permission of John Wiley & Sons, Inc.; illustration, p. 257 from International Centre for Diffraction Data; illustrations from John M. Newsam/Molecular Simulations Inc.; photo, p. 85 by Steven Suib

Printed in the United States of America
10 9 8 7 6 5 4 3 2 1

ISBN 0-13-841909-4

Prentice-Hall International (UK) Limited, London
Prentice-Hall of Australia Pty. Limited, Sydney
Prentice-Hall Canada Inc., Toronto
Prentice-Hall Hispanoamericana, S.A., Mexico
Prentice-Hall of India Private Limited, New Delhi
Prentice-Hall of Japan, Inc., Tokyo
Pearson Education Asia Pte. Ltd., Singapore
Editoria Prentice-Hall do Brasil, Ltda., Rio De Janeiro

TABLE OF CONTENTS

Part I Background Material 1

Part II Laboratory Exercises 121

PREFACE

Laboratory Experiments

The basic philosophy of the laboratory is based on the models experiments. In our long association with A. F. Wells, a conclusion was reached that the concept of inorganic structures could not be adequately taught from a text supplemented by lectures. Wells himself experienced a certain degree of frustration as he tried to give seminars on structural inorganic chemistry. It was under his tutelage that the models laboratory was developed. Because it was felt that a constant review and rethinking about structures were needed to truly learn structural concepts, the practical laboratories were developed to accomplish this review. In addition, a secondary, but not less important, goal of our laboratory was to introduce techniques which are not introduced in other undergraduate courses. High temperature techniques, vacuum line operations, ampoule sealing, solid-state reactions, controlled atmosphere reactions, reactions with gases, and inorganic characterization are some of these concepts.

The models laboratories are best performed on desktops. If an appropriate setting such as a seminar room is not available, they can be done in a laboratory. However, it will not be as comfortable for the students to stand for three hours while pondering the spatial relationships using the models as visual aids. (Unfortunately, an arm table classroom type chair does not provide a large enough working area for model construction.) A black or whiteboard is also necessary while teaching the models laboratories.

Because the laboratory is based on the models experiments, the first two models experiments are scheduled at the beginning of the semester. Some students find model building to be "uninteresting," therefore, several laboratory experiments are scheduled before the final models laboratory. If the instructor's preference is to decrease the time spent on models, the third models laboratory can be eliminated. It is, however, imperative that the first two models laboratories be done.

Many students find the models laboratory interesting and informative. The degree of interest of the students in the models laboratory is probably directly related to the enthusiasm of the instructor. Students obtain insights into structures that are virtually impossible to obtain by way of a printed text or by the arm waving of a lecturer. The students who find the models laboratory "uninteresting" are the ones who have difficulty visualizing relationships in three dimensions. It is ironic that it is these very students who need the exercises the most.

Essays

The essays found in this manual are not meant to be comprehensive treatments of the subjects covered; books have been written on each of the essay subjects. They are geared more towards giving students material that might be included in a single lecture on a topic. A bibliography to help the students fill in the gaps for each subject is provided at the end of the essay so that the students can read further on any single subject.

The essay on the importance of the solid state does not make a lot of sense to the students at the beginning of the semester. It has been found that this is a good way to conclude the course. The instructor in a lecture session or two can "walk" the students through the concepts discussed in this essay.

Characterization Experiments

Some of the characterization techniques presented in Part III may have been covered in Analytical Chemistry or Physical Chemistry. For example, some analytical courses cover conductivity experiments. Some of the physical chemistry laboratory courses cover X-rays. The limited number of characterization techniques is offered with the philosophy that repetition, if it occurs, is useful in teaching. A second important pedagogic tool is to relate a characterization technique to the compounds prepared in the laboratory. X-ray powder patterns are to inorganic chemistry as melting points are to organic chemistry. Because a powder diffractometer is not as readily available as a melting point apparatus, the process of obtaining X-ray data is introduced as "dry labs." If an X-ray diffractometer is available to an undergraduate class, the diffractometer experiment in Part III can be used. Vacuum lines are also of limited availability to an undergraduate class. For this reason, vacuum techniques are taught in an abbreviated way in experiments such as the tin(IV) chloride experiment. If vacuum lines are available, and the class size is small, the experiments in Part III can be done. Magnetic susceptibility is useful in "counting" the number of unpaired electrons. Once again, there is the problem of the limit to the number of magnetic susceptibility balances. The conductivity experiment is one that is perhaps easier to do; most institutions have more than one conductivity bridge.

Even with two laboratory periods per week, there is not enough time to do all that one would like. Recognizing the above limitations to the characterization experiments, one strategy is to cover the characterization techniques in the recitation section and to reserve the laboratory periods for the preparative experiments. The solution as to what balance to strike always rests with the teacher and the equipment available.

Acknowledgements

The authors would like to thank all of the members of the inorganic division of the University of Connecticut who helped with various aspects of these laboratories. Prof. Ronald Krause contributed the experiment on the synthesis of ethylenediamine complexes of cobalt and nickel and much of the crystal growth in gel experiment. Professors Jeffrey Bocarsly and Carl David contributed the chapters on computational chemistry. In addition, the experiment on metal complex used for DNA cleavage is based on papers published by Prof. Bocarsly. Prof. Ulrich Mueller-Westerhoff gave advice on the experiment on ferrocene. As already indicated, the late Prof. A. F. Wells was the primary instigator and source of inspiration for this laboratory course.

The authors would also like to thank all of the graduate students who acted as teaching assistants for the course using these experiments, especially Shawn P. Davis, Charles A. Detmer III, Bryan Gavini, Athanasios Kostapapas, Stephen Leach, Kerry C. McMahon, Dimitri Psaras, and Beth Vileno.

The following have given permission to use material in their copyright resources. Clarendon Press has given us permission to use some of the figures and tables from A. F. Wells, "Structural Inorganic Chemistry." Physical Electronics has given permission to use figures from page 82, 83, 86, and 87 of the book "Handbook of X-ray Photoelectron Spectroscopy" by Moulder, J. F.; Stickle, W. F.; Sobol, P. E.; and Bomben, K. D. The International Centre for Diffraction Data has given permission to reproduce the powder diffraction data from their Powder Diffraction File (PDF). John Wiley & Sons, Inc. has given permission to reproduce the magnetic susceptibility data from Lewis/"Modern Coordination Chemistry". Richard P. Muldoon of the Lucent Technologies/Bell Labs has given permission to use the structure of a superconductor as it appeared in page 8 of Chemical and Engineering News dated May 11, 1987. We thank John M. Newsam for considerable help with the Molecular Simulation, Inc. software.

Part I   BACKGROUND MATERIAL

SAFETY IN THE CHEMICAL LABORATORY

Safety in the laboratory evolves around knowledge and prevention. Knowledge not only includes an understanding of the nature of the chemicals being used and the process being undertaken, but it also involves an understanding of how to deal with emergencies. Prevention involves safe practices and above all the use of common sense.

With the number of new regulations being issued daily, the best way to keep up with the safety scene is to access the updates on the web. For example, the Materials Safety Data Sheets that are essential to the safe operation with any chemical are available at http://www.pdc.cornell.edu/issearch/msdssrch.htm. Many organizations have safety pages on the web. For example, The University of Connecticut has a 25 page Chemical Hygiene Plan at http://vm.uconn.edu/-wwwehs/chemplan.html. Also available from the University of Connecticut home page are the Laboratory Safety Inspection (policies and procedures), Laboratory Hazardous Waste Management (policies and procedures), a Quick Guide to Chemical Waste Disposal, How to Survive an OSHA and EPA/DEP Compliance Inspection, Laboratory Accident and First Aid Information, and a Chemical Resistant Glove Guide. If your institution has similar data on the web, one should be familiar with it and to access it periodically to keep posted on the updates. Of course, material from any other institution can be accessed and downloaded from the web.

1. Familiarize yourself with the hazards associated with the chemicals being used in the experiment. Do not rely on labels alone. Familiarize yourself with the Materials Safety Data Sheets. Know how to use the reference manuals such as those listed at the end of this write-up. Know the type and extent of the hazard associated with the material. Merely being afraid of chemicals does not maximize safety.

2. Learn to recognize the principles of safe operating conditions. For example, high pressure cylinders should be clamped or otherwise secured. Noxious fumes should not be vented into the lab. Open flames should not be used to heat flammable liquids.

3. Practice personal safety. Always wear eye protection in the laboratory. Wear adequate shoes, not sandals or slippers. Wear appropriate clothing. Lab aprons or lab coats are desirable and are recommended.

4. Common sense prohibits horse-play or practical jokes. Working alone is discouraged. Common sense says that one should not eat or smoke while in the laboratory. Common sense does include learning the location of safety equipment such as fire extinguishers, eye-wash fountains, fire blankets, and emergency showers. Common sense says that one should know how to use each of these items. Common sense further says that one should not be afraid or embarrassed to ask how to use each of these safety items if one is not certain how it is used. Common sense also includes the necessity for prior orientation and reading before initiating a laboratory experiment.

5. One should know some general rules for emergencies. The first thought in all chemical spills or splashes is to inundate the portion of the anatomy involved with lots of water. Acid spills should be neutralized with bicarbonate. If the instructor orders the lab evacuated, leave at once.

6. As with the familiar litany in commercial aviation about seat belts, oxygen masks, and emergency exits, the lab instructor is required (as are the stewardesses) to caution the class about the safety hazards at the beginning of each laboratory session. Although there are good reasons for this requirement, it could compromise laboratory safety if the students come to use these warnings as the sole source of safety information or even worse start to look on it as a boring exercise. Each student should do the safety part of the laboratory preparation independently as if an announcement were not to be made. The announcement should be used as a supplementary source of information and a reminder of the safety practices to be exercised. It is only with a good understanding of the hazards involved that safety in the laboratory can be maximized.

Helpful Background Literature

1. DiNardi, S. R. "The Occupational Environment – Its Evolution and Control", American Industrial Hygiene Association: Fairfax, Virginia, 1997. Chapter 48, Laboratory Health and Safety, pp. 1225-1240.

2. "Prudent Practice in the Laboratory (Handling and Disposal of Chemicals)"; National Academy Press: Washington, D. C., 1995.

3. Leach, Robert E., Editor "The Sigma-Aldrich Library of Chemical Safety Data"; Sigma-Aldrich Corp., 1985.

4. Mohn, Wm. J. "Fundamentals of Laboratory Safety"; Van Nostrand Reinhold: New York, 1991.

5. Kaufman, James A. "Waste Disposal in Academic Institutions"; Lewis Publishers: Chelsea, Michigan, 1990.

6. Pipitone, David A. "Safe Storage of Laboratory Chemicals, 2nd Ed."; John Wiley: New York, 1991.

7. American Chemical Society "Safety in Academic Chemistry Laboratories"; American Chemical Society: Washington, D. C., 1985. 69 pp.

8. ASTM; "Guide to the Safe Handling of Hazardous Materials Accidents"; ASTM STP 825, 1983. 64 pp.

9. Bennett, G. F.; Feates, F. S.; Wilder, I. "Hazardous Materials Spills Handbook"; McGraw-Hill: New York, 1982. 704 pp.

10. Bretherick, L. "The Handbook of Reactive Chemical Hazards", 4th ed.; Butterworths: Stoneham, MA, 1990. 2000 pp.

11. Clayton, G. D.; Clayton, F. E. "Patty's Industrial Hygiene & Toxicology" Vol. I. General Principles, 3rd ed.; Wiley: New York, 1978. 1466 pp. Vol. II, Part A. Toxicology, 3rd ed.; Wiley: New York, 1980. 1430 pp. Vol. II, Part B. Toxicology, 3rd ed.; Wiley: New York, 1981. 955 pp. Vol. II, Part C. Toxicology, 3rd ed.; Wiley: New York, 1982. 1312 pp. Vol. III. Theory and Rationale of Industrial Hygiene Practice; Wiley: New York, 1979. 752 pp.

12. Commission Report "Maximum Concentrations at the Workplace and Biological Tolerance Values for Working Materials 1987"; VCH Publishers, Inc.: New York, NY, 1988. 87 pp.

13. Compressed Gas Association, Inc. "Handbook of Compressed Gases"; 2nd ed.; 1980. 916 pp.

14. Dux, J.; Stalzer, R. "Managing Safety in the Chemical Laboratory"; Van Nostrand Reinhold: New York, 1988. 144 pp.

15. Fawcett, H. H. Ed. "Hazardous and Toxic Materials", 2nd ed.; John Wiley & Sons, Inc.: Somerset, N.J., 1988. 600 pp.

16. Fire, F. L. "The Common Sense Approach to Hazardous Materials"; Fire Engineering: New York, 1986. 369 pp.

17. Freeman, Harry M. "Standard Handbook of Hazardous Waste Treatment and Disposal"; McGraw-Hill: Hightstown, N.J., 1990. 992 pp.

18. Gosselin, R. E. et al. "Clinical Toxicology of Commercial Products", 5th ed.; Williams and Wilkins: Baltimore, MD, 1984.

19. Hampel, C. A.; Hawley, G. G. "Glossary of Chemical Terms", 2nd ed.; Van Nostrand Reinhold: New York, 1982. 306 pp.

20. Hodgson, E.; Mailman, R.; Chambers, J. "Dictionary of Toxicology"; Reinhold: New York, 1988. 395 pp.

21. Kaigai, G; Shiryo, K. "Toxic and Hazardous Industrial Chemicals Safety Manual"; The International Technical Information Institute: Tokyo, 1978. 700 pp.

22. Keith, L. H.; Walters, D. B. "Compendium of Safety Data Sheets for Research and Industrial Chemicals, Parts I-III"; VCH Publishers: New York, NY, 1985. 1862 pp.

23. Keith, L. H.; Walters, D. B. "Compendium of Safety Data Sheets for Research and Industrial Chemicals, Parts IV-VI"; VCH Publishers: New York, NY, 1987. 1696 pp.

24. Lefevre, M. J.; Conibear, S. "First Aid Manual for Chemical Accidents", Second Edition; Van Nostrand Reinhold: New York, 1982. 250 pp.

25. Meyer, E. "Chemistry of Hazardous Materials"; Prentice Hall: Englewood Cliffs, NJ, 1977. 370 pp.

26. Meyer, R. "Explosives" 2nd ed.; Verlag Chemie: Deerfield Beach, Fla., 1981. 440 pp.

27. NFPA "Flash Point Index of Trade Name Liquids", 9th ed.; National Fire Protection Association: Boston, MA, 1978. 308 pp.

28. National Research Council "Prudent Practices for Handling Hazardous Chemicals in Laboratories"; National Academy of Science: Washington, D. C., 1981. 291 pp.

29. Parmeggiani, L. "Encyclopedia of Occupational Health & Safety", 3rd ed.; International Labour Office: Geneva, 1983.

30. Phifer, R. W.; McTigue, W. R. Jr. "Handbook of Hazardous Waste Management for Small Quantity Generators"; Lewis Publishers, Inc.: Chelsea, MI, 1988. 200 pp.

31. Pitt, M. J.; Pitt, E. "Handbook of Laboratory Waste Disposal"; Halsted Press: New York, 1985. 360 pp.

32. Sax, N. Irving; Lewis, R. "Dangerous Properties of Industrial Materials", 7th ed.; Van Nostrand Reinhold: New York, 1988. 4000 pp.

33. Sax, N. Irving; Lewis, R. "Hazardous Chemicals Desk Reference"; Van Nostrand Reinhold: New York, 1987. 1095 pp.

34. Sax, N. Irving; Lewis, R. "Rapid Guide to Hazardous Chemicals in the Workplace"; Van Nostrand Reinhold: New York, 1987. 256 pp.

35. Schwope, A. D. et al. "Guidelines for the Selection of Chemical Protective Clothing", 2nd ed.; American Conference of Governmental Industrial Hygienists: Cincinnati, Ohio, 1985.

36. Sittig, M. "Handbook of Toxic and Hazardous Chemicals and Carcinogens", 2nd ed.; Noyes Publications: Park Ridge, NJ, 1985. 950 pp.

37. Steere, N. V. "Handbook of Laboratory Safety", 2nd ed.; CRC Press: Boca Raton, FL, 1972. 854 pp.

38. Stricoff, R. S.; Walters, D. B. "Laboratory Health and Safety Handbook", John Wiley & Sons, Inc: Somerset, NJ, 1990. 336 pp.

39. Weiss, G. "Hazardous Chemicals Data Book"; Noyes Publications: Park Ridge, NJ, 1986. 1093 pp.

40. Young, J. A. "Improving Safety in the Chemical Laboratory: A Practical Guide"; John Wiley & Sons: New York, NY, 1987. 350 pp.

LABORATORY NOTEBOOKS

Laboratory notebooks fulfill a technical and legal need. Learning to keep a good laboratory notebook is an on-going process. Even experienced researchers sometimes find that when it comes time to write a paper, a crucial bit of data has not been recorded. Almost everyone has had the unpleasant job of repeating an experiment because some little point important for the write-up was not recorded. For this reason, it is important to strive to keep a good laboratory notebook whenever laboratory work is done. Good habits are not learned instantly.

The basic requirements for a laboratory notebook are as follows:

1. Notebooks must be bound.

2. Write with an indelible writing instrument.

3. Entries must be dated and signed.

4. Writing should be readable and well organized.

5. Entries should be periodically witnessed.

Bound notebooks are a must. If a notebook is involved in an important patent litigation, loose-leaf notebooks or spiral bound notebooks will not stand up in court. It is important to enter into the notebook directly. The worst habit one can acquire is to make entries on scraps of paper with intentions of later entry so the notebook can be kept ″neat.″ Making entries on scraps of paper will not be tolerated in this class!

Notebook entries must be made using an indelible writing tool. A pencil is not acceptable. If original entries can be erased or otherwise easily modified, it cannot be considered to be sound legal evidence. It is important to use a waterproof ink which does not fade in light. Inks since the days of the fountain pen inks have been either of the permanent type or water soluble type. If water is accidentally splashed on a notebook, it is desirable to be able to still read the entry. The greatest problem these days is with the flair or felt tip pens. Those with permanent ink are generally so marked at the store. If in doubt, write on a scrap of paper and hold it under a faucet. If the ink washes off, don′t use the pen. Thought should also be given to photochemical fading. Red inks of any kind are most susceptible to fading. Blue ball-point inks also fade in light. Black inks are the best. The blue-black permanent fountain pen ink is good. The iron compound in the ink oxidizes with time to a water insoluble compound.

Entries made into the notebook should always be dated. Generally start off the day′s entry with a date. If a preliminary lab write-up is done, it should be dated with the date the write-up is actually done. When entries are made in the laboratory on experimental observations, it should have the date the observations were made. At the conclusion of the lab, once again note the date and sign your name.

The writing should be readable and well organized. The notebook is not judged for its beauty, but it is a record which other people should be able to read and understand. The criterion for good organization is that of being able to convey to others what has happened in a unmistakable fashion without need for your help in interpretation. If an error is made, cross out the error with a single line. It sometimes happens that what was at one time thought to be an error is in fact a valuable observation. One should be able to read the original entry even when crossed out. Some prefer to use a laboratory notebook with crosshatched lines. This is so that graphs and charts can be entered directly into the notebook. Some prefer to use a notebook where all the notes are taken on the right page (odd numbered pages) of the book with the left page a graph paper for entry of drawings and graphs. Either type of laboratory notebook is satisfactory.

Preparations or characterizations should be systematically identified. This is especially important in recording research results. One technique is to code a preparation or characterization in some way to a page in the laboratory notebook. For example, a compound preparation described on page 34 of notebook IV might be labeled IV-34. If more than one compound preparation is described on page 34, the samples might be labeled IV-34-1, IV-34-2, etc. A system involving notebook number and page number allows one to easily find the notebook location where a particular compound, characterization or analysis is described. It is particularly frustrating to encounter a compound or an infrared spectra and not be able to identify its source.

In addition to indicating where in the notebook the preparation of a product is described, the label on the sample vial should also include the tare weight. The tare weight is the empty vial which includes the label and lid. The label should, of course, have net weight of the product, your name, and date.

Notebook entries should be periodically witnessed. In a class, the laboratory instructor periodically looks at the notebook. In a research lab, the group leader will want to make sure he or she understands your research results. Get in the habit of having your notebook witnessed. The questions which are asked will help you improve your notebook-keeping technique.

Helpful Background Literature

1. H. M. Kanare, "Writing the Laboratory Notebook"; American Chemical Society: Washington, D. C., 1985.

LABORATORY GLASS AND LABORATORY GLASSBLOWING

Introduction

One wonders what the state of chemistry would be without glass. Common laboratory ware such as beakers and flasks are made of glass. Special apparatus such as those used for separations are fabricated from glass.

In the first section, the types of glass used in the laboratory and their properties are discussed. In addition, some of the characteristics of laboratory glassware such as joints and stopcocks are presented.

In the second section, the rudiments of fabricating laboratory glassware is presented, not so much to transform the uninitiated into a glassblower in one easy lesson, but to enable an experimental chemist to intelligently design and to order special glassware.

Laboratory Glass

Depending on the use to which a glass apparatus is to be put, glass with the desired property is selected. For ordinary laboratory use, a glass which has resistance to thermal shock is desirable. For other uses, glass of certain spectral properties is needed. For some procedures, a glass which is resistant to alkali is indicated. For glass-to-metal seals, glass with the correct thermal expansion coefficient is necessary. There are a large number of glass compositions for a range of different uses, e.g. for optical glass in making telescopes, but this section will deal only with the glass compositions which might be used for chemical applications.

Borosilicate Glass

Before borosilicate glass became a common commercial item, laboratory glass consisted of a variety of glasses with compositions to meet specific needs. Because these different glass compositions had different coefficients of expansion, one could not be sealed to another. In order to identify which glass was which, Jena Glass in Germany came up with a scheme of putting a fine hairline of colored glass in its glassware. Thus an Erlenmeyer flask with a brown hairline could be sealed to another apparatus with a brown hairline marker, but not to some piece with a red hairline identifier.

While most chemistry laboratories in the world were using European glass equipment, Corning Glass in the U. S. was developing a glass which would withstand thermal shock. This effort was in response to a need for a reliable railroad signalmen's lattern, the only means available at the time for communicating with the locomotive engineer. The glass lantern globe would often break when the globe was thermally stressed by the flame of the lantern and the cold rain or snow on the outside. With the failure of the lantern, the signalman had no way to warn the locomotive engineer of a track blockage ahead. When this led to disasters, the need for an improved lantern globe became acute. By 1908, the Corning researchers had come up with a borosilicate glass which had the desired thermal resistivity. However, the initial borosilicate glass was not sufficiently moisture resistant. By 1912, a satisfactory borosilicate glass had been developed which solved the lantern globe problem. This borosilicate glass was called Nonex. The story is told that one of the researchers cut the bottom off a Nonex jar and brought it home for his wife to bake a cake. The success of this experiment led to the line of glass cooking ware.

By 1915, laboratory apparatus made of borosilicate glass was produced with a Pyrex label. Glassware made in the period 1915 – 1919 can be identified with a green double ring with Pyrex and Corning on the donut part of the double ring. With the German glassware cut off during the First World War, the use of the new Pyrex ware took off. Glassware produced between 1919 and 1923 can be identified with the same double ring except that it has the term "Pat. 5-27-19" outside of the double ring. Pyrex glass produced between 1923 and 1925 has the same double ring with "Pat. 5-27-23" and "Made in USA" outside of the double ring. Glass produced between 1925 and 1931 has the same double ring with the terms "TM Reg US Pat Off" and "Made in USA" surrounding the double ring. After 1931, the label indicated Pyrex within a single ring. Because of the superiority of Pyrex as a laboratory glass, it rapidly became the standard worldwide. When the Pyrex patent expired, other manufacturers started to offer a similar borosilicate product. Kimble Glass Company called their product KG-33. Schott Glass in Germany called their product Duran. Because in these days when corporations are buying and selling divisions, the designation for borosilicate glass may be very different in the next few years. The term Pyrex is used for historic reasons, and the assumption should be made that when the term Pyrex 7740 is used, it should be understood that the material is Pyrex 7740 *or the equivalent.* Two of the equivalent glasses are, of course, Duran and KG-33. Schott, the owner of the Duran tradename, bought the Pyrex tradename in 1995. However, glass will continue to be sold under the tradename Pyrex as well as Duran.

People often assume that Pyrex is a specifically identifiable product. In fact, the term Pyrex is applied to a range of glass compositions. The Pyrex used for most laboratory ware is Pyrex 7740. Some other Pyrex glasses with different compositions and properties are: Pyrex 7052 for Kovar sealing, Pyrex 7070 for low loss electrical use, Pyrex 7720 for tungsten sealing, and Pyrex 7250 for baking ware.

Pyrex 7740 has a coefficient of expansion of 33×10^{-7}cm/cm/^oC. This compares with the thermal expansion of 92×10^{-7}cm/cm/^oC for soda lime or soft glass. The upper normal service temperature for Pyrex 7740 is 230^oC and the upper useful temperature is 490^oC. The softening temperature is 820^oC and the working temperature is 1245^oC. The annealing temperature for Pyrex 7740 is 565^oC.

The composition of Pyrex 7740 is 80.5% SiO_2, 12.9% B_2O_3, 3.8% Na_2O, 0.4% K_2O, and 2.2% Al_2O_3.

Soda Lime Glass

Soda lime glass, or soft glass, was at one time a staple in chemistry laboratories. Bending glass tubing to make wash bottles, or whatever, was at one time a laboratory exercise for freshman chemistry. It is still useful in the laboratory for making traps or gas purification trains.

The thermal coefficient of expansion for soda lime glass is almost three times that of Pyrex 7740 and the upper use temperature is only 110°C. The softening temperature is 696°C and the working point is 1000°C. The lower working temperature for soda lime glass indicates that it can be worked in a Bunsen or Meker flame whereas Pyrex with its higher working temperature is virtually impossible to work in a Bunsen flame.

Glass for Special Uses

Kovar is a cobalt-nickel-iron alloy developed by the Stupakoff Ceramic and Manufacturing Co. of Latrobe, Pennsylvania in 1934 for glass to metal seals. Kovar "A" and either Corning Glass 7052 or Corning Glass 7040 or its equivalent have thermal expansion curves which almost identically follow each other from room temperature to 500°C. Large glass to metal seals can be made with this system. Kovar works similarly to stainless steel. The glass which seals to Kovar is a hard glass. If it is desired to attach the metal system to borosilicate glass (Pyrex 7740), a graded seal must be used. Usually two different glasses with coefficient of expansion intermediate between the Kovar sealing glass and Pyrex are selected. A strain will be formed where each glass in the graded seal are attached, but the strain will be small enough so that it will withstand temperature changes.

Nonex, Pyrex 7720, or its equivalent is a borosilicate glass containing lead. Its coefficient of expansion is 36×10^{-7} which is fairly close to the ordinary laboratory borosilicate (Pyrex 7740). Its primary use is in sealing tungsten into glass, a procedure needed, for example, to make an electrical lead into a glass apparatus. Although small tungsten wires can be sealed into ordinary laboratory borosilicate glass, larger diameter tungsten requires Nonex. The coefficient of expansion of Nonex does not exactly match that of tungsten, but it is close enough for most uses. A rule of thumb is that tungsten wire larger than 1.2 mm (50 mil) should be sealed in Nonex rather than in laboratory borosilicate glass. Smaller diameter tungsten wire can be sealed in Pyrex 7740.

In addition to tungsten sealing, Nonex, because of it coefficient of expansion which is close to that of laboratory borosilicate glass is used as one of the glasses in graded seals.

Uranium glass (3320) is also used for sealing larger diameter tungsten wire. It is easily recognized because it has a light green color. Uranium glass is also useful in making graded seals. It has good working characteristics and can be easily sealed to Pyrex 7740. It is usually the first step in making a graded seal to soft glass.

Vycor

Vycor originated from a chance observation at Corning Glass that borosilicate glass which contains fairly large amounts of boron oxide is rapidly leached with acids. It was also noted that the rate of leaching was increased if the high borosilicate glass was cooled slowly. If an object is made from the high borosilicate glass and slowly cooled followed by leaching in hot dilute acid, the boron oxide and soda are leached away leaving a porous skeleton of almost pure silica. When this skeleton is fired at 1200°C, a clear structure some 14% smaller than the original object is formed.

Vycor is essentially identical to fused quartz in its properties. The coefficient of expansion is $8 \times 10^{-7}/^{o}C$. Its softening temperature is 1500 – 1530°C, and the normal working temperature is generally quoted as being 800°C. However, it can be used as a furnace tube in a tube furnace at 1000°C for short periods of time, e.g. half an hour, because the thicker wall withstands deformation for a short time.

Fused Quartz

Fused quartz is prepared from any pure quartz source. One source is pure quartz crystals which can be found in nature. Fused quartz has a coefficient of expansion of $5.5 \times 10^{-7}/^{o}C$ and a softening temperature of 1580°C. Quartz has a compressive strength of 13,350 kg/sq cm (190,000 psi) and a tensile strength of 492 kg/sq cm (7000 psi). It has almost perfect elasticity and for this reason quartz fiber is used for torsion balances.

A word on the reasons for the mechanical strength of quartz and the high softening point might be appropriate. The material SiO_2 exists in three forms. Quartz is stable to 870°C, tridymite from 870°C to 1470°C, and cristobalite from 1470°C to 1710°C, the melting point for SiO_2. All three forms of SiO_2 are tetrahedra sharing all four corners. That is, if the center of the tetrahedra indicate positions of the silicon atom and the corners of the tetrahedra indicate positions of the oxygen atom, each oxygen atom is shared between two silicons. The transition from one form to another is very sluggish and all three forms are found as minerals in nature.

The tridymite and cristobalite structures are analogous to the wurtzite and zinc-blende structures mentioned at the end of the first structures exercise. An idealized structure can be visualized by making rings of six using triangles (to indicate tetrahedra pointing down into the table top) alternating with tetrahedra pointing up. If the tip of the tetrahedron pointing up is connected to a corner of another tetrahedron so that the two tetrahedra are in the eclipsed conformation, then this pattern repeated infinitely in three dimensions is that of tridymite. If the tetrahedra pointing up is connected in a linear fashion with other tetrahedra in such a way that each pair of tetrahedra are in a staggered configuration, and the pattern is imagined as extending in three dimensional space, then the structure is that of cristobalite.

Quartz also has the silicate tetrahedra connected to other tetrahedra at all four corners. The tetrahedra are in a spiral configuration in either a left hand or right hand spiral. Because of this asymmetry, quartz is optically active being either levorotatory or dextrarotatory. See Fig. 1.

All three forms of SiO_2 exhibit the α and β forms. There are only minor changes in atomic positions from one form to the other. This is indicated by the observation that when, for example, the left hand spiral quartz undergoes transformation from the α to the β form, the left hand spiral pattern is not changed.

Since quartz is a four connected three dimensional net, one would expect that it would demonstrate great mechanical strength. Since melting involves the free movement of one part of the structure with respect to another, such movement would not be possible without some bond cleavages. This explains the high melting point of quartz.

Soda lime glass, which has the composition 72% SiO_2, 15% Na_2O, 9% CaO, 3% MgO, and 1% Al_2O_3, has many of the Si-O-Si bonds broken and replaced by $Si\text{-}O^-\ Na^+$, $Si\text{-}O^-\ Ca^{++}$, and $Si\text{-}O^-\ Mg^{++}$ type bonds. Because this destroys some of the Si-O-Si bonds which serve to give quartz its rigid three dimensional four connected network structure, the temperature needed to move part of the material relative to another is expected to be less. As previously noted, the softening temperature for soda lime glass is 695°C as opposed to 1580°C for quartz.

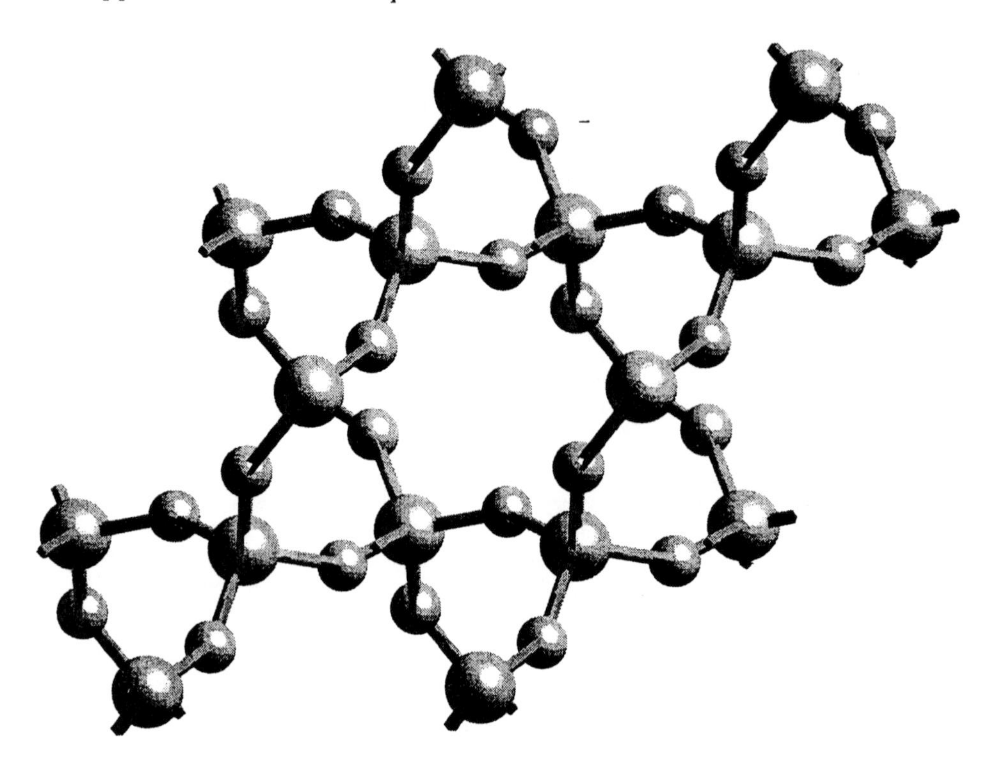

Fig. 1. Structure of β Quartz

Glass Tubing, Joints, and Stopcocks

Laboratory Borosilicate Stock

Laboratory glass is fabricated from glass tubing. The tubing comes in four foot lengths and is always designated in millimeters outside diameter. Larger diameter tubing is considered to be glass pipe and, because it is used by chemical engineers, is designated in inches (outside diameter). The standard tubing comes in standard wall, medium wall, and heavy wall. The wall thickness and sizes are listed in the standard catalogs.

In addition to tubing, capillary tubing is also avilable. Capillary tubing is designated by the diameter of the capillary. Because capillary tubing is made by pulling glass, the inner diameter is approximate and sometimes not uniform from one end of a capillary to the other. If a uniform bore is needed, a batch of several capillaries should be checked. One trick is to put a short string of mercury into the capillary and move it down the length of the capillary by gently blowing on the end of the capillary. If the length of the mercury column does not change, then the diameter of the capillary can be assumed to be constant.

Glass rod is also available. It can be obtained in diameters ranging from 3 mm to 19 mm.

Standard Taper Joints

In 1929, the glass industry, in conjunction with the Bureau of Standards, standardized all interchangeable joints, stopcocks, and stoppers to a taper of 1 mm ± 0.006 mm/cm. Standard taper joints are designated by two numbers. For example, a 24/40 joint measures 24 mm from grind surface to grind surface on the large end of the joint. The 40 indicates that the shank length is 40 mm long. A standard taper joint can come in several lengths. For example a joint with the large end of 19 mm diameter can be 19/38 or 19/22. If a short shank joint of another length is desired, the male joint can be cut with an abrasive saw to the desired length. The following table indicates the typical sizes of the full size joints.

Table I

Standard Taper Joints

7/25	34/45
10/30	40/50
12/30	45/50
14/35	50/50
19/38	55/50
24/40	60/50
29/42	71/60

In addition to the typical joints seen in the lab, through joints are available in the above sizes. Through joints come with a tube of smaller diameter than the small end of the joint sealed onto the small end. These through joints are useful in designing laboratory equipment. For example, a drip extension might be desired at the end of a reflux column. The drip extension can be fabricated without losing part of the grind of the joint by using a through joint.

Ball and Socket Joints

Ball and socket joints are spherical in shape. Because of the spherical nature of the joint, it can adjust to changes in the geometry of the equipment setup. It is, however, not self supporting as is the case with standard taper joints. Ball and socket joints are also designated by two numbers. The first number is the diameter of the ball and the second number is the diameter of the hole in the ball or socket in millimeters. An alternate definition of the second number is that it is the inner diameter of the connecting tube.

Table II

Common Sizes of Ball and Socket Joints

12/2	28/12
12/5	28/15
18/7	35/20
18/9	35/25

"O" Ring Joints

"O" ring joints have a groove in a flat surface which will accept a Buna-N or Viton "O" ring. These joints typically come in a size designated by the clamp size of the type used for the spherical joints.

Table III

"O" Ring Joints

Approx. id	Clamp Size
7.5 mm	18/7
15.5 mm	28/15
20 mm	35/20
41 mm	65/40

Stopcocks

Stopcocks can be of two types. One is the classical glass stopcock of the type which has been found on dropping funnels and burets for years. A second is one with a Teflon stem fitted into a glass cylinder with "O" rings and which makes a closure by being screwed down such that an "O" ring at the tip comes in contact with the seat.

The classical glass stopcock can be divided further into ones used for fluid control and ones used for high vacuum systems. The stopcocks used for fluid control or for low vacuum systems generally have straight through bores. For use in dropping funnels or burets where stopcock grease may be a problem, Teflon plugs are used. The standard taper discussed for joints above make the interchangeablility feasible. The major problem in interchanging plugs in a stopcock is the position of the hole. Because it does not always line up with the inlet and outlet tubes, the absolute interchangeability is somewhat limited.

For high vacuum systems, stopcocks have slanted holes rather than straight through holes. This is to reduce the chance that a leak might be caused by a striation in the grease as the stopcock is turned. Vacuum stopcocks also frequently have a vacuum cup which acts to hold the plug firmly in the barrel. Because the plug has one hole which evacuates the vacuum cup, the stopcock must be installed so that the tube leading to this hole is on the pump side of the stopcock.

Vacuum stopcocks come in different configurations depending on the company making the stopcocks. One configuration is a stopcock with right angle sidearms. This design is useful in that a sizeable opening can be created by removing the stopcock plug. Such an arrangement is useful if the vacuum apparatus is to be cleaned. Cleaning solution can be introduced and removed through this opening whereas it is virtually impossible to do so through a convoluted small opening such as found on the slant bore stopcock or the Teflon stem stopcock.

Vacuum stopcocks are generally sized by the diameter of the opening in the stopock plug. A 2 mm stopcock has a 2 mm bore in the plug. For high vacuum work, it is best to use at least a 4 mm stopcock so that diffusion of the gas molecules can take place relatively rapidly.

The plugs for vacuum stopcocks are generally hollow. This means that any dimensional effects caused by temperature changes will be similar for the plug and the barrel. Vacuum stopcocks are also individually ground. The first step is to grind the stopcock using the standard taper mandrels. Finally, the plug is individually ground into the barrel by hand lapping. For this reason, vacuum stopcocks will have a number scratched to the barrel and the same number scratched on the plug. The plug and barrels for vacuum stopcocks are not interchangeable.

The grease used for lubricating high vacuum stopcocks is important. Two types of grease are commonly used, silicone and hydrocarbon. The silicone grease consists of silicone oil which is jelled with finely divided silica. The vapor pressure for the silicone grease is around 10^{-6} torr. One disadvantage to silicone grease is that if some is left on the glass, heating to blow a modification or repair will cause a thermal decomposition with formation of silica which burns into the glass causing a white flare. Careful washing with chloroform or mineral spirits (available in hardware stores) will remove the silicone for glassblowing purposes, but the disadvantage is that the absence of silicone on the glass surface is difficult to detect. Lowry in J. Chem. Ed., 74, No. 7, July 1997, p. 841 recommends alcoholic NaOH for silicone removal.

The best grease to use for vacuum stopcocks and joints is a hydrocarbon grease sold under the tradename of Apiezon. The vapor pressure for the various Apiezon greases varies from 10^{-9} torr to 10^{-11} torr at 20°C. The common Apiezon greases are L, M, N, and T. Apiezon N is a good general purpose stopcock grease. Apiezon T is a dark colored grease which can be used at higher temperatures than Apiezon N, but it tends to be stiffer and the stopcock greased with T is more difficult to turn. Apiezon M is a good choice for a student laboratory because it is a fraction of the cost of the other Apiezon greases. Apiezon L, N, and T run about $100 for a 25 g tube.

One advantage to Apiezon greases is that no oxidation or polymerization takes place because there is no olefinic content. A stopcock greased many years before can still be rotated although it might require slight warming with a hot air gun.

The Teflon stem stopcocks have become very popular, one reason being that there is no grease involved. The "O" rings on the stem which provide the vacuum seal are very effective. Because the stem is screwed up and down, fine adjustments in flow are easy. This is especially true in controlling liquid flow. There is no problem with "grease set" which makes a glass stopcock difficult to turn if it has not been used for a period of time. Disadvantages with these Teflon stem stopcocks include slower on and off and more limited access for cleaning purposes.

Laboratory Glass Fabrication

Laboratory glassblowing is conceptually very simple. It consists of about eight seals some of which are variations of one another. The difficulty in becoming an accomplished glassblower is in learning to rotate glass and to get a feeling for viscous glass. These two skills are sufficiently difficult to master so that most people do not have the patience or the time necessary to invest in acquiring these two skills.

It is certainly not possible to teach glassblowing in one laboratory period. At best, one can get a feel for the difficulty of rotation and of the handling of viscous glass.

In the laboratory experiments in this laboratory manual, the feeling for glass is introduced in learning to seal ampoules. Students will be asked to practice sealing a 6 mm tube under vacuum and, when confident of this procedure, to seal off ampoules containing the samples prepared in the laboratory. The two instances where this is done are the Stannic Chloride and the Cuprous Chloride experiments.

The reason for outlining the seals used in glassblowing is to enable a practicing chemist to intelligently design special apparatus which might be needed in the course of one's work. The question to be asked when designing a piece of apparatus is: Can this apparatus be made by a series of steps using the seals described? In advising graduate students, a research director soon realizes that a piece very difficult to make is proposed where a piece much simpler to make which will provide the same functionality is also a possibility. Professional glassblowers can volunteer advice on apparatus design, but the chemist will have to be able to ask the proper questions to maximize the design from the point of functionality and ease of construction.

Rotating Glass

Glass can be rotated in two different ways. One is an overhand stance with both hands. The other is to hold the glass in the left hand using the third, fourth, and fifth fingers as the bearing surface on which the glass rotates, and the thumb and forefinger acting to rotate the glass. The palm of the left hand would be down during this rotation process. The right hand should pick up the glass much as one picks up a pencil. The hand is then turned palm up. The glass rotates on the bearing surface formed by the third, fourth, and fifth fingers (or the fourth and fifth fingers (depending on the physiology of the glassblower) and the notch of the hand between the forefinger and the thumb. The forefinger and the thumb provide the rotation of the glass. The advantage to holding the glass in the right hand in the latter fashion is that the two pieces of glass being sealed can be swung to bring the right hand end of the apparatus up to the mouth for blowing. Also, if the glass is being held in this way, the analogy to holding a pencil indicates that the fingers give good manipulative control of the end of the tube. Whatever the method used, the trick is to hold the pieces such that the glass in the right hand does not wave about in relation to that in the left hand. The rate of rotation of the glass in the right hand should also equal the rate of rotation of the glass in the left hand. The right hand does the manipulation of the glass in addition to the rotation.

Breaking Glass Tubing

The smaller diameter glass tubing (20 mm or less) as well as rod and capillary are best "cut" or broken by scratching, wetting, and breaking. Scratching, or scoring the glass tubing, is done by use of a carbide glass scoring knife or with the corner of a file. Tradition in chemistry laboratories seems to be to use a triangle file. However, these soon get dull and will not make a good scratch. It is better to get a six inch single cut flat file with reasonably coarse teeth and to use the four corners. (A single or double cut bastard or mill file can be used. This writer happens to prefer the single cut bastard file.) When the corners of these files get dull, the file is ground on the narrow edge on a grinding wheel. Care must be taken to not overheat the file. The file is held flat on the guide and given an even smooth pass. After about two such passes, the file should be placed in water to cool. As soon as new corners are established, the grinding can be stopped. Usually, two or three passes are sufficient. Care must be taken to make one, and only one, scratch. The scratch is then moistened, usually with a finger moistened with saliva or water, and with the thumbs placed under the side of the tubing away from the scratch, the glass is pulled down such that the micro crack initiated by the scratch can be enlarged to break the tube.

For glass larger than about 20 mm, A single scratch is made as above and the end of a red hot Pyrex rod is touched to the scratch. The thermal stress from the hot rod will initiate a macro crack which usually runs around the tube to make a clean cut. If the glass is scratched more than once, several cracks will initiate giving a break which is not clean.

For very large diameter tubing, a piece of Nichrome wire can be wrapped once around the tube and heated electrically until it is red hot. Several drops of water dripped on the hot glass will initiate a clean break.

Sealing Glass

There are eight fundamental seals to be mastered in making laboratory apparatus. All but one require the rotation skill if a smooth workmanship job is to be done. These eight seals are described.

T-seal. The T-seal is the only seal which does not require rotation. The spot on the apparatus where a side arm is required is heated and blown out. Experience dictates the area to be heated so that the hole blown in the apparatus is of the same diameter as the tube to be attached. The lip of the hole and the end of the tube are simultaneously heated and the viscous glass pressed together and slightly pulled out so that a big wad of glass is not formed at the junction. Part of the skill in making a T-seal is to make sure that the viscous glass is touched evenly all around the point of attachment. If this is not done, a hole results which must be closed up in one way or another. Perhaps it is easiest for a beginner to use a 2 or 3 mm "fill rod" to patch in the hole. This is probably a good place to make the observation that glassblowing is easy if done right in the first place. It is only a very good glassblower who can get into a problem situation and pull off a good seal. The area of attachment is then worked by heating with a flame and blowing the glass in and out in order to smooth it out. If the work can be moved, it is advantageous to turn the seal up or down as needed to have the viscous glass flow in the direction desired by gravity. If a T-seal is being made to a fixed system, care must be taken to not heat the upper part to the extent that the viscous glass flows toward the bottom by gravity leaving the upper part thinner than desired.

Test Tube End. The test tube end is made by rotating the glass while the end to be closed is in the flame. A glass rod is used to pull the hot edges together and to gradually pull the end out. It is not important that closure of the end be made at this point. However, the points of the opening should be symmetrically attached to the glass rod such that the hot end of the tube to be closed can be pulled out evenly. By moving the flame back a bit from the end attached to the rod, the tube can be allowed to collapse while pulling out sufficiently to prevent the glass from becoming too thick. Care must be taken not to pull a long "point." The short closed end is then heated while rotating and blowing in or out as needed. The blowing in and out causes the glass to "flutter" and flow so that the thickness of the closed end is uniform.

Straight Seal. A straight seal is to attach two pieces of tubing of the same diameter so that it becomes in essence one piece. The ability to rotate is critical in making this seal. The two ends of the tubes to be attached are heated until viscous and then pushed together so that they make contact around the circumference. Once again, as in the T-seal, it is desirable to pull the two tubes slightly after the initial contact to diminish the lump of viscous glass which might result on making the contact. While rotating evenly, the glassblower will blow in and out to "flutter" the glass so that the glass at the seal is as close to the thickness of the tubing as possible.

Big-Little Seal. This seal is similar to the straight seal except that a large diameter tube and a smaller diameter tube are to be joined. The technique is to make a test tube end on the large diameter tube. A hole is then blown in the test tube end so it is of the same diameter as the smaller tube. If this hole is too small, it can be enlarged by using a carbon rod to open up the hole to the desired diameter while rotating. If the hole is too large, the small diameter tube can be flared out using the carbon rod. The flaring will create a small funnel shape at the end of the small diameter tube. After the hole and the small diameter tube are made to match, the big-little seal is made in a somewhat similar way to a straight seal. Skillful rotation is important. The flame is directed at the shoulder. The actual location of the joint is usually blown into the shoulder to give a pleasing uniform seal.

Ring Seal A ring seal is most familiar as that connecting the inner tube and the outer cooling jacket of a condenser. A test tube end is made in the larger tube and the inner tube is pushed into the larger tube after being wrapped with high temperature ceramic glassblowing tape. These tapes are available as substitutes for asbestos tape which was used for many years. If the inner tube is long, it may be desirable to support the inner tube with two wrappings of tape. With the inner tube pressed against the test tube end, the outer part of the test tube end is heated until it can be "fluttered" against the inner tube. When a good seal is made to the inner tube, the heating is ceased and, after short cooling, the portion of the glass within the inner diameter of the centered tube is heated. This hot portion is then blown out. The tube to continue the inner tube is then attached. If a condenser or a similar piece of apparatus is being constructed, it is useful to attach the water inlet tube immediately. Once the entire end is constructed, the ring seal and the T-seal should be thoroughly hand annealed.

Insert Seal. An insert seal is similar to the ring seal except that the portion inside the larger tube is too short to support with tape. An insert seal can be made by making a hole in the test tube end on the larger tube and then inserting the inner tube into this hole. The seal of the outer tube to the inner tube can be made more easily if a "maria" is blown on the tube to be inserted. A "maria," also sometimes called an olive or button, is the name given to a bump created in the tubing by heating the tube evenly with a sharp flame and then pushing the tube together slightly. A use for "marias" other than for insert seals is for water inlet or outlet tubes. A rubber hose can be slipped over the enlarged portion of the tube and be held more securely than by a tube without an enlargement. In making the insert seal, the inner tube is pushed into the hole in the test tube end until it comes to a stop because of the "maria." The contact of the "maria" with the hole is then heated until the glass fuses. The insert seal is then worked by heating the seal. If the short inner portion of the insert seal flops to one side or the other, the shoulder of the large tube can be heated and the inner extension centered. If this causes the outer part of the tube to not be aligned with the larger tube, the portion of the insert seal just outside the seal itself can be heated and the outer tube aligned. Perfect rotation will, of course, minimize the flopping of the inner tube from one side to the other.

Dewar Seal. A Dewar seal connects two concentric tubes annularly. It is the type of seal found at the top of a Dewar flask. A Dewar seal is made in much the same way as a ring seal. After the inner tube is sealed to the test tube end of the larger tube, the center portion of the larger tube is blown out. By blowing and by shaping with carbon tools, the Dewar seal is made. If excess glass remains after the center is blown out, the excess glass is peeled off with a glass rod.

If an actual Dewar flask is being made, there are two approaches. In one, a small hole can be left in the test tube end of the inner tube. After the center portion has been blown out, the small hole can be temporarily sealed. After the Dewar seal is finished, the small hole can be permanently sealed. A second approach is to seal the finished inner tube into the larger tube. The ring seal is started and the center portion which starts to bulge out due to the hot enclosed gas is pulled out with a glass rod. The remainder of the excess glass is picked off using the glass rod. The trick in this second approach is to make sure that there is a seal around the entire circumference before the center portion is opened.

Bend. Because in the traditional freshman chemistry laboratory bending glass was historically one of the first experiments, many feel that the bend is the easiest operation in glassblowing. Contrary to popular belief, many glassblowers consider the bend to be the most difficult of the eight seals. The trick is to heat a fairly long piece of glass evenly before attempting the bend. In order to understand the physics of this situation, take a piece of standard wall rubber tubing. If the tubing is held so that there is only a short piece between the two hands, the tubing on bending will kink. One finds that only by moving the hands apart will it be possible to bend the tubing uniformly without kinking. The trick then in bending glass tubing is to heat a fairly long portion evenly. This requires good skills in rotation and the ability to move the glass in the burner to obtain even heating. Bending small diameter tubing is relatively easy. Bending larger diameter tubing, say 20 mm or larger is more difficult. It is easy to get one part of the glass hotter than another part. It is also easy to have one hand rotate at a slightly different rate from the other so that the glass starts to twist.

Glass to Metal Seals

There are a number of glass to metal seals. Some of these, such as the Kovar seal previously mentioned and the Housekeeper seal to copper, are so sophisticated that they should only be attempted by professionals. The two most common wires to be sealed into glass are platinum and tungsten. Small platinum wires can be sealed into soda lime glass. Small tungsten wires can be sealed into Pyrex 7740. For larger diameter tungsten wire, Nonex or uranium glass is used. Nonex contains lead. In order to work glass containing lead, an oxidizing flame is needed. Unless the flame is very rich in oxygen (sometimes called a hissing flame), the lead will reduce and come out as a black coating on the glass.

A good glass to metal seal is formed when the coefficient of the glass is close to that of the metal and the glass "wets" the metal or its oxide. Tungsten makes a vacuum tight seal in laboratory borosilicate glass when it is evenly oxidized so that the glass can effectively "wet" the oxide. The tungsten wire is first cleaned with a stick of sodium nitrite. When hot tungsten is rubbed with sodium nitrite, a reaction takes place. The tungsten will turn red hot even after being removed from the flame due to this reaction. The bright, clean tungsten wire is rinsed in distilled water and, when dry, heated evenly by a steady pass through a gas-oxygen flame such that a smooth even oxide coating is formed. A close fitting glass sleeve of 0.5 mm to 1 mm is slipped over the oxidized tungsten. This sleeve is heated from one end to the other shrinking it onto the oxide surface. If the procedure is done properly, a golden yellow or reddish-copper color results. This glass sleeve is then sealed into the apparatus where the lead is desired. If more than one lead is desired a press seal is used. Two, three, or four of the tungsten leads coated with glass sleeves are placed in the end of a tube which is prepared with a rectangular opening. This rectangular opening is then heated and the hot glass pressed down on the glass sleeves. A large tweezer modified with small squares of steel welded on the tips is useful for pressing the glass to make this type of seal.

Graded Seals

Two kinds of graded seals which might be desired are from laboratory borosilicate glass to soda lime glass and from laboratory borosilicate glass to quartz. Because of the difference in the coefficient of expansions between laboratory borosilicate and soda lime glass, about seven different glasses with intermediate coefficient of expansions are needed. In order to make the transition from soda lime glass with a coefficient of expansion of 90×10^{-7} to laboratory borosilicate with a coefficient of expansion of

32×10^{-7}, each intermediate glass differs from the next by about 10×10^{-7}. For a graded seal from laboratory borosilicate to quartz, three different borosilicate glasses are used. The glass sealed to quartz is high in silica and relatively low in boric acid. The boric acid content is gradually increased over the next two grades in the seal.

Annealing Glass

If laboratory borosilicate glass is to be hand annealed, one should plan to spend about as much time annealing as in the construction of the apparatus. One school of thought says that the piece should be blackened by a non-oxygen containing flame and then heated with a strong gas-air flame until the carbon is burned off. A hand held propane torch of the type available in any hardware store is also useful for annealing. Make sure that it is not a "turbo" burner or one which uses MAAP gas. The latter two give a flame which is too hot. Hand annealing can also be done with a bushy gas-oxygen flame, but the conditions for this are somewhat of an art. One wants to anneal without introducing new strains. A polarized light source to examine for strains is useful in following the progress of the annealing. (A piece of glass examined between cross polarized light will show white flares indicating the position of the strain. If the white strain area is sharp, there is danger of the work cracking on cooling.)

If a glass annealing oven is available, the glass apparatus constructed can be annealed in the oven or lehr. Hand annealing is only needed to ensure that the apparatus survives intact until it is oven annealed. A general rule of thumb is to run the oven up to 550°C and hold for half an hour, or to run it up to 600°C and then to turn off the oven.

Helpful Background Literature

1. Wheeler, E. L. "Scientific Glassblowing," Interscience: New York, 1958.

2. Corning Glass Works "Laboratory Glass Blowing with Corning's Glasses," 30 page booklet, Copyright 1969 by Corning Glass Works: Corning, NY.

3. Corning Glass Works "Properties of Selected Commercial Glasses," 16 page booklet, Copyright 1963 by Corning Glass Works: Corning, NY.

4. Corning Glass Works "Glass in Science and Industry," 23 page booklet, Bulletin GSI-10.

5. Bethlehem Apparatus Company "Manual of Simplified Glassblowing," 57 page booklet; Bethlehem Apparatus Company: Hellertown, PA.

6. Hammesfahr, J. E.; Stong, C. L. "Creative Glassblowing," W. H. Freeman: San Francisco, 1968.

7. Pfaender, H. G., Revised by Schroeder, H. "Schott Guide to Glass," Van Nostrand Reinhold Co.: New York, 1983.

HEATING AND COOLING

HEATING

The methods of heating encountered in the usual undergraduate experience include burners, hotplates, heating mantles, baths, ovens and furnaces. It will be the purpose of this discussion to evaluate each of these methods.

Burners

The usual burners used in the laboratory are Bunsen, Tirrill and Meker burners. All burn a mixture of gas and air. Although it is possible to reach maximum temperatures of 1000°C with a Tirrill and 1200°C with a Meker when heating small objects such as a platinum crucible, the usual temperatures obtained are much lower. Consider that the softening point of Pyrex glass is 800°C (with loss of dimensional integrity starting at 600°C) and the working point (for glassblowing) is 1200°C. A dry Pyrex apparatus can be heated with a Bunsen or Tirrill burner with essentially no distortion. This indicates a glass temperature of less than 800°C. Even under the full heat of a Meker burner, a Pyrex combustion tube only distorts slowly. Higher temperatures can be obtained by using a blast lamp in which gas (methane, propane, or butane) is burned with compressed air or oxygen. Flame temperatures of 1800°C to 2000°C are possible with these devices. The principle limitation of burners is the difficulty in precisely controlling the temperature.

Hot Plates

The laboratory cousin of the household electric range comes in several forms. They all share in common the feature that some flat surface is heated with resistance wire. The lower temperature varies from slightly above room temperature for some of the stepless control models to 200°C for the step control models. The upper temperature for most models is about 500°C although there are some models which are incapable of going much above 300°C. The hot plate is superior to a burner because the temperature of the surface can be controlled fairly well. There is also a safety factor in not having an open flame.

Heating Mantles, Tapes and Immersion Heaters

These items are variations of the hot plate with a heating surface which is not flat. Heating mantles and heating tapes are resistance wires fabricated into a flexible or semiflexible unit with glass cloth, asbestos and/or silicone rubber. The upper temperature limit is determined by the material of construction and ranges from 400°C to 500°C. Immersion heaters are often made by imbedding a resistance wire in mineral insulation (MgO, Al_2O_3, asbestos) inside a copper, brass, Monel or stainless jacket. Many immersion heaters are made exactly like the Cal-Rod heating elements of the household electric range. These consist of a straight piece of Chromel or Nichrome resistance wire surrounded by magnesium oxide and swaged into a metal casing. The carefully purified MgO acts as an electrical insulator which has reasonable thermal conductivity.

Heat Transfer Baths

Steam Baths

Steam baths are an old favorite for heating reactions or to drive off solvent. Many chemistry buildings have steam piped into the laboratory. Some hoods have a steam table installed in the hood. The advantage to steam baths is that there is little danger associated with them except for the thermal hazard. The disadvantage is that the temperature is not variable and is around 100°C.

Paraffin Oil Baths

Oil baths are heated by a burner, hot plate or immersion heater. The temperatures achievable are variable with an upper limit for a paraffin oil bath of around 250°C. Sometimes the uninitiated will use vegetable oil instead of paraffin oil. Vegetable oils are unsaturated triglycerides and will polymerize forming a gummy substance when heated in air. They also tend to smoke at temperatures lower than that of paraffin oil.

Silicone Oil Baths

Silicone oils with a viscosity higher than 30 centistokes can be used as a heat transfer fluid. The advantage to silicones is that, because they are not as combustible as the hydrocarbon fluids, the fire hazard is less. The lower molecular weight siloxanes are cyclic and are volatile. The higher viscosity oils have the cyclic compounds stripped out. The very high viscosity oils have no boiling point whereas the lower viscosity oils have boiling points.

The siloxanes can be divided into four broad classes. (1) The silicon has two methyl groups substituted on it; (2) the silicon has both methyl and phenyl groups substituted on it; (3) the silicon has only phenyl groups substituted on it; and (4) the silicon is substituted with trifluoropropyl groups. Dow Corning calls the polydimethylsiloxane the 200 series. The viscosities available vary from 0.65 centistokes to 2,500,000 centistokes. The common polymethylphenylsiloxanes are labeled as either 510 or 550. These are resistant to oxidation and gumming and are slightly more thermally stable than the 200 series polymers, but they are more expensive. The Dow Corning 550 fluid is a clear 115-centistoke polyphenylmethylsiloxane fluid which is serviceable as a heating bath fluid up to 232°C in air. In closed systems, it is serviceable to 315°C. The less expensive dimethyl compound with a viscosity greater than 50 centistokes is also recommended as a heat transfer fluid. It is serviceable to 200°C. Although the 200 series dimethyl polysiloxanes can be used as a heat transfer fluid, it should be understood that it is a general purpose fluid and is used for insulating, lubricating, damping, coating, defoaming, and releasing. It is also used as a component in cosmetics and polishes.

Silicones are not completely stable entities. There is slow gel formation between 180°C and 200°C. At 250°C the gel formation rate becomes reasonable rapid. Gel formation is inhibited in the absence of oxygen.

One disadvantage to silicones is that the silicone becomes adsorbed on glass and is hard to remove. Chloroform will dissolve much of it off. An oxidizing hot acid bath will also remove the silicone.

Other Heat Transfer Fluids

Concentrated sulfuric acid is sometimes used as a heat transfer fluid. It is not as thermally stable as some of the other fluids and tends to start fuming at temperatures above 200°C. Another disadvantage is that it is very reactive. In general, concentrated sulfuric acid is not recommended as a heat transfer fluid.

Dibutyl phthalate is sometimes used as a heat transfer fluid. However, there are no large advantages over paraffin oil. The cost is slightly greater than for paraffin oil.

Temperatures higher than 250°C can be obtained by using a paraffin wax. The disadvantage to the wax is that it is a solid at room temperature, and the flask to be heated cannot be immersed in the heating medium until the wax melts.

Molten Metal Baths

Wood's metal is a low melting alloy. With a composition of (50% bismuth, 25% lead, 12.5% tin, 12.5% cadmium), it melts at 66°C and can be used as a bath liquid to 500°C.

Molten lead can be used up to 700°C, but does not melt until 328°C.

Molten Salt Baths

The sodium nitrate-potassium nitrate eutectic occurs at 50 mole percent and melts at 222°C. The nitrate can be safely used to 500°C; there is a danger of decomposition at higher temperatures. A caution with molten salts is that carbonaceous material can be (not will be) explosively oxidized by the hot nitrate and the accidental addition of water can cause violent splattering.

Furnaces

Laboratory ovens and furnaces, whether they are the low temperature drying oven type or the very high temperature furnaces, are almost always electrically heated. Nickel-chromium alloy heating wire is generally used for temperatures up to 1000°C. For example, Chromel-A is an 80-20 alloy of nickel-chromium and Chromel-C a 60-16-24 nickel-chromium-iron alloy. Heating elements are generally used in wire or ribbon form and can be coiled or bent to compress a length of wire of the proper resistance into the area desired. The resistance for Chromel-A remains fairly constant with rise in temperature. The change in resistance vs. temperature is greater for smaller diameter wires. Nickel-chromium alloys develop a protective oxide coating with use which helps prolong their useful life at temperatures below 1000°C. During furnace construction the wire should not be stretched since the higher resistance at the smaller wire diameter will cause a "hot spot" to form which eventually results in a failure.

For temperatures above 1000°C silicon carbide rods can be used. Silicon carbide rods (Hot Rods) are made of highly purified silicon carbide self bonded by recrystallization at high temperatures. The typical rods which are commercially available range in diameter from 3/8″ to 1 3/4″ and lengths from 12″ to 63″. The resistance of a rod will vary from 0.5 to 3 ohms depending on the length and thickness. Hot Rods are generally used for furnaces operating in the range 1000°C to 1700°C. No single resistance wire can be used over this range. An iron-chromium-aluminum alloy is useful from 1000°C to 1300°C. For the range 1300°C to 1600°C platinum wire is used. Molybdenum and tungsten wire can be used to 1800°C, but only in a reducing atmosphere.

Furnace Construction Materials

A consideration for high temperature work is the type of materials which can be used for thermal insulation, furnace lining and reaction receptacles. The upper limit for borosilicate or Pyrex type glass is 490°C. The upper limit for fused quartz or 96% silica (Vycor) is 1100°C. Asbestos is serviceable to 900°C. Alundum, which is fused alumina powdered and bonded into a ceramic, is useful from 1400°C to 1800°C depending on the alumina content and method used for bonding. Mullite, an aluminum silicate ($3\ Al_2O_3 \cdot 2\ SiO_2$) is also bonded to form refractories and is useful to 1800°C. Zirconia (ZrO_2) is the highest temperature commercial refractory commonly available and is useful to 2500°C. The above materials can be formed into cores or tubes for furnaces. They can also be made in brick form from which furnaces are constructed with or without mortar. A convenient thermal insulation for use to 1200°C is felt or paper made from a ceramic fiber (Fiberfrax). A representative composition of these fibers is 51.7% Al_2O_3, 47.6% SiO_2, 0.15% B_2O_3, 0.30% Na_2O. Refractory materials are also obtainable as castable solids. These materials are mixed with water and worked like ordinary Portland cement.

Crucible materials are determined as much by the chemical nature of the reactants and the conditions of the reaction (oxidizing or reducing atmosphere) as by their mechanical or thermal properties. Various metal reaction vessels are used. Platinum, iron, molybdenum and tungsten boats or crucible are examples. Of the ceramic types, alundum (Al_2O_3), silicon carbide (SiC), fused magnesia (MgO) and fused zirconia (ZrO_2) are some of the materials used. Choices are based on observations such as the following. Silicon carbide starts to oxidize slowly around 760°C but is stable under reducing conditions to 2100°C; iron is easily oxidized whereas platinum is not; alkali oxides, hydroxides, nitrates, nitrites and cyanides attack platinum; all but the cyanides are unreactive toward iron.

Temperature Measurement

The International Practical Temperature Scale - 1968 (IPTS 68) is defined to agree with the Thermodynamic Temperature Scale within limits of error by interpolating between fixed points by different interpolating instruments. Table I indicates the heating range under discussion.

Although a number of other interpolating instruments are known, the most commonly used are the resistance thermometer, thermocouple and pyrometer.

Table I

Temperature °C	Fixed Point	Interpolating Instrument
100°	B.P. of H_2O	Platinum Resistance Thermometer
231.9681°	F.P. of Sn	Platinum Resistance Thermometer
419.58°	F.P. of Zn	Platinum Resistance Thermometer
630.74°	F.P. of Sb	Pt vs. Pt/10% Rh Thermocouple
961.93°	F.P. of Ag	Pt vs. Pt/10% Rh Thermocouple
1064.43°	F.P. of Au	Pyrometer

The resistance thermometer makes use of the change of resistance of some materials with temperature. Nickel, nickel-iron, and platinum wires have a positive temperature coefficient. For example, the resistance of platinum wire changes about 0.4% per degree Centigrade, that of nickel-iron 0.45%, and that of nickel wire 0.6%. In order to interpolate temperature, a length of wire with the desired resistance is wound into a convenient shape and connected as one arm of a Wheatstone bridge. The advantages of wire resistance thermometers are good sensitivities, high accuracy, high reproducibility, good linearity and wide range. The platinum resistance thermometer is *the* interpolating instrument of choice from the triple point of hydrogen at − 259.35°C (13.81 K) to the freezing point of antimony at 630.74°C. Care must be taken to avoid straining the wire since this affects the resistance. The single greatest disadvantage is the high cost of platinum wire.

Thermistors are semi-conductors of ceramic material made from mixtures of oxides. Some of the oxides used are those of manganese, nickel, cobalt, copper and uranium. The semi-conductors exhibit large negative temperature coefficients of resistance; the resistance decreases with an increase in temperature. The resistance of typical thermistors will vary 3% to 5% per degree Centigrade. The change, however, is far from linear. They also lack long term stability. Their advantages are their small size and low cost. If a specific temperature needs to be detected or small temperature changes need to be accurately determined, the thermistor may be the sensor of choice. The useful range for a thermistor is from 4 K to 200°C although no single thermistor offers linearity over this range.

Thermocouples consist of two dissimilar wires twisted or fused together. The Seebeck effect which results is a potential difference caused by free electrons escaping more easily from one material than from the other due to the different energy levels in the two materials. Since energy levels vary with temperature, observable changes can be detected in the potential difference. An operating thermocouple consists of junctions at each end of dissimilar wires. If one junction is hot and the other cold, the net potential in the circuit is proportional to the difference in temperature between the hot and cold junction. The cold junction is kept either at room temperature or at ice temperature (0°C) and is called the reference junction. Table II on the following page lists the common types of thermocouples used.

Table II

Type	Usable Temperature Range ^oC	EMF/Degree C
J, Iron/Constantan	-190° to 870°	0.034 to 0.0855 mv/deg C
K, Chromel/Alumel	-190° to 1370°	0.02 to 0.03 mv/deg C
T, Copper/Constantan	-190° to 395°	0.018 to 0.062 mv/deg C
E, Chromel/Constantan	-184° to 980°	0.46 to 1.0 mv/deg C
S, Platinum/10%Rh-Pt	-18° to 1760°	0.005 to 0.012 mv/deg C
R, Platinum/13%Rh-Pt	-18° to 1700°	0.0068 to 0.018 mv/deg C
B, Platinum/30%Rh-Pt	$+150^o$ to 1810°	0.0024 to 0.017 mv/deg C

The advantages of the thermocouple are its applicability at high temperatures, a selection of size to fit the job and ease of fabrication at low cost. The disadvantages are several. Cold junction compensation is a nuisance. For accurate work the temperature of the reference junction must be carefully controlled. For most of the high temperature work, room temperature is assumed to be constant and compensation is made with a device built into many of the measuring bridges. The output of thermocouples is low. As can be seen from Table II, the change in output per degree is in the microvolt range. In contrast, the output from a wire resistance sensor is over 100 times this magnitude. Thermocouples are characterized by limited accuracy. Not only are there problems related to precise measurements of small outputs, but there are also inherent variations in the potential produced by the junction. Since there are not only the hot and cold junctions but also junctions to permit connection to a measuring bridge, the possible errors from these variations can be additive. Long lead wires may serve to increase resistance. Although the thermocouple itself is cheap, the measuring bridge may be expensive. Despite the disadvantages, the thermocouple remains the sensor of choice for high temperatures and for rough temperature measurements.

Although tungsten rhenium alloy thermocouples are useful to 2300°C, temperatures above 1700°C are generally measured with a pyrometer. The pyrometer is an instrument which relates radiant energy emitted by a hot body to temperature. It will be recalled that the intensity of the radiant energy is a function of the black body temperature and wavelength.

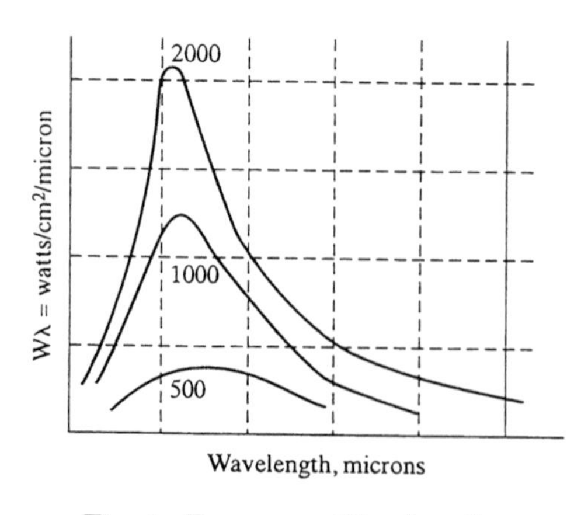

Fig. 1. Energy vs. Wavelength

Some pyrometers measure all or nearly all the energy under the curve and are called total radiation pyrometers. If only a certain wavelength is measured, the device is called a spectrally selective pyrometer. If more than one selected wavelength is measured, the device is called a multicolor pyrometer. The sensing agent in a pyrometer can be a series of thermocouples, silicon photodiodes, other solid state sensors, or an optical matching device. In the latter, the incandescence of a tungsten filament is varied with applied voltage and matched to the radiant output.

Other Methods

This discussion has not attempted to be exhaustive. Rather common heating devices such as hot air guns, hot vapor jackets and sand baths have not been mentioned. Reference is made to Dodd and Robinson (1) for these techniques. High temperature techniques such as induction melting, levitation melting, image furnaces, shock waves and lasers are discussed by Bautista and Margrave (2).

COOLING

In the modern chemistry laboratory, thermally unstable compounds are stored in refrigerators or freezers. Laboratory refrigerators and freezers differ from their corresponding household appliances in that there is no internal light bulb and the controls, if any, are outside the box. These steps are taken to insure that there will be no spark to set off an explosion of chemical fumes which might accumulate in the cold compartment. Small electrical and electronic cooling devices are also on the market that are adapted to cooling traps. They, however, are more expensive than various cooling baths, especially for intermittent use. Cooling baths are also useful for conducting low temperature reactions. Some of the cooling baths used in chemical laboratories are discussed in the sections below.

Ice-Salt Mixtures

Ice-salt mixtures have been used for a number of years to achieve low temperatures. Some of the readers may be familiar with ice cream made at home in the crank type ice cream maker. The can holding the ice cream ingredients is cooled with a crushed ice and salt mixture. The ice cream mixture is stirred to help it achieve thermal equilibrium with the cooling mixture and to prevent large ice crystals from forming and thus achieving a smooth ice cream product. The ice and salt mixture is cold enough to freeze the mixture, a temperature colder than that required to freeze pure water.

Another historical use of an ice-salt mixture was by Fahrenheit who devised the temperature scale still widely used in the U.S. The Fahrenheit scale is currently defined as 32^oF for the freezing point of water and 212^oF for the boiling point of water. However, Gabriel D. Fahrenheit (1686 – 1739), who invented the mercury in glass thermometer, devised his scale by selecting as zero the lowest temperature he could achieve with an ice-salt mixture and 100 as the body temperature of a man.

An ice and salt mixture consists of three phases: ice, salt, and a liquid phase consisting of salt in water. There are two components, salt and water. According to the phase rule, $F = C - P + 2$, where F is the number of degrees of freedom, C the number of components, and P the number of phases. For a three phase system with two components with pressure held constant at one atmosphere, the system becomes invarient with no change in temperature as long as the three phases exist in equilibrium. The system is said to be a eutectic with the ice melting at a minimum temperature. Another way of looking at this is that a constant temperature bath is obtained when all three phases are present. To achieve this, a proper ratio of ice to salt must be used. The following table gives the temperature for various ice salt systems (from Int. Crit. Tables Vol. I, pp. 63 – 65, 1926.)

Table III

Ice-Salt Mixtures

Mixture	Temperature
23% NaCl and 77% ice	– 22.4°C
37% $NaNO_3$ and 63% ice	– 18.5°C
30% $Na_2S_2O_3$ and 70% ice	– 11°C
19% NH_4Cl and 81% ice	– 15.8°C
43% NH_4NO_3 and 57% ice	– 17.5°C
59% $CaCl_2.6H_2O$ and 41% ice	– 54.9°C
25% HCl and 75% ice	– 86°C

Dry Ice

A similar phase diagram argument to the above can be made for systems that involve dry ice (solid carbon dioxide) and an organic solvent. This system is frequently used in cooling traps to condense out volatiles. The temperatures achieved with dry ice and various solvents are shown in the table below:

Table IV

Dry Ice and Solvent Systems

System	Temperature
dry ice and ethyl alcohol	– 72°C
dry ice and chloroform	– 77°C
dry ice and ether	– 77°C
dry ice and liquid SO_2	– 82°C
dry ice and acetone	– 85°C

Liquid Nitrogen and Liquid Air

Probably the most convenient coolant for traps on a vacuum line is liquid nitrogen. Liquid nitrogen boils at – 196°C. Unlike dry ice and organic liquid coolant, the cooling Dewar can easily be removed and replaced. Liquid nitrogen is easier to replenish than powdered dry ice. There is no fire hazard and there is no disposal problem. With the dry ice and organic liquid system, the organic liquid will either have to be stored for re-use or disposed of properly.

Liquid air or liquid oxygen (bp – 183°C) can also be used in the place of liquid nitrogen. However, the nitrogen will boil off preferentially to oxygen from liquid air. (Note the difference in boiling points.) Whenever there is a liquified gas high in oxygen, there is a possibility of a vigorous reaction because oxygen is a strong oxidizer. The safety factor makes nitrogen the preferred cooling agent as opposed to liquid air or liquid oxygen.

Slush Baths

In addition to the two component systems discussed above, a one component two phase system will give a constant temperature at atmospheric pressure. The classic example of this is the ice and water system. As long as there is some solid ice phase in equilibrium with a liquid water phase, the temperature will be held constant at 0°C. Various slush baths can be prepared that give a constant temperature as long as a solid phase is in equilibrium with the liquid phase. For example, if solid chloroform is in equilibrium with liquid chloroform, the temperature will remain constant at – 63.5°C. (Justify how the phase rule predicts this phenomenon.)

A slush bath is prepared by placing an appropriate organic liquid in a Dewar flask and lowering the temperature by adding a small amount of nitrogen while stirring with a spatula. Care must be taken not to pour too much liquid nitrogen at one time into the Dewar containing the organic liquid as the nitrogen will form a second phase on top of the organic liquid. This nitrogen layer will freeze a crust on the top of the organic phase faster than stirring can achieve thermal equilibrium. The liquid under this crust will be at a higher temperature than the frozen crust above; and, when the crust is broken, the warmer liquid below will spew out of the Dewar.

A well made slush will consist of small solid particles in a liquid matrix. A stiff slush, with more of the solid phase, will retain the constant temperature for a longer time than a thin slush. Some slush temperatures are listed in the table below.

Table V

Organic Liquid and Its Slush Temperature (°C)

Organic Liquid	Slush Temperature (°C)
cyclohexane	+6.5
nitrobenzene	+5.7
benzene	+5.5
eucalyptol	+2.0
aniline	– 6.6
methyl salicylate	– 9.0
tert-amyl alcohol	– 12.0
benzaldehyde	– 14.0
benzyl alcohol	– 15.3
butyl benzoate	– 20.0
carbon tetrachloride	– 22.9
bromobenzene	– 30.6
1,2-dichloroethane	– 35.6
anisole	– 37.3
3-pentanone	– 42.0
tetrachloroethane	– 43.8
chlorobenzene	– 45.2
1-hexanol	– 48.0
ethyl malonate	– 50.0
trichloroacetal	– 57.5
chloroform	– 63.5
ethyl acetate	– 83.6
toluene	– 95.0
methylene chloride	– 96.7

methanol	– 97.8
carbon disulphide	– 111.6
1-bromobutane	– 112.4
2-chloropropane	– 117.0
1-chloro-3-methylbutane	– 117.2
ethyl bromide	– 119.0
1-chloropropane	– 122.8
1-chlorobutane	– 123.1
methylcyclohexane	– 126.3
1-propanol	– 127.0
allyl alcohol	– 129.0
1-chloro-2-methylpropane	– 131.2
allyl chloride	– 136.7
chloroethane	– 138.7
2-methylbutane	– 160.5

The temperature of the slush will, of course, be affected by the purity of the solvent. The slushes made from carbon tetrachloride, chlorobenzene, chloroform, ethylacetate, carbon disulphide, and methylcyclohexane generally give reliable temperatures, and these slushes are considered to be low temperature secondary standards.

Helpful Background Literature

1. Dodd, R. E.; Robinson, P. L. "Experimental Inorganic Chemistry"; Elsevier: Amsterdam, 1957.

2. Bautista, R. G.; Margrave, J. L. in Jonassen and Weissberger, "Technique of Inorganic Chemistry, Vol. IV", Interscience: New York, 1965, p. 65.

HANDLING AIR AND MOISTURE SENSITIVE COMPOUNDS

Introduction

There are different degrees of air sensitivity. Some compounds are unstable in oxygen, some in air, some in the presence of water vapor, and some are influenced by more than one of these species. There are many ways to handle air sensitive compounds. For gaseous species, vacuum lines are available. For solid and liquid samples, several types of apparatus are involved.

Glove Bags

For compounds that are relatively stable in air such as $Ni[(PO(C_2H_5)_3]_4$ or $KCuCl_3$, glove bags can be used. The glove bag consists of a plastic material with indentations for gloves and an opening for introducing materials, as well as an exit for introducing gases and for evacuating the bag. In most cases N_2 can be used. However, the polyethylene from which the bag is constructed is not especially good for keeping out O_2 and H_2O, especially over extended periods. Balances, tools, etc. can be placed inside the bag if necessary. Very often a wood framework made of for example tinker toy parts can be used to provide some structure.

Glove Boxes

If compounds are very unstable in air or in contact with O_2 or H_2O, then two basic approaches are commonly used. One approach involves use of a glove box. The other approach concerns use of Schlenkware.

A glove box (or dry box) is a metal container that has various parts including an entry port, a window panel, latex gloves built into the window, a gas handling system, getter materials for adsorbing trace levels of O_2 and H_2O, a recirculation pump, and a pump for evacuation purposes. Heated copper particles or copper wool can be used to adsorb oxygen whereas alumina, zeolites, and other materials can be used to adsorb H_2O. In time these getters need to be regenerated by heating these materials in various gaseous atmospheres.

There are several other potential components of a glove box including electrical outlets, gas introduction, and evacuation ports. In addition there may be components permanently stored and used inside the glove box including refrigeration units, heaters, stirrers, balances, tools, and other materials.

A basic procedure involves putting reagents and other materials into a port which is attached to the dry box followed by evacuation of the port. After filling the port with inert gas, the inside door of the port can be opened to put these materials into the glove box to carry out reactions. A diagram of a glove box is given below.

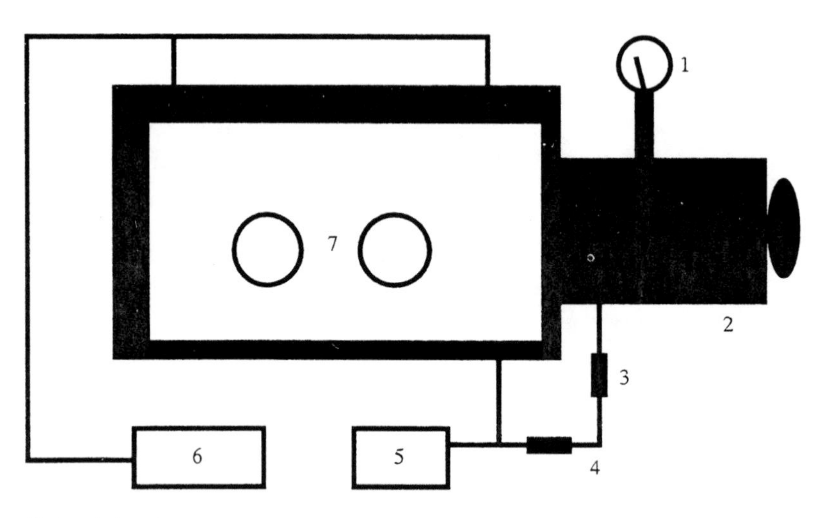

Parts of glove box:
1 - Gauge, 2 - Port, 3 - Trap (H_2O), 4 - Trap (O_2), 5 - Pump, 6 - Recirculation Pump, 7 - Gloves

Fig. 1. Diagram of Glove Box

Schlenkware

The other mechanism for synthesizing and treating oxygen and water sensitive compounds involves use of Schlenkware. Schlenkware is a series of glass apparatus that have high quality evacuation stopcocks, sintered glass frits to separate solid and liquid components, and entrance and exit adapters for attachments to vacuum lines, pumps, tubing, gas tanks and other apparatus. Schlenkware can be used to synthesize air sensitive materials via introduction of liquids through syringes, gases or with solids. All manipulations are done in the glassware under inert atmosphere.

Special procedures need to be developed for using glove boxes and Schlenkware. For example, solvents need to be distilled and stored in bottles so that water and oxygen are not introduced. Solid samples brought in and out of a glove box need to be placed into vials under inert atmosphere, or the bottles need to be evacuated in the port of the glove box to remove adsorbed oxygen. At times other open vessels of desiccants like P_2O_5 and NaK alloy are used in glove boxes to getter small quantities of oxygen and water. Such procedures have been described in detail elsewhere (1, 2).

Helpful Background Literature

1. Angelici, R. J. "Synthesis and Technique in Inorganic Chemistry"; Saunders: Philadelphia, 1977.

2. Shriver D. F.; Drezdon M. A. "The Manipulation of Air Sensitive Compounds"; John Wiley: NY, 1986.

CHROMATOGRAPHY

Introduction

Chromatography involves a series of methods for separation and identification of components of a mixture. There are both stationary and mobile phase in chromatographic experiments. The components in a mixture are transported through the stationary phase. Different components dissolved in the mobile phase migrate at different rates and can be separated in this way. Very often this migration time or retention time is used to calibrate and identify unknown components of a mixture.

The different types of chromatographic methods are distinguished by the nature of the stationary phase and the types of interactions involved in the chromatographic separation. In liquid chromatography (LC), a liquid is used as the mobile phase. In gas chromatography (GC), a gas is the mobile phase.

Liquid Chromatography

A diagram of the general process showing a packed column is given below.

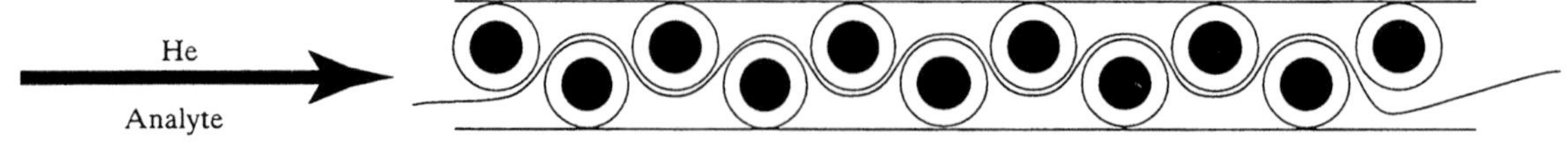

Packed column

Fig. 1. Path of Analyte through Packing

In the above figure, the analyte is the mobile phase, the small filled circles are the support, the outer circle is the stationary phase, and the line shows the path through the column of the analyte which travels between the particles.

In LC methods, there are various types of interactions including liquid-liquid or partition, liquid-solid, ion exchange, and gel permeation with stationary phases that are liquid adsorbed on solid, a solid phase, an ion exchange resin or zeolite type material, and liquid trapped in a solid polymer, respectively. In LC there is an equilibrium established between the mobile and stationary phase which can be described by A_m, the activity of a species in the mobile phase, and A_s, the activity in the stationary phase. The following equations apply:

$$A_m \Leftrightarrow A_s \qquad K_D = \frac{[A_s]}{[A_m]} -$$

Gaussian peaks require a linear isotherm and both $[A_S]$ and $[A_M]$ need to be small. In other words, the column should not be overloaded.

Column chromatography is an LC method where the mixture to be separated is placed on a column and eluted with solvents. In this case a solid component is moved through a column by washing with several batches of solvent. The added solvent or eluent moves the dissolved part of the sample down the column creating a partitioning of the different components in time. The different rates of elution give rise to bands of different components along the length of the column. If enough eluent is used, the various components are separated in time and collected at the bottom of the column. This method is often used in the separation of a complex mixture of components such as some metal phthalocyanine complexes. If a detector such as an infrared or a UV-visible instrument is used, then a chromatographic experiment can be done via column chromatography by monitoring for the products with these spectroscopic detectors.

High performance liquid chromatography (HPLC) was developed throughout the 1960s from the basis of LC column chromatography which is used primarily for synthetic separations. In HPLC, small particles on the order of 5μ are used which can markedly enhance the efficiency of the column. In order to obtain usable flow rates, pumps are used at hundreds of atmospheres. An efficient column is one that gives good separation which produces sharp peaks. The longer a material stays on a column, the broader the peak. The relative sharpness is related to the formula:

$$\text{relative sharpness} = \frac{W_b}{V_r}$$

where W_b is a normalized peak width (with respect to the time on the column) and V_r is the volume the material passes through the column which is related to the time on the column, t_r. The number of theoretical plates (n) should be large.

Normal phase involves use of a polar stationary phase and a nonpolar mobile phase whereas reverse phase is the opposite. To minimize band broadening, a large stationary phase volume, a thin film of stationary phase, smooth particles for prevention of solvent traps, and particles of uniform size are needed. A silica, SiO_2, support which has been surface treated with silane coupling agents containing R groups such as $C_{18}H_{37}$ (octadecyl) might be used for the nonpolar stationary phase.

There are several types of detectors that might be used. The following table summarizes some of the common types of detectors and what is measured.

Table I

Detector Type	Relation to Concentration	Comment
UV	$A \propto c$	Analyte must absorb
Fluorescence	$F \propto c$	High sensitivity, very selective
Electrochemistry	$i \propto c$	Must be electroactive
Refractive Index	Measure Δ RI	Inexpensive, general, sensitivity

Various advantages and requirements are outlined in the table for the different types of detectors. In all of these detectors it is important to minimize the dead volume. An overall diagram of an HPLC apparatus is given below.

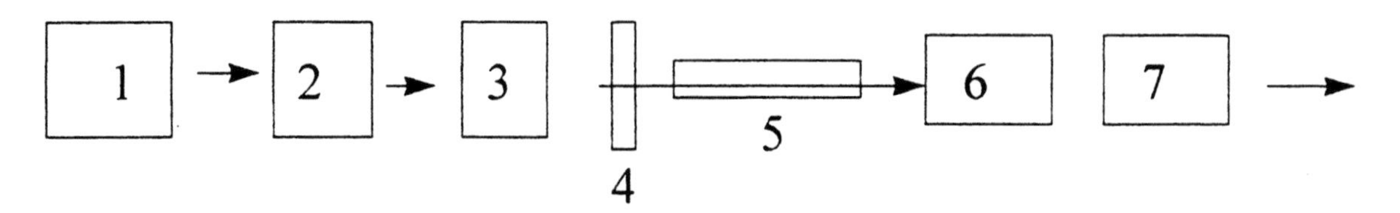

Fig. 2. Schematic Diagram of HPLC

The various pieces of apparatus include a solvent reservoir (1), a pump (2), an injector (3), a pre-column (4), a column (5), a detector (6), and a pulse damper (7) with an outflow. Further discussion of liquid chromatography and HPLC can be found elsewhere (1-3).

Gas Chromatography

In gas chromatography, the gas is the mobile phase which contains species to be separated. The stationary phase can be a liquid phase adsorbed on a solid, an appropriate solid phase, or can be some organic species coordinated to a solid surface. In this case, a carrier gas such as helium is used in conjunction with pressure regulators, and flow meters. The column is contained inside an oven which can be run isothermally or via a temperature program depending on the nature of the separation. The detector is attached to the outflow of the reactor. The sample injection port can have a septum though which a syringe is injected, or a gas sampling valve can be used to inject a pulse of gas through a known sample volume loop. If large volumes are used, it is possible to used packed columns, typically 1/4 or 1/8 inch o.d. stainless steel. The column contains the stationary phase. If rapid analyses of small amounts of sample and rigorous separation which require a high number of theoretical plates are needed, then capillary columns can be used. These capillary columns have diameters on the order of 0.25 mm and lengths of 50 m.

Two major types of detectors are used in GC experiments. Thermal conductivity detectors (TCD) are electrically heated where the temperature depends on the thermal conductivity of the gas around it. Helium or hydrogen can be used as carrier gases. The several advantages of a TCD are its low cost, its simplicity, its ability to respond to most chemical species, and, perhaps most importantly, its rugged character. The sensitivity, however, is not very good as compared to other detectors.

Flame ionization detectors (FID) are considerably more sensitive than TCD systems. When materials burn in a flame, ionic species are produced. The combusted species are measured. A major disadvantage is that the sample is destroyed. Very often both a TCD and an FID are used in complicated analyses for cross-checking.

Finally, there are advantages of coupling chromatographic methods with other methods such as infrared and mass spectrometry methods. In such instruments such as a GCMS, a column is used to separate the individual component of a mixture and a mass spectrometer detector is used. Such GCMS and GCIR instruments offer tremendous advantages over singular instruments and can be used to clearly identify peaks in a chromatogram if standards are not available, if unknowns are being studied, and for other purposes.

Helpful Background Literature

1. Skoog, D. A.; West, D. A. "Analytical Chemistry: An Introduction"; Saunders Golden Sunburst Series: Philadelphia, PA, 1986, Chapter 18, 492 – 527.

2. Skoog, D. A.; Leary, J. J. "Principles of Instrumental Analysis"; Saunders College Publishing: Fort Worth, TX, 1992, 579 – 667.

3. Sibilia, J. P. "A Guide to Materials Characterization and Chemical Analysis"; VCH: NY, 1988, 61 – 84.

THERMAL ANALYSIS

Instrumentation

Thermal analysis refers to a number of methods in which changes occurring during a temperature rise or fall are detected. The range of temperatures can be from ambient to 1500°C and the rate of rise or fall can be controlled over a range of "hold" to 200°C per minute. Properties as varied as weight loss to viscoelastic properties can be measured as a function of the temperature change. Because determination of properties such as thermal mechanical analysis are more useful for polymer chemistry than for inorganic chemistry, the discussion here will be confined to thermogravimetric analysis (TGA), differential thermal analysis (DTA), and differential scanning calorimetry (DSC).

Thermogravimetric Analysis

Thermogravimetric analysis involves weighing a material with a sensitive balance as the temperature is varied in a controlled manner. The signal from a thermocouple measuring the sample temperature is fed into one axis of an X-Y recorder. In a TGA instrument with an optical sensor, any change in the beam position is detected optically by a photomultiplier tube. This signal is transmitted to an amplifier which provides a current to a feedback coil which acts to maintain the position of the balance beam. The amplifier, in addition to providing the current necessary to maintain the position of the balance beam also sends a signal to the other axis of the X-Y recorder. Any weight changes which takes place during the heating process are thus recorded as a function of the temperature.

Several examples of useful chemical data which can be obtained by TGA are listed below.

As discussed in the strontium zincate(II) experiment, the temperature at which water is lost from strontium acetate hemihydrate can be determined by TGA. If it is desired to know at which temperature all of the water of hydration is driven off, TGA can be used to provide this information.

The various hydrates of calcium bromide have been studied by TGA. The weight vs. temperature curve shows inflections indicating $CaBr_2.H_2O$ and $CaBr_2.2\ H_2O$. It is difficult to identify $CaBr_2.3\ H_2O$ since $CaBr_2.6\ H_2O$ melts in its own water of crystallization at 33°C.

Silica gel is used as a desiccant. The sorption of water by silica gel can be studied by TGA.

Alumina can be chlorinated to increase its acidity. Its catalyst efficiency for isomerizing butane depends on the acidity. The extent of the acidity introduced by chlorination has been determined by thermal desorption of pyridine which is reacted with the acid sites.

Thermal analysis has also been used to follow the hydration process during the setting of concrete.

Differential Thermal Analysis

In differential thermal analysis, the temperature of the sample being studied is compared to a thermally inert reference material as the temperature of both are changed in a controlled manner. The thermally inert material is a substance like alumina or glass beads. If some transition takes place in the sample such as melting or phase change, the heat absorbed or emitted during the change will give rise to a temperature of the sample different from that of the reference. The thermocouple outputs from the sample and the reference are amplified about 1000 times with an amplifier appropriate for low voltage signals. The differences in temperatures indicated by these voltages are recorded as a function of the controlled temperature change. The direction of the difference will indicate whether the transformation is endothermic or exothermic.

Modern DTA instrumentation can be used to detect temperature changes with very small samples. The sample and reference are placed in small aluminum pans which are then positioned above sensitive thermocouples. Sample sizes of 0.1 mg to 100 mg are common.

In the strontium zincate(II) experiment, the fusion of the strontium acetate and of the tetrazinc(II) monooxoacetate (zinc oxyacetate) and zinc acetate can be detected by DTA. For zinc acetate, DTA gives evidence that some of the zinc acetate decomposes to the tetrazinc(II) monooxoacetate and that these two substances, when they fuse, give rise to the endotherms at 252°C and 258°C.

The differential thermogram of calcium oxalate hydrate is reproducible under certain conditions and is often used to calibrate the thermal instrument. At around 210°C the hydrate decomposes and becomes cooler (endothermic reaction). At about 480°C the oxalate decomposes to the carbonate and carbon dioxide. This oxidation reaction makes the sample warmer than the reference (exothermic reaction). Finally, at about 815°C the sample once again becomes cooler than the reference because of the endothermic reaction of calcium carbonate decomposing to calcium oxide and carbon dioxide. The exact temperatures of the transitions vary a bit depending on factors such as the crystal size.

Another use of TGA is in the study of the dehydration of $CuSO_4.5\ H_2O$. The DTA clearly substantiates the character of the waters in $CuCO_4.5\ H_2O$. The coordination about the cupric ion is by four waters in a square planar geometry with the oxygens from two different sulfate ions at the remaining two octahedral sites. The fifth water is bridging between sulfate oxygens and some of the square planar water oxygens. An alternate description is to visualize chains of octahedral copper atoms bridged by sulfate with the fifth water between the chains. The DTA of $CuSO_4.5\ H_2O$ depends on the particle size. Part of the reason is that it takes time for water to diffuse to the surface of larger particles. For fine particles, two endothermic peaks are observed as would be expected from the two structurally different waters in the compound.

DTA has been used to study the decomposition of superconducting precursor materials to understand the nature of the decomposition which leads to the superconducting compound. Other studies such as the phase transitions of $YBa_2Cu_3O_x$ with change in the value of x have been carried out by varying the partial pressure of oxygen during a DTA determination.

Differential Scanning Calorimetry

Differential scanning calorimetry is similar to differential thermal analysis. Instead of measuring the difference in temperature between the sample and a reference material as is done with DTA, DSC measures the heat flow into the sample and the reference. Whereas DTA measures differences in temperature, DSC measures differences in energy.

Two different approaches are used to obtain DSC information. One is the power compensated DSC and the other is heat flux DSC. In the former, the sample and the reference are heated by two separate heaters which are controlled such that the temperatures of the sample and reference are kept constant while the instrument temperature is increased or decreased linearly. The difference in the energy flow necessary to keep the sample and reference at constant temperature is, therefore, determined. In a heat flux DSC, the heat flow into the sample and reference is measured directly. Chromel disks are placed on the bottom of the sample and reference pans. These pans are then placed on a constantan platform. The chromel disk and the constantan disk form a sheet thermocouple. The differential heat flow is proportional to the difference in output from the two sheet thermocouples.

DSC has been used to determine the melting point and enthalpies of fusion of eutectic salt mixtures. It has also been used to determine the eutectic of binary alloy systems.

The changes in quartz and dolomite in cement manufacture has been studied using DSC.

Most of the experiments which give results using DTA can also be determined using DSC. Because of the information obtained by DSC, it has become the method of choice for much of thermal analysis.

Helpful Backgound Literature

1. Willard, H. H.; Merritt, L. L. Jr.; Dean, J. A.; Settle, F. A. Jr. "Instrumental Methods of Analysis," 6th Ed.; D. Van Nostrand Company: New York, 1981, pp. 606 – 619.

2. Skoog, D. A.; Leary, J. J. "Principles of Instrumental Analysis," 4th Ed.; Saunders College Publishing: Ft. Worth, TX, 1992, pp. 568 – 577.

3. Wendlandt, W. W. "Thermal Analysis," 3rd Ed.; John Wiley & Sons: New York, 1986.

4. Dellimore, D. *Anal. Chem.* 1990, *62*, 44R.

5. Dellimore, D. *Anal. Chem.* 1988, *60*, 274R.

6. Wendlandt, W. W. *Anal. Chem.* 1986, *58*, 1R.

7. Wendlandt, W. W. *Anal. Chem.* 1984, *56*, 250R.

INFRARED SPECTROSCOPY

Introduction

Electromagnetic radiation in the infrared region interacts with molecules by interaction of the oscillating electric field component of the radiation with oscillating electric dipole moments in the molecule. For this reason, there must be a change in the dipole moment of the molecule as it vibrates in order for the molecule to absorb infrared radiation. A second selection rule suggests that the only transitions between vibrational states from v_0 to v_1 (the change in $v = 1$) are allowed. This is called a fundamental frequency.

The above model is based on a harmonic oscillator. Since this model is not correct for a real molecule, overtone transitions (the change in $v = 2, 3$) can also occur. These processes are shown in Fig. 1.

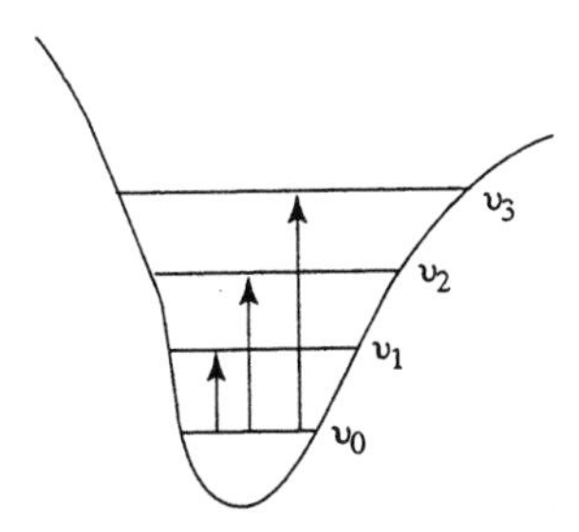

Fig. 1. Fundamental and Overtone Transitions.

The number of independent vibrations of a polyatomic molecule is given by the 3n − 6 rule. A way of understanding the 3n − 6 rule is as follows: If the position of each atom is designated by Cartesian coordinates, then the positions of n atoms is described by 3n Cartesian coordinates. For a molecule consisting of n atoms, the atoms are bonded together and each atom in the molecule cannot independently take up any position. This does not mean that there are less than 3n degrees freedom, but some are reclassified because of the bonding which occurs. Because the atoms are bonded together in a molecule, there will be three degrees of freedom needed to describe the translational motion of the molecule and three degrees of freedom needed to describe the rotational motion of the molecule. If these six degrees of freedom are subtracted from the 3n degrees of freedom, then the 3n − 6 degrees of freedom describe the vibration of the atoms in the molecule.

Note that for linear molecules, there can only be two rotational modes, that about the two axes which are perpendicular to the molecular axis. If these two rotational modes and the three translational modes are subtracted from 3n, there will be 3n − 5 degrees of freedom internal for a linear molecule.

There are symmetric and asymmetric stretching vibrations as well as bending modes. Combination and difference bands are also observed.

The type and number of infrared vibrations can be rigorously determined with group theory methods. A general procedure is to determine the point group of the molecule, to construct an x, y, z coordinate vector system for each atom, to determine the total representations for the molecule, to use the decomposition formula to determine the symmetry of the 3n modes, and to subtract out the translational and rotational modes. The vibrational modes (3n − 6 for a nonlinear molecule) are left. These fundamental modes of vibration will be infrared active if the irreducible representations correspond to the x, y, or z vectors which can be determined by referring to a character table.

The infrared active vibrational modes are useful for characterizing molecular inorganic materials. Gaseous inorganic materials are subject to detailed analysis. The infrared frequencies observed can often be assigned to specific vibrational modes. For binary compounds such as HCl, CO, or NO, the 3n − 5 rule (the diatomic molecular is linear) indicates that, since n = 2, 3n − 5 = 1. Thus there is one stretching force constant for a diatomic molecule. The frequency which should be observed for this stretching force constant can be calculated from a harmonic oscillator approximation. For a simple polyatomic molecule like SO_2, the 3n − 6 rule (n = 3) is 3n − 6 = 3. One would expect three fundamental modes of vibration for SO_2. These fundamental modes are the symmetrical stretching (ν_1), in-plane wagging (ν_2), and asymmetric stretch (ν_3) illustrated below.

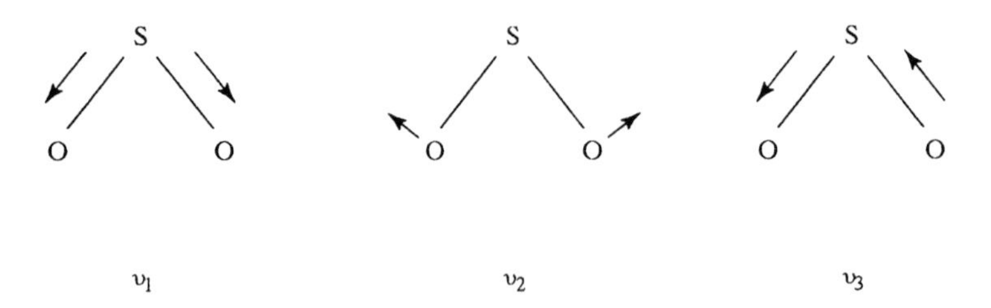

Fig. 2. Three Fundamental Modes for SO_2

The arrows only indicate vibration in one direction. For example, for ν_1 the two SO bonds simultaneously stretch and then contract. For ν_2, the O-S-O bond angle gets larger then smaller. For ν_3, one oxygen stretches while the other contracts. This is followed by the reverse process. These three modes are predicted by the 3n − 6 rule. However, the infrared spectrum for SO_2 shows more than three bands. These are assigned to overtones or combination bands. The $2\nu_1$ band is an overtone of the ν_1 band. The combination bands occur when absorption of infrared energy results in the simultaneous excitation of two different vibrational modes.

Table I
Infrared Spectrum of SO_2

ν (cm^{-1})	Assignment
519	ν_2
606	$\nu_1 - \nu_2$
1151	ν_1
1361	ν_3
1871	$\nu_2 + \nu_3$
2305	$2\nu_1$
2499	$\nu_1 + \nu_3$

The infrared spectra of molecular inorganic compounds obtained in the gas phase exhibit an absorption accompanied by fine structure on either side of it. This is due to a combination of vibrational and rotational transitions. These fine structures are not seen in the infrared spectra of liquids and solids because the rotation of molecules is hindered in these condensed phases. In vibration rotation spectra, the center line is called the Q branch, the family of peaks at the larger wave number is called the R branch and the family of peaks at the lower wave number is called the P branch. The Q branch is due to the vibration alone without rotation. This branch may or may not appear depending on the selection rules (Σ states, L $=$ 0 exhibit no Q branch). The R branch is due to the vibration plus the various rotational quantum numbers (v_r) and the P branch is due to the vibration minus the various rotational quantum numbers.

The infrared frequencies observed for some common inorganic species are shown in Table II. The values shown are for the most abundant isotope. For example, ν_1, ν_2, ν_3 shown in the table for $HC^{12}N$ are 3311, 712, and 2097 cm^{-1}. Because of the difference in the mass of the isotopic carbon, the ν_1 and ν_2 bands for $HC^{13}N$ occur at 3295 and 706 cm^{-1}. There are some bands which are indicated by two values. These are due to inversion doubling. For example, ammonia is a pyramidal molecule which shows the following vibrational modes:

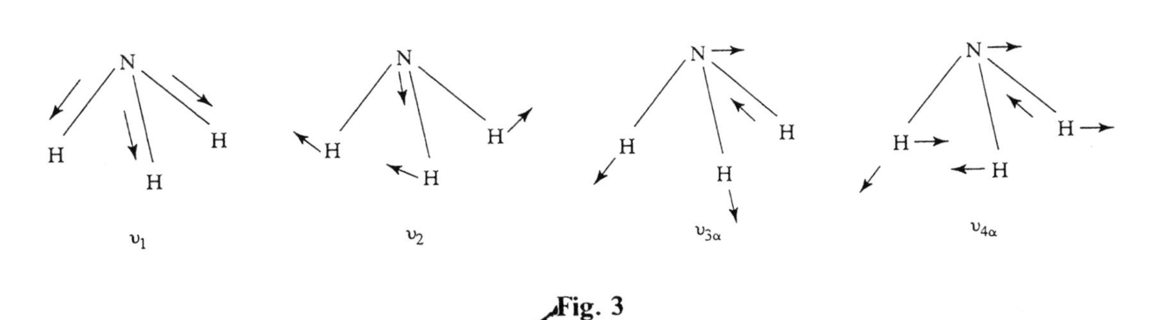

Fig. 3

It can be seen from the table that ν_1 has two values and ν_2 has two values. This is due to inversion doubling. The ammonia inversion is shown in Fig. 4.

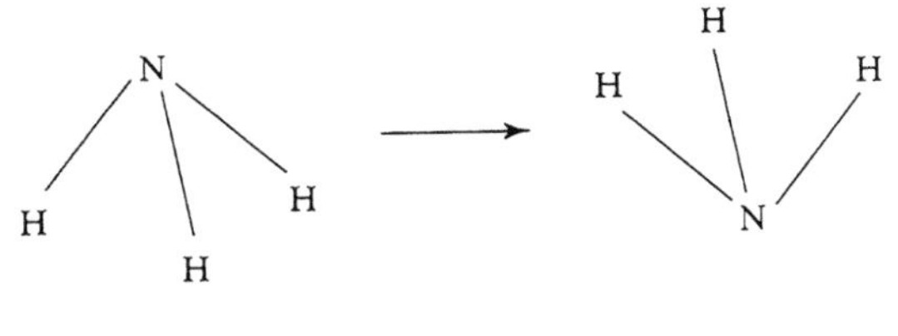

Fig. 4.

The potential barrier for this interconversion is small giving rise to an easy conversion of one form to the other. These two forms give rise to splits in the vibrational energy level, one being positive and the other being negative. Transitions between levels of different sign are allowed in infrared and transitions between levels of the same sign in Raman. Thus the two infrared bands arise from a transition from a negative $v = 0$ to a postive $v = 1$ and a positive $v = 0$ to a negative $v = 1$.

Table II

Infrared Frequencies for Some Common Inorganic Species

Substance	ν_1	ν_2	ν_3	ν_4
H_2 (g)	4161.13			
HF (g)	3962			
HCl (g)	2886			
HBr (g)	2558			
HI (g)	2230			
HCN (g)	3311	712	2097	
N_2O (g)	2224	589	1286	
NO_2 (g)	1318	750	1618	
H_2O (g)	3657	1595	3756	
H_2O (l)	3219	1627	3445	
H_2S (g)	2615	1183	(2627)	
SO_2 (g)	1151	518	1362	
NH_3 (g)	3336	932	3414	1628
	3338	968		
PH_3 (g)	2327	990	2421	1121
		992		
NF_3	1032	647	905	493
PCl_3	507	260	494	189
ClO_3^- (s)	910	617	960	493
SO_3^{2-} (s)	1010	633	961	496
BF_3	888	718	1505	482
BCl_3	471	480	995	243
$CaCO_3$ (calcite)		879	1429 – 1492	706
KNO_3		831	1405	692
			974	
$NaNO_3$		831	1405	692
NH_4NO_3	1050	830	1350	715
NH_4^+	3040	1680	3145	1400
SiF_4	800	268	1010	390
$SiCl_4$	424	150	608	221
$TiCl_4$	388	119	490	139
$SnCl_4$	368	106	403	131
PO_4^{3-}	938	420	1017	567
SO_4^{2-}	983	450	1105	611
ClO_4^-	928	459	1119	625
CrO_4^{2-}	830	(330)	765	330
MnO_4^-	845	(355)	910	395

The infrared spectra of coordination compounds involve the vibrational frequency of the ligands modified by the coordination to the metal atoms. In addition, the bonding of the metal to a ligand atom will give rise to an infrared frequency.

Another way of understanding infrared data for coordination compounds is to do a normal coordinate analysis of a simplified model. For example, an ammine can be modeled as a tetrahedral nitrogen with a metal atom and three hydrogens at the vertices of a tetrahedron. This model gives rise to the following normal modes of vibration.

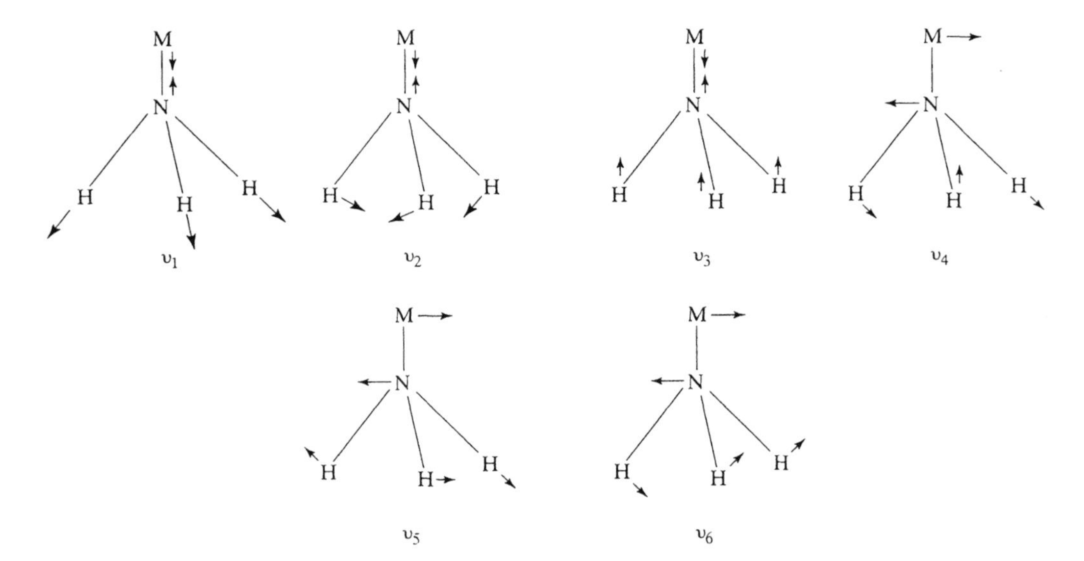

Fig. 5. Ammine as Tetrahedral Nitrogen

The simple halogen ligands provide a relatively straight forward vibrational frequency. Bridging halogens can be identified from the infrared spectrum. Linkage isomerism (nitro/nitrito, cyanato/isocyanato, cyano/isocyano, etc.) can be characterized by the infrared spectra. Infrared methods are also useful in identifying isomerism in the organic ligand, and to determine whether a species is coordinated or an ionic counter ion. Careful analysis leads to a number of characteristics of coordination compounds, however, since most ligands are polyatomic, the analyses are not always easy. See references such as Nakamoto (5) for a more complete treatment.

Group theory can be used to describe the symmetry of molecules. It is possible to use Group Theory to predict the number of infrared and Raman bands in a material. There are several excellent discussions of how this can be done (3, 4, 8, 9, 10).

Techniques for Obtaining IR Spectra

The infrared spectra of solids can be obtained by preparing a salt pellet or by making a nujol mull. The sample to be determined is mixed with infrared pure KBr or CsI. These salts when pure have minimal absorption in the infrared. The mixture of the sample and the matrix salt is pressed in a pellet press. The pellet thus prepared is placed in the sample chamber of an infrared spectrometer. The amount of sample is guaged to provide a sample which indicates the absorption clearly. If the concentration is too high, the peaks will not be readily visible because of the high background. If the sample concentration is too low, even the strong peaks will be small and the weaker peaks may not be readily visible. A second method of determining the infrared spectra of solids is to make a nujol mull. Nujol is a highly purified hydrocarbon oil which is rather viscous. If the solid is made into a mull with this oil, all the sample peaks other than those hidden by the C-H peaks of the nujol will be visible. The nujol mull can either be painted on a salt plate or, if too runny, can be sandwiched between two salt plates.

Liquid samples can be studied by IR or FTIR spectroscopy in a variety of ways. In this case a cell with IR transparent windows such as NaCl, CsI, or other such materials needs to be used. Glass and quartz absorb IR radiation so must be avoided. Nonaqueous solvents are necessary with the salt cells used for liquid samples in order to avoid dissolution of the cell by water as well as to avoid the broad resonances due to water. Sealed liquid cells are available from several manufacturers. It is useful to subtract the absorption of the solvent from the liquid sample and computer programs are available for this. A wise choice of solvent can be made by referral to IR optical absorption data to pick the best materials with minimal absorption in the regions of absorption of the sample.

The infrared spectrum of a gas sample can be determined by condensing the sample into a gas infrared cell. This process is described in the Vacuum Line experiment. As indicated in the experiment on the Characterization of a Gaseous Unknown, pressures of 20 to 100 mm are sufficient. Once again, the infrared spectrum obtained will indicate whether or not the concentration (pressure) is too high or too low.

Fourier Transform Infrared Spectroscopy (FTIR) is more an instrumental capability rather than a technique. A Fourier Transform instrument is capable of scanning the sample thousands of times. Very weak signals can thus be separated from the noise by a built-in computer. Part of the power of the FTIR instrument is that it is able to scan the whole spectral output without the use of a slit to examine one small wavelength increment at a time. In addition to its ability to improve the signal to noise ratio by multiple scanning, the optics are inherently simpler leading to better signals compared to the older instruments. FTIR is useful in studying a minor component present as an impurity, or formed as a by-product, or formed as an unfavorable component in an equilibrium process.

Helpful Background Literature

1. Bellamy, L. J. "The Infra-red Spectra of Complex Molecules", Methuen & Co. Ltd: London, 1954.

2. Cotton, F. A. "Chemical Applications of Group Theory", Wiley-Interscience: New York, 1971.

3. Drago, R. S. "Physical Methods for Chemists", 2nd ed., Saunders: Ft. Worth, 1977; Chapter 1, 2, and 6.

4. Ebsworth, E. A. V.; Rankin, D. W. H.; Cradock, S. "Structural Methods in Inorganic Chemistry", CRC Press: Boca Raton, 1991, Chapters 4 and 5.

5. Nakamoto, K. "Infrared Spectra of Inorganic and Coordination Compounds", Wiley-Interscience: New York, 1970.

6. Silverstein, R. M.; Bassler, G. C.; Morrill, T. C. "Spectrometric Identification of Organic Compounds", John Wiley & Sons: New York, 1981, Chapter 3.

7. NIST/EPA Gas-Phase Infrared Database, http://webbook.nist.gov

8. Huheey, J. E.; Keiter, E. A.; Keiter, R. L. "Inorganic Chemistry"; 4th ed., Harper Collins: New York, 1993, Chapter 3.

9. Douglas, B.; McDaniel, D. H.; Alexander, J. J. "Concepts and Models of Inorganic Chemistry"; 2nd ed., John Wiley & Sons: New York, 1983, Chapter 3.

10. Harris, D. C.; Bertolucci, M. D. "Symmetry and Spectroscopy"; Oxford University Press: New York, 1978, Chapters 1 and 3.

MAGNETIC SUSCEPTIBILITY

Introduction

Magnetic properties have been used to characterize a wide variety of materials ranging from elemental oxygen and ozone to complex systems such as glass and metallic alloys. One of its most extensive applications has been its use in the characterization of coordination complexes of the transition metals. For the first row transition elements, magnetic measurements often give a straighforward estimate of the number of unpaired electrons. However, the correlation of the magnetic properties of the second and third row transition elements is considerably more complex.

Instrumentation

The classical magnetic susceptibility definition is in terms of a linear displacing force in a non-uniform magnetic field with a gradient $\delta H/\delta s$ where H is magnetic field and s is the direction of the linear displacing force. The magnetic moment acquired by a body in a magnetic field is given by

$$f = \kappa v H \frac{\delta H}{\delta s}$$

where f is the linear displacing force, kappa is susceptibility per unit volume, v is the volume of the body, H is the field, and the gradient as defined above.

There are several techniques used to determine the magnetic susceptibility. Two of the classical techniques are the Gouy method and the Faraday method. In addition, there are modern instrumentation which use methods which are unlike either of the classical methods.

Gouy Method

In the Gouy method, a long cylinderical sample is suspended between the poles of a magnet such that the sample is positioned with one end at a high field strength and the other end in a region of negligible field. The change in force observed in the presence of and in the absence of a magnetic field can be treated by the theory which follows to arrive at a magnetic susceptibility.

A Gouy balance for instructional purposes can be easily constructed by using instrumentation generally found in a chemistry laboratory. An old two pan analytical balance can be used to determine the change in force. If a hole is drilled in the table top holding the balance, a chain can be attached to the left hand pan leading to a hook. A non-magnetic costume jewelry chain can be used for this purpose. A compensating Gouy tube can be made by anyone capable of making a straight seal in 5 or 6 mm glass tubing. A permanent magnet with a magnetic field from 5000 to 15,000 oersteds should be mounted on an aluminum plate with "skids" which slide in grooves on an aluminum plate fixed to the table or bench. The magnet can then be slid back and forth to subject the Gouy tube to a magnetic field or to remove the magnetic field. A slide device is needed in that it is important to be able to reposition the magnet in the same way each time.

Samples can be powdered, liquid, or solid. Powdered samples must be packed in a reproducible way. Solids are ground to a fine powder in an agate mortar and put into the tube with tamping. This is done by dropping the tube through a larger tube onto a wooden block. Liquid sampling is easy since a packing problem does not exist. They are placed in the same Gouy tube as the solid powders. Solids such as metals can be cast or machined into the same shape as the Gouy tube.

Temperature control is needed for careful magnetic susceptibility studies. The susceptibility of paramagnetic compounds change with temperature, but the susceptibility of diamagnetic compounds does not change with temperature. In order to hold the temperature constant, a Dewar with a small lower tip is constructed. The lower tip must fit into the cavity between the two magnet pole faces. Heaters can also be constructed to heat the sample to any desired temperature. Once again, the construction of the heater must be such that the heating can be done between the two magnet pole faces.

One source of error in the Gouy determination is in the possible preferential orientation of magnetically anisotropic crystals. For example, powdered graphite can give different magnetic susceptibilities if the crystals are randomly oriented or are somewhat crystallographically oriented. Another source of error is in the temperature control. For samples exhibiting small changes in force, the change in the buoyancy of the surrounding air can contribute to serious errors in the experimentally determined force. Because of this change of buoyancy with temperature, the temperature should be controlled to 0.1°C if small changes in force are expected. For forces measurable on a standard analytical balance, this buoyancy change is not too critical.

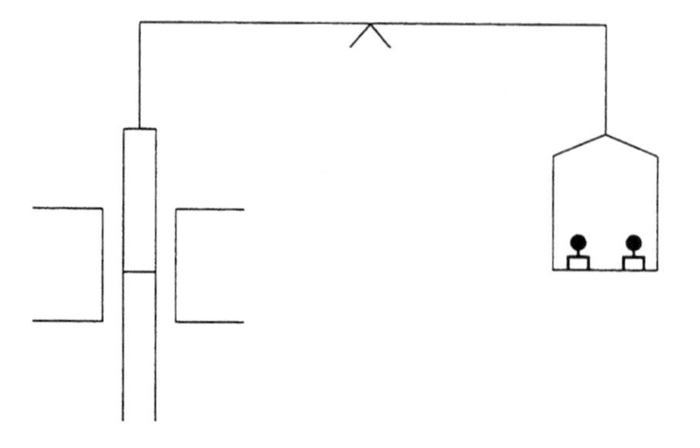

Fig. 1. Schematic of Gouy Balance

Faraday Balance

The Faraday method differs from the Gouy method in that the magnet poles are inclined toward each other or are shaped such that the part of the pole faces near the sample are not parallel to each other. A non-homogeneous field is created with an axis of symmetry. If the sample is placed in a position such that the strength of the field changes with movement along the axis of symmetry, then the sample will be subjected to a force along the axis. The advantages to the Faraday method are that only a small amount of sample is required, and the determination of density is not needed. A requirement is that a sensitive microbalance is needed. Various other force indicators such as silica springs or torsion bars have been used.

Evans Method

The Evans method is a modification of the Gouy method. Instead of measuring the force exerted on a sample by a magnet, the method uses a stationary sample and moving magnets. Two pairs of magnets are mounted on opposite ends of a beam. If a sample is placed between the poles of one pair of magnets, a deflection of the beam results. There is a coil between the poles of the other pair of magnets. A current through this coil restores the beam to its equilibrium position. The current needed to restore this equilibrium is proportional to the force exerted by the sample and this electrical property is used to measure the magnetic susceptibility. Commercial instruments of this type are more compact and allow for easier measurements of magnetic susceptibility.

Other Instrumental Methods

More sophisticated equipment such as a vibrating sample magnetometer can be used for these types of experiments. For further discussion of magnetic apparatus, measurements and theory refer to Mulay (2), Mabbs and Machin (3), and Smart (4).

Basic Principles

In order to interpret magnetic properties of the first row transition metal complexes in terms of unpaired electrons several areas need exploration. (1) The various units used to express magnetic behavior need to be defined. (2) The various types of magnetic phenomena and their causes should be understood. (3) The relationships used to convert experimental measurements to susceptibilities need to be outlined. (4) The equations used for relating susceptibility to the number of unpaired electron need to be derived.

Definition of Units

The degree to which a substance acquires a magnetic moment in a field is expressed in several ways. The magnetic susceptibility per unit volume is represented by the symbol κ.

$$\kappa = \frac{I}{H}$$

The quantity I is the intensity of magnetization induced in the sample by the external magnetic field. This induction may be pictured as being caused by alignment of dipoles and/or by the formation of charge polarization in the sample. The extent of I, the magnetic induction, depends on the sample. H is the field strength of the external magnetic field and is usually measured in oersteds (oe). If it is assumed that I and H have the same units, then κ is dimensionless.

Because it is more convenient to use weight units for solids, the specific susceptibility or gram susceptibility is defined as

$$\chi = \frac{\kappa}{d}$$

where d is the density of the solid.

The molar susceptibility is defined as the gram susceptibility times molecular weight and the atomic susceptibility as gram susceptibility times atomic weight.

$$\chi_M = \chi(\mathrm{MW})$$

$$\chi_A = \chi(\mathrm{at.\ wt.})$$

All these terms, κ, χ, χ_M, and χ_A are measures of the magnetic moment exhibited by different substances in magnetic fields. For purposes of the following discussion κ will be assumed to be dimensionless and χ, χ_M, and χ_A to have the dimensions of reciprocal density.

Types of Magnetic Effects

The magnetic effects observed can be classified as diamagnetic, paramagnetic, ferromagnetic, antiferromagnetic and ferrimagnetic. The two with which we will be most concerned is diamagnetism and paramagnetism.

Diamagnetism is exhibited by elements or compounds in which all electrons are paired. A diamagnetic substance will be weakly pushed out of an inhomogeneous magnetic field. Quantum mechanical analyses of atomic and molecular diamagnetism have been done. However, an intuitive feel for the diamagnetic effect can probably be most easily obtained by considering the classical treatment of atomic diamagnetism. An electron in an orbit is considered equivalent to a current in a wire. As a current in a wire creates a magnetic field, the electron in an orbit also creates a magnetic field. An external magnetic field H on an atom will cause a Larmor precession of this atomic field whose frequency is opposite in sign and proportional in magnitude to that of the external field. This suggests the repulsion of a diamagnetic substance from an external magnetic field and the independence of the susceptibility to field. Another way of getting a feel for the latter is to consider that the change in magnetic moment as expressed by the classical model is:

$$\Delta\mu_A = -\frac{\mathrm{Ne}^2\mathrm{H}}{6\mathrm{mC}^2}\sum_n \overline{(r)}^2$$

The negative sign indicates that the induced Larmor precession is opposite in sign to the applied field. The H in the numerator expresses the fact that the magnitude of the change in magnetic moment is proportional to the field in which the substance finds itself. Recalling that susceptibility is defined as a ratio of the induced magnetism to the magnetic field, and noting that the expression for the induced magnetic moment above contains the term H, it can quickly be concluded that the H terms cancel out and the susceptibility will be field independent.

Diamagnetism is essentially unaffected by temperature since the effect arises from paired electrons. Moderate temperature changes are not capable of significantly changing the induced moment of electrons in filled orbitals. The tendency for these electrons to remain paired is too great.

Paramagnetism is exhibited by compounds in which there are one or more unpaired electrons. A further condition is that there must be no interaction of the dipole represented by one atom or ion and another. This condition is fulfilled by the ligands surrounding the central metal atom which effectively shield the metal atom from its neighbor. Such systems are said to be magnetically dilute.

The spin and orbital angular momentum of the unpaired electron creates a permanent magnetic moment. In a magnetic field, these permanent moments are aligned and attracted into the field. The effect of paramagnetism then is that the sample is drawn into an inhomogeneous field. A paramagnetic material is also diamagnetic since there will always be a closed shell of electrons along with the unpaired electrons. However, the effect of paramagnetism is generally at least 10 times greater than diamagnetism so that a differentiation between a paramagnetic compound and a diamagnetic compound can almost always be made by noting whether a sample is attracted or repelled from an inhomogeneous magnetic field.

Paramagnetic susceptibility, like diamagnetic susceptibility, is independent of field. The susceptibility is a function of the ratio of the force exerted on the sample to the square of the field strength. Since the degree of alignment of the moments which determines the force of attraction into the field is directly related to the strength of the field, the ratio remains constant with changes in field strength.

Unlike diamagnetism, paramagnetism is dependent on temperature. Since thermal motions oppose the tendency of the permanent moments of the unpaired electrons to align with the field, the susceptibility varies inversely as the absolute temperature. This is expressed by the Curie Law:

$$\chi \propto \frac{1}{T}$$

In most cases, the Curie Law does not quite hold. When a correction factor is added, the relationship is called the Curie-Weiss Law:

$$\chi \propto \frac{1}{T + \Delta}$$

The correction factor Δ can sometimes be justified by alignments of moments, but on the whole cannot be explained. It is best considered an empirical correction factor which expresses the extent of deviation from ideality as expressed by the Curie law.

Ferromagnetic materials have unpaired electrons like paramagnetic materials but instead of being magnetically dilute as in the paramagnetic substances there is positive interaction of the moments. This interaction tends to align electron spins parallel and leads to susceptibility values 4 to 10 orders of magnitude greater than diamagnetic materials. The ferromagnetic material is attracted into inhomogeneous magnetic fields and this attraction is found to be field dependent.

Antiferromagnetic materials differ from ferromagnetic materials in that there is a negative interaction of the permanent moments. The atoms in antiferromagnetic materials tend to align in such a way that spins cancel one another. The magnitude of the susceptibility is about the same as that for paramagnetism.

Ferrimagnetism is exhibited by substances in which there is a simultaneous positive and negative interaction of the permanent moments. This is brought about by the presence of two magnetic atoms in compounds such as ternary oxides where ferromagnetic interactions can occur between atoms in one type of site and antiferromagnetic interactions can occur in atoms in another type of site.

Conversion of Experimental Data to Susceptibilities

The direct experimental observation made in determining susceptibilities is the force exerted on a sample in a magnetic field. The relation of this force to susceptibility is expressed by:

$$f = \kappa v H \frac{\delta H}{\delta s}$$

The magnetic moment acquired by a substance in a field of strength H is proportional to κvH where κ is the susceptibility per unit volume, v is the volume of the sample and H is the field strength. If the external magnetic field were uniform at all parts of the sample, there would be no resultant force on the sample. However in an inhomogeneous field, the part of the sample in the stronger part of the field is affected differently from the part in the weaker field. Therefore, if the field gradient is expressed by $\delta H/\delta s$ where s is the linear direction perpendicular to the magnetic force fields, the above expression indicates the force on a sample of volume susceptibility κ in an inhomogeneous magnetic field.

In the Gouy method, a long cylindrically shaped sample is hung in a magnetic field such that one end of the cylinder is placed at the center of the pole gap and the other end extended out to where the field is zero or very small. The force acting on the sample is given by integrating all area cross sections of the cylinder from the end of the sample at maximum field to the other end at zero or minimum field. Integration of

$$f = \kappa v H \frac{\delta H}{\delta s}$$

gives

$$f = \frac{1}{2} \kappa H^2 A$$

where κ is the volume susceptibility, H is the external field strength and A is the cross-sectional area of the cylindrical sample. For susceptibilities determined in air, a correction will have to be made for the volume susceptibility of the air displaced. This is because the oxygen in air is paramagnetic. If a diamagnetic gas like nitrogen is used, the correction is small enough so that it can be safely ignored. The equation corrected for air is

$$f = \frac{1}{2}(\kappa - \kappa_o)H^2A$$

where κ_o is the volume susceptibility of the air displaced by the sample. If the end of the sample not in the pole gap is not in a zero field, a suitable correction must be made.

$$f = \frac{1}{2}(\kappa - \kappa_o)(H^2 - H_o^2)A$$

where H_o is the minimum field experienced by the end not in the pole gap. If the sample is a powder or liquid and contained in a glass or plastic tube, a diamagnetic correction for the tube material needs to be made. The equation then becomes:

$$f = \frac{1}{2}(\kappa - \kappa_o)(H^2 - H_o^2)A - \delta$$

where δ is the diamagnetic correction for the tube material. Compensating tubes can be built where there is as much diamagnetic material on one side of the field as on the other. The force exerted on the diamagnetic material by the inhomogeneous field on one side of the magnet will cancel the force exerted on the diamagnetic material by the inhomogeneous field on the other side of the magnet. Such compensating tubes should be checked by weighing in and out of a magnetic field to make sure that the observed weight is the same in both cases. Only in such cases can δ be set equal to zero.

The above equation can be expressed in terms of gram susceptibiltiy χ in the following way:

$$\text{Set} \quad \frac{H^2 - H_o^2}{2} = \frac{1}{C}$$

This term will be constant for a fixed magnetic field and for samples of constant length or one which is long enough so $H_o = 0$.

$$F = \frac{A}{C}(\kappa - \kappa_o) + \delta$$

$$\kappa = \frac{C}{A}F - \frac{C}{A}\delta + \kappa_o$$

To convert to χ, remember that

$$\chi = \frac{\kappa}{d} \qquad \kappa = \chi d$$

$$\chi d = \frac{C}{A}(F - \delta) + \kappa_o$$

$$d = \frac{wt}{vol}$$

$$\chi = \frac{\frac{C\ vol}{A}(F - \delta)}{wt} \pm \frac{\kappa_o vol}{wt}$$

$$\frac{vol}{A} = \text{length in cm}$$

$$\chi = \frac{C\ell(F - \delta)}{wt} + \frac{\kappa_o\ vol}{wt}$$

By convention,

F, δ expressed in mg. (For a diamagnetic material the value of F or δ will be negative)

wt expressed in g

vol expressed in cc

$\kappa_o = +0.029 \times 10^{-6}$ cgs units/cc

The second term in this expression is the correction for the air displaced. κ_o is the volume susceptibility of air and the value for vol is the volume of the sample. The entire term can be neglected if the susceptibility is determined in a nitrogen atmosphere. It can often be neglected if the volume of the sample is small. The first term contains δ, the diamagnetic correction for the tube. This can be neglected if a compensating Gouy tube is used. The constant C which contains the magnetic field term is usually determined by measuring the force on a material whose susceptibility is accurately known. One must always keep in mind whether or not H_o is a significant value or not. The length of sample necessary to make H_o equal to zero can be determined by determining C for varying amounts of a calibrating substance. If this is not done, χ can be determined by using exactly the same length of calibrating substance as the unknown material. If the same tube is used for all measurements, C becomes a constant. Since this is a constant for a particular tube in a given magnetic field, it is often called a tube constant.

Relating Susceptibility to Unpaired Electrons

The result of the quantum mechanical treatment of the relationship between the applied magnetic field and the moments leading to diamagnetic and paramagnetic susceptibilities along with contributions from effects such as Van Vleck paramagnetism can be stated in terms of the effective magnetic moment, μ_{eff}

$$\mu_{eff} = \sqrt{\frac{3k\chi'_A T}{N\beta^2}} = 2.84\sqrt{\chi'_A T}\ \ B.M.$$

B.M., the Bohr magneton, is the unit of magnetic moment and is equal to $eh/4\pi mc = 9.27 \times 10^{-21}$ erg/gauss. χ'_A is the atomic susceptibility corrected for diamagnetic components of ligands and associated ions. In the following table, the diamagnetic corrections are listed for anions, cations, and individual atoms found in ligands. These latter are labeled Pascal's constants.

Table I
(from Figgis and Lewis)

Diamagnetic Corrections
(All values x 10^{-5}/g atom)

Cations		Anions	
Li^+	1.0	F^-	9.1
Na^+	6.8	Cl^-	23.4
K^+	14.9	Br^-	34.6
Rb^+	22.5	I^-	50.6
Cs^+	35.0	NO_3^-	18.9
Tl^+	35.7	ClO_3^-	30.2
NH_4^+	13.3	ClO_4^-	32.0
Hg^{2+}	40	BrO_3^-	38.8
Mg^{2+}	5	IO_4^-	51.9
Zn^{2+}	15	IO_3^-	51.4
Pb^{2+}	32	CN^-	13.0
Ca^{2+}	10.4	CNS^-	31.0
Fe^{2+}	12.8	SO_4^{2-}	40.1
Cu^{2+}	12.8	CO_3^{2-}	29.5
Co^{2+}	12.8	OH^-	12.0
Ni^{2+}	12.8		

Pascal's Constants

H	2.93	P	26.3
C	6.00	As(V)	43.0
N ring	4.61	As(III)	20.9
N open chain	5.57	Sb(III)	74.0
N mono amide	1.54	Li	4.2
N diamide imide	2.11	Na	9.2
O ether alcohol	4.61	K	18.5
O ketone aldehyde	− 1.73	Mg	10.0
O carboxyl	3.36	Ca	15.9
F	6.3	Al	13
Cl	20.1	Zn	13.5
Br	30.6	Hg(II)	33
I	44.6	Si	20
S	15.0	Sn(IV)	30
Se	23	Pb	46
Te	37.3		

Constitutive Corrections

C=C	−5.5
C=C−C=C	−10.6
C≡C	−0.8
N=N	−1.8
C=N−R	−8.2
C in benzene ring	−0.24
C−Cl	−3.1
C−Br	−4.1
C−I	−4.1

Ligand Corrections

Dipyridyl	105
Phenanthroline	128
Water	13
ortho-phenylene	194

If spin-orbit coupling is negligible, L and S interact independently and

$$\mu_{eff} = \sqrt{4S(S+1) + L(L+1)} \text{ B.M.}$$

Experiment has shown that the orbital contribution for many of the first row transition elements can be completely ignored.

$$\mu_{eff} = \sqrt{4S(S+1)} \ B.M.$$

Since S = n/2 where n is the number of unpaired electrons,

$$\mu_{eff} = \sqrt{n(n+2)} \text{ B.M.}$$

Equating n, the number of unpaired electrons to χ'_A

$$\sqrt{n(n+2)} = 2.84\sqrt{\chi'_A T}$$

$$n(n+2) = (2.84)^2 \chi'_A T$$

$$n^2 + 2n - (2.84^2 \chi'_A T) = 0$$

$$n = -1 + \sqrt{1 + 8.0656\chi'_A T}$$

This shows how the number of unpaired electrons, n, can be estimated from the atomic susceptibility and the Kelvin temperature for the first row transition metals.

Helpful Background Literature

1. Figgis, B. N.; Lewis, J. in Lewis, J; Wilkins, R. G. "Modern Coordination Chemistry"; Interscience Publishers Inc.: New York, 1960, pp. 400 – 454.

2. Mulay, L. N. "Magnetic Susceptibility"; Interscience Publishers Inc.: New York, 1963.

3. Mabbs, F. E.; Machin, D. J. "Magnetism and Transition Metal Complexes"; John Wiley and Sons, Inc.: New York, 1973.

4. Smart, J. S. "Effective Field Theories of Magnetism"; W. B. Saunders Co.: Philadelphia, 1966.

MASS SPECTROMETRY

Instrumentation

Basics

The workings of a mass spectrometer will first be described in a very simplistic way. Once an understanding of the basic principle has been obtained, the variations on the basic technique will be discussed.

In a mass spectrometer, ions are created which are then sorted by mass. Because ions would not exist for long in the presence of large numbers of gas molecules due to collisional processes, a high vacuum is necessary. Since introduction of a relatively high pressure of a sample will destroy this high vacuum, various means are employed to control the rate at which molecules enter the mass spectrometer. In one such technique, the sample is expanded into a bulb which then leaks a small gas stream into the mass spectrometer through an appropriate pin hole leak. In this way, a vacuum of 10^{-7} torr or better can be maintained in the instrument.

The sample molecules which have been leaked into the instrument are then ionized in various ways. In one, energetic electrons emitted from a hot filament are accelerated to about 70 eV by an appropriate potential difference between the filament and a cathode which then impact on a sample molecule, cleaving the bonds and forming ions and radicals.

The positive ions which are formed are pushed toward the accelerator plates by a small repeller electrode. The other molecular fragments formed by electron impact are removed from the instrument by the vacuum pump maintaining the necessary vacuum in the instrument. As the positive ion passes through the hole in the first accelerator plate, it encounters a large potential difference set up between the first accelerator plate and the second accelerator plate. This produces an accelerated beam of positive ions. The beam of positive ions can be collimated in much the same way as a light beam can be collimated. However, instead of using lenses as is done with light, the positive ion beam is focused using focusing plates on which appropriate potentials are applied.

The beam of positive ions is then mass analyzed by a number of techniques. A common one is by magnetic deflection. In a single focusing instrument, the strength of the magnetic field is changed by varying the current through the electromagnet. The ions of varying mass over charge ratios (m/e) are deflected differently with different magnetic fields. The different m/e ions are detected as the ions of the separated ion beam passes through an ion exit slit and fall on the ion collector. Using electronic techniques, it is possible to detect a single ion falling on the collector by amplifying the signal with electron multipliers.

Instrumental Variations

The ion beam can be separated, or focused, by a variety of methods in addition to the magnetic sector deflection. These are time-of-flight, cyclotron resonance, cycloidal focusing, and radio frequency gating (quadrupole). One of the common tools at present is the latter (quadrupole). Because a quadrupole mass analyzer is relative inexpensive and affords very fast scanning over a given range of masses, it is extensively used in GC/MS instruments. As a peak comes through the gas chromatograph, the mass scan is fast enough to provide a mass spectrum of the peak. The disadvantage to a quadrupole mass spectrometer is that the resolution is not high. In order to achieve high resolution, double focusing magnetic sector instruments are used. In these instruments, the beam is focused twice, first by an electrostatic field then by a magnetic field. High mass accuracy can be obtained by these instruments. Accuracies of five parts per million are claimed with possibilities of one part per million by computer massaging of the data. Practically, a millimass unit is routine. Using the characteristic mass defects of the elements, elemental mapping of the mass peaks is possible. A good example of instrumentation and techniques can be found in Beynon (3).

There are other methods for sample introduction in addition to the reservoir bulb and molecular leak technique. A solid probe can be used to introduce samples which do not have the vapor pressures necessary for the vapor phase sample introduction. Fast scanning techniques coupled with high speed recording of the fast mass sweep reduce the flow time necessary to get a mass spectrum. Among other advantages, this allows for very small samples to be analyzed. For example, coupling of a mass spectrometer to a gas chromatograph would not be possible without this feature. In a GC/MS, the mass spectrometer not only acts as a gas chromatography detector, but it also gives a mass spectrum of each peak. This makes it possible to not only characterize the peaks by the retention time, but to also get far more information by enabling, in many cases, the specific identification of the compound causing the peak by the analysis of the mass data for the peak.

Intrepretation of Data

In interpreting mass spectrometer data, it should be noted that the relative abundance of an ion is a function of relative bond strengths and the way in which a molecule cleaves when a high energy electron impacts on it. A useful species which is very often formed is a molecule from which only one electron has been lost. This positive species is called the molecular ion peak. Depending on the molecule, this peak can be very large to non-existent. Fortunately, the number of doubly charged peaks is small. This enables the interpretation of most of the peaks to be the mass since the charge, e, is equal to one. The relative abundances of the various ion peaks is characteristic of the molecule and can be measured to 0.01% with accuracy. The most abundant peak is assigned a value of 100% and is called the parent ion peak. The other ion peaks are listed as a percentage of the parent ion peak. Finally, the isotopic abundances of the elements play an important part in the interpretation of the data produced.

Molecular Ion Peak and Fragmentation Patterns

In order to illustrate the molecular ion peak and the fragmentation pattern of a molecule, the carbon dioxide mass spectrum will be examined.

Table I

Mass Spectrum of Carbon Dioxide

m/e	Relative Abundance
12	8.7
16	9.6
22	- 1.9
28	9.8
29	0.13
30	0.02
44	100.
45	1.2
46	0.42

Note that in the mass spectrum of carbon dioxide shown above, the parent ion peak is the molecular ion peak. Carbon + oxygen + oxygen is 12 + 16 + 16 = 44. The C^+ is 8.7 % of the parent ion peak and the O^+ peak is 9.6% of the parent ion peak. The CO^+ peak (12 + 16 = 28) is 9.8% of the parent ion peak. The instances when the molecular ion peak is also the parent peak is the exception rather than the rule. Some types of molecules do not display a molecular ion peak at all. A case of a compound where the molecular ion peak is almost insignifant is that of trifluoroacetonitrile.

Table II

Mass Spectrum of Trifluoroacetonitrile

m/e	Relative Abundance
12	13.
14	2.1
19	2.0
24	2.7
26	11.
27	0.11
31	22.
32	0.28
38	6.2
50	25.
51	0.31
69	100.
70	1.08
76	46.
77	1.0
95	2.4
96	0.06

Note that the molecular ion peak at m/e of 95 is only 2.4% of the parent peak at m/e = 69. A quick consideration of atomic weights and combination of atomic weights indicates that the m/e peak of 12 is due to C^+, the m/e peak of 14 is due to N^+, the m/e peak at 26 is due to CN^+, the m/e peak at 31 is due to CF^+, the m/e at 50 is due to $CF_2{}^+$, the m/e peak at 69 is due to $CF_3{}^+$, and the m/e peak at 76 is due to C_2F_2N. The $CF_3{}^+$ is the most abundant positive ion formed by the fragmentation of trifluoroacetonitrile by electron impact. Because the bonds cleave easily, the molecular ion peak at 95 which consists of three fluorines, two carbons and one nitrogen (19 + 19 + 19 + 12 + 12 + 14 = 95) is only 2.4% of the largest peak in the spectrum.

Doubly Charged Peaks

It is generally emphasized that the peaks observed in mass spectra data are mass over charge peaks. Thus the same mass peak will appear at half the value for a singly charged species if it has two positive charges. Fortunately, the doubly charged species are not abundant. The triply charged peaks can essentially be ignored. The doubly charged peaks can be examined in the spectrum of carbon tetrachloride.

Table III

Mass Spectrum of Carbon Tetrachloride

m/e	Relative Abundance	Ion Peak
35	41.	Cl^+
36	0.71	
37	13.	
38	0.23	
41	1.2	
42	0.80	
43	0.15	
47	40.	CCl^+
48	0.49	
49	13.	
50	0.16	
58.5	4.7	CCl_3^{++}
59.5	4.5	
60.5	1.4	
61.5	0.16	
70	1.4	
72	0.92	
74	0.16	
82	29.	CCl_2^+
83	0.77	
84	19.	
85	0.46	
86	2.9	
87	0.07	
117	100.	CCl_3^+
118	0.57	
119	96.	
120	0.46	
121	30.	
122	0.11	
123	3.1	

If the doubly charged species are of odd mass numbers, they are easily discernible in the mass spectrum because they will show up with non-integral mass numbers. Thus the 58.5 peak is the same species as the 117 peak except with a double charge. Likewise, the 59.5 peak is the doubly charged 119, the 60.5 is the doubly charged 121, and the 61.5 is the doubly charged 123. It should be noted that the doubly charged peaks are relatively small. The doubly charged peaks of species with even mass numbers are not as easy to identify. However, it might be assumed that the 41 peak is the doubly charged 82 species.

In general, the doubly charged peaks are found only for highly stabilized species. For example, polycyclic aromatics, organo-metallic compounds, and inorganic species such as mercury exhibit multiply charged species. Because the doubly charged species are not commonly formed and because they are relatively weak peaks, interpretations which involve the major peaks can safely ignore the possibilities that the peaks may be doubly charged.

Isotopic Peaks

Whereas chemists deal with the natural mixture of isotopes found in nature, thereby using the atomic weights of 1.008 for hydrogen or 35.5 for chlorine, the mass spectrometer will separate and give data for each of the isotopic species. Thus in interpreting mass spectra, it is important to know the natural isotopic abundances of the elements. These are shown in the table below.

Table IV

Abundances of Common Isotopes

Isotope	Mass	%	Mass	%	Mass	%
H	1	100.	2	0.0156		
B	10	22.9	11	100.	12	25.
C	12	100.	13	1.120		
N	14	100.	15	0.36		
O	16	100.	17	0.04	18	0.20
F	19	100.				
Si	28	100.	29	5.07	30	3.31
P	31	100.				
S	32	100.	33	0.78	34	4.39
Cl	35	100.	37	32.7		
Ga	69	100.	71	66.7		
Ge	70	56.0	72	75.0	74	100. etc.
As	75	100.				
Se	76	18.2	78	47.4	80	100. etc.
Br	79	100.	81	97.5		
Sn	116	43.6	118	72.8	120	100. etc.
Sb	121	100.	123	74.7		
I	127	100.				
Hg	199	56.9	200	77.7	202	100. etc.
Pb	206	50.9	207	40.4	208	100. etc.

Those marked etc. in the third column have more than three isotopes existing naturally. The three largest isotopes have been listed in order to keep the table simple. Tin for example has the isotopes:

Table V

Isotopes of Tin and Abundances

112	3.07
114	2.08
115	1.07
116	43.6
117	23.4
118	72.8
119	26.5
120	100.
122	14.5
124	18.3

Thus, interpretation of mass spectra containing any of the ions marked "etc." in the third column should be done after looking up the natural isotopic ratios in an appropriate reference source (2).

The isotopic peaks explain why there are small peaks above the molecular ion peak in the spectrum of carbon dioxide above. In addition to the $C^{12}O^{16}O^{16}$ peak at m/e = 44, there will also be $C^{13}O^{16}O^{16}$, $C^{12}O^{18}O^{16}$, as well as other combinations of the C^{12}, C^{13}, O^{16}, and O^{18} Because C^{13} is 1.12% of the C^{12} abundance, it would be expected that the m/e 45 peak would be 1.12% of the m/e 44 peak. In fact, the data shows that it is 1.2% of the m/e 44 peak. This is because of a small contribution from the $C^{12}O^{17}O^{16}$ ion. The abundance of O^{18} is 0.20%. If there were only one oxygen in carbon dioxide, one might expect the parent-plus-2 peak at m/e 46 to be 0.20% of the parent peak. However, there are two oxygens in carbon dioxide. There is a chance that either might be O^{18}. Because of this, one would expect the parent-plus-2 peak to be twice the 0.20% or 0.40%. The actual data shows 0.42%.

It should be noted that the molecular ion peak is always chosen as the combination of the most common isotopes. In the case of carbon dioxide, the m/e 45 and m/e 46 are both ions formed by loss of an electron from the otherwise intact molecule. However, it is the m/e 44 peak which is called the molecular ion peak and the 45 and 46 peaks are referred to as isotope peaks.

In many cases, the isotopes appear in substantial amounts rather than in small amounts as in the case with carbon and oxygen. Chlorine, for example, has an isotopic ratio of Cl^{35} to Cl^{37} of about 3 to 1. It can be seen from the mass spectrum of carbon tetrachloride shown above that the Cl^{35} peak at m/e of 35 is about three times that of the Cl^{37} peak at m/e of 37. This same ratio holds for the CCl^{+} peaks at m/e 47 and 49.

If an ion species contains two chlorine atoms as is the case for the $CCl_2{}^{+}$ peaks, then the possible combination of isotopes are $C^{12}Cl^{35}Cl^{37}$, $C^{12}Cl^{37}Cl^{35}$, and $C^{12}Cl^{37}Cl^{37}$. Because the chlorine isotopes contribute substantially to the peaks formed, the small contribution from C^{13} is being ignored. Whereas with only one chlorine, the Cl^{35} to Cl^{37} ratio is 3 to 1, with two chlorines, there is twice the probability of having a Cl^{37} in the ion. Thus it would be expected that the $CCl_2{}^{+}$ peak at m/e of 82 would have a peak 3/2 the size at m/e of 84. The peak intensities of 29 and 19 are indeed in the approximate ratio of 3 to 2. In addition, the $C^{12}Cl^{37}Cl^{37}$ peak would be expected at m/e 86. As expected, this peak is not large.

If there are three chlorines as in the $CCl_3{}^{+}$ peak, the possibilities of finding a Cl^{37} isotope increases. If the Cl^{35} to Cl^{37} is in the ratio of 3 to 1 for a single chlorine and the Cl^{35} to Cl^{37} is in the ratio of 3 to 2 for two chlorines, then the ratio of Cl^{35} to Cl^{37} is in the ratio of 3 to 3 or 1 to 1 for three chlorines. The 117 peak at 100% and the 119 peak at 96% are in this approximate ratio. Also, the presence of three chlorines increases the possibility of having two Cl^{37} in the ion fragment. This is indicated by the 30% peak at m/e of 121. The m/e 123 peak indicative of three Cl^{37} isotopes in the ion peak is small as expected.

The calculation of the ratios of isotope peaks which might be expected for different numbers of atoms in the ion peak are shown in Beynon (3).

Examples of the use of mass spectrometry in inorganic chemistry can be found in reference (4).

Helpful Background Literature

1. McLafferty, F. W. "Interpretation of Mass Spectra", W. A. Benjamin, Inc.: New York, 1967.

2. McLafferty, F. W. "Mass Spectral Correlations", Advances in Chemistry Series #40, American Chemical Society: Washington, D. C., 1963, p. 5

3. Beynon, J. H. "Mass Spectrometry and Its Applications to Organic Chemistry", Elsevier: Amsterdam, 1960.

4. Margrave, J. L. (Symposium Chair) "Mass Spectrometry in Inorganic Chemistry", Advances in Chemistry Series #72, American Chemical Society: Washington, D. C., 1968.

5. NIST/EPA/NIH Mass Spectral Database, http://webbook.nist.gov

NUCLEAR MAGNETIC RESONANCE

Introduction

Nuclei have magnetic properties. Nuclear Magnetic Resonance is a technique for probing the magnetic property of a nucleus to determine useful chemical information about a chemical compound.

Detailed quantum mechanical background will not be attempted in this short essay. A number of excellent references exist that present this material (1-5). Briefly, the magnetic quantum number m has values ranging from I, I − 1, I − 2, . . . − I + 1, − I where I is the spin quantum number. The nuclear spins of the elements can be characterized by the mass number A and the charge number Z (number of protons). If A is an odd value, I will be half integral, i.e. 1/2, 3/2, 5/2 etc. If A and Z are both even, the spin is zero and there is no magnetic resonance spectrum. If A is even and Z is odd, the spin is integral, i.e. I = 1, 2, 3 . . . A complete table of isotopes and their nuclear properties is found in Ref. 1, Appendix A and on the inside front cover of Ref. 2. A few examples are listed below:

Table I

Isotope	NMR frequency in Mc for a 10 kilogauss field	Spin I, in multiple of $h/2\pi$
^{1}H	42.577	1/2
^{7}Li	16.547	3/2
^{10}B	4.575	3
^{25}Mg	2.606	5/2
^{49}Ti	2.401	7/2
^{50}V	4.245	6

Examples of atoms possessing nucleus with A and Z both even are ^{16}O, ^{12}C, and ^{32}S.

Most students who have had organic chemistry will be familiar with proton NMR. For ^{1}H, I = 1/2 and the nuclear angular momentum quantum number, m_I, is ± 1/2. These two states are degenerate in the absence of a magnetic field but become split in a magnetic field. The m_I = + 1/2 is lower in energy than the m_I = − 1/2 and the lower energy state is slightly more populated by nuclei with a lower energy magnetic moment. If the nucleus with a magnetic moment is treated as a bar magnet, the magnet can be visualized to precess in the applied field as shown below.

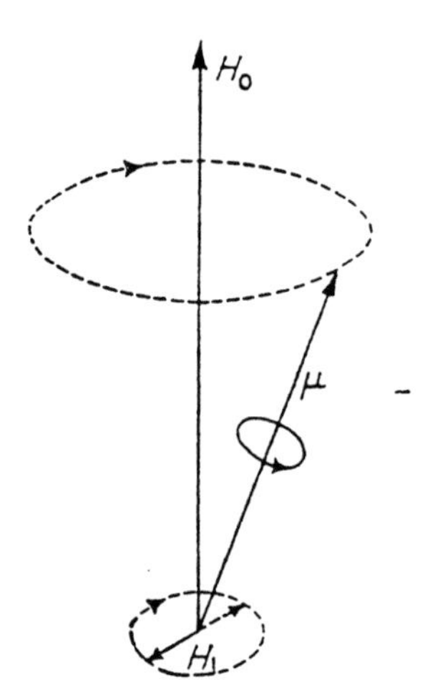

Fig. 1. Magnetic Fields vs. Nuclear Precession

If a magnetic field H, smaller than H_0 is introduced to be perpendicular to H_0, and this field is made to rotate in a plane perpendicular to the direction of the field, it will affect the precessing nuclei. If the rotation becomes that of the Larmor frequency itself, the precessing nuclei will be promoted to the higher energy $m_I = -1/2$ and a nuclear magnetic resonance is observed. Because the Larmor precession is governed by the environment in which the nucleus finds itself, the magnetic resonance observed will give information on the type of proton in the organic molecule.

The NMR frequency required to promote a nucleus from a lower to a higher quantum state varies with magnetic field and nucleus. Table I indicates the frequency needed at 10,000 gauss for the several isotopes listed.

Experimentally, the NMR spectrometer accomplishes the above as follows. The sample to be characterized is placed between two poles of a strong magnet. A radio frequency transmitter transmits a signal through an antenna placed in such a way that the signal is perpendicular to the field of the magnet. In this way the rf signal is exerting a rotating field H as shown in the figure above which is perpendicular to the H_0 of the large magnet and capable of interacting with the Larmor precession. When the rotating field of the signal is in synchronization with the Larmor frequency, the two are said to be in resonance. The nucleus in the lower energy state is excited to the higher energy state. This transition is picked up by a detector antenna arranged to be perpendicular to the exciting antenna and also perpendicular to the direction of H_0.

Because the $2I + 1$ orientations of the spin-I nucleus is not split into sufficiently different energy levels by the external magnetic field, both the lower energy levels and the higher energy levels are populated by nuclei. Thermal energies do a good job of causing transitions from one level to the other. The lower energy level is, however, populated by a slightly higher number of nuclei than the higher energy level. If the same number of nuclei undergo transition from one energy level to the other, no net absorption will be seen. Because there will be a slightly larger number of transitions from the lower energy level to the higher energy level, a nuclear magnetic transition will be seen. However, the number of upward transitions exceeds the number of downward transitions by one in about 10^4 to 10^6. For this reason, the NMR produces weak signals in comparison to some other spectroscopic techniques.

The extent of the split of the energy levels of the nuclear orientations is governed by the strength of magnetic field H_0, the gyromagnetic ratio γ, and the z component of the angular momentum. The gyromagnetic ratio is the ratio relating the magnetic moment μ to the spin quantum number I. Thus $\mu = \gamma I$. The actual energy difference between the nuclear orientations brought about by the magnetic field is given by:

$$\Delta E = h\nu \qquad [1]$$

where h is Planck's constant and

$$\nu = \frac{\gamma H_O}{2\pi} \qquad [2]$$

The selection rules for NMR is $\Delta m = \pm 1$. The transitions are, therefore, allowed only between adjacent energy states. For example, for $I = 1$, the magnetic quantum numbers are $m = 1, 0$, and -1. Transitions between $m = 1$ to $m = 0$ or from $m = 0$ to $m = -1$ are allowed, but not a transition from $m = +1$ to $m = -1$.

The NMR for nuclei with $I = 1/2$ are easier than for those with $I = 1$ or larger. Since there are only two energy states for $I = 1/2$, this leads to simpler multiplets. Also, the nuclear charge distribution is not spherical for nuclei with $I > 1/2$. This leads to a quadrupole moment which in turn leads to rapid relaxation and hence to line broadening.

Chemical Shifts

The nucleus in an atom or molecule is surrounded by electrons. In a magnetic field, these electrons acquire a diamagnetic moment. (See diamagnetism in the essay on magnetic susceptibilities.) The moving electrons create a small magnetic field which effectively diminishes the magnetic field seen by the nucleus and thereby decreases the energy difference of the nuclear energy levels. This electronic effect on the nucleus is called the screening constant. Since the energy levels are brought closer together, the NMR will be observed at a lower frequency.

The classical example quoted for this chemical shift is that of ethyl alcohol. The OH proton, the CH_2 proton, and the CH_3 proton, each in its own unique electronic environment, exhibited three distinct peaks with intensities in the ratio of 1 : 2 : 3 indicating the number of protons of each type.

Whereas the chemical shift for protons in organic compounds range over about 12 ppm, the chemical shift in inorganic compounds can be large. For example, the shift for hydrogen in transition metal hydrides is about 70 ppm, and the shift for hydrogen in paramagnetic compounds can be as much as 1000 ppm. The chemical shifts for other nuclei are also very large. For example, the ^{19}F exhibits shifts over a range of about 1000 ppm. Compared to fluorine in $CFCl_3$ (shift of zero), CH_3F has a shift of -272 ppm, HF has a shift of -203 ppm, XeF_6 has a shift of $+550$ ppm, and UF_6 has a shift of $+746$ ppm. The shifts for ^{13}C in organic compounds range over about 220 ppm with acids, aldehydes, and ketones showing the largest shifts. The ^{13}C for metal carbon bonds show a larger range of shifts. The shifts for ^{31}P show similarly large shifts, with phosphorous in various phosphines showing shifts ranging from zero to 400 ppm.

Spin-spin Coupling

In addition to the chemical shift, high resolution NMR gives further information due to spin-spin coupling. Because both this and chemical shifts are covered in organic chemistry, only a cursory description will be given here.

Spin-spin splitting occurs because neighboring nuclei shield or deshield the nucleus in question. The high resolution NMR of ethanol shows that the OH proton is split into a triplet, the CH_2 protons are split into a quartet, and the CH_3 protons are split into a triplet. This is graphically illustrated by the following:

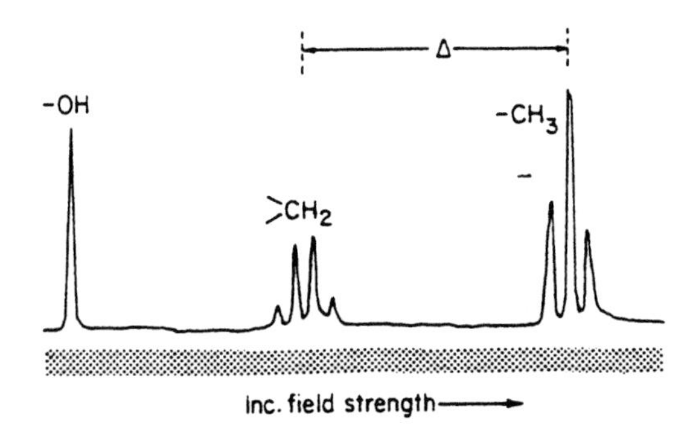

Fig. 2. Representation of High Resolution NMR of Ethanol

If there are two neighboring nuclei with two nuclear configurations, + 1/2 and − 1/2, the possible orientations of these two nuclei can be schematically represented as follows:

+ 1/2 + 1/2

+ 1/2 − 1/2   − 1/2 + 1/2

− 1/2 − 1/2

The sum of the magnetic configurations of the neighboring nuclei will split the NMR of the nucleus being observed with a given chemical shift because they cause a slight change in the magnetic field being experienced by the nuclei. The amount of the shift will be determined by the probability of a particular combinations of nuclear configurations giving rise to a given magnetic field. It can be seen that the Σm_I for the first row configuration of the two nuclei is + 1. For the two nuclei with the second row configurations, the Σm_I is 0. For the the two third row nuclei, the Σm_I is − 1. The nuclear spins with the first row configuration gives rise to the low field peak of the CH_3 protons. The second row nuclear spins give rise to the middle CH_3 peak, and the third row spin configuration gives rise to the high field peak.

If there are three neighboring nuclei with two nuclear configurations, + 1/2 and − 1/2, the possible orientations of these two nuclei can be schematically represented using + and − only as:

+ + +

+ + −   + − +   − + +

− − +   − + −   + − −

− − −

Fig. 3.

Similarly to the CH_3 group on ethanol "seeing" different arrangements of the two adjacent hydrogens in the CH_2 group, there will be four different magnetic fields as shown above for three protons. If the three adjacent protons have the nuclear configuration in the first row, the Σm_I will be $+3/2$. The nuclei in the second row with the configurations as shown will have Σm_I of $+1/2$. The nuclei in the third row will have Σm_I of $-1/2$, and, if the three nuclei have the configuration as shown in the fourth row, the Σm_I will be $-3/2$. The four different total net spins will give peaks in the ratio 1 : 3 : 3 : 1.

The splittings and the relative intensities observed for the number of neighboring nuclei for nuclei with spin 1/2 are tabulated below:

Equivalent I = 1/2 Nuclei	Relative Intensities
0	1
1	1 : 1
2	1 : 2 : 1
3	1 : 3 : 3 : 1
4	1 : 4 : 6 : 4 : 1
5	1 : 5 : 10 : 10 : 5 : 1
6	1 : 6 : 15 : 20 : 15 : 6 : 1

Fig. 4.

A question might be raised as to why the protons on the CH_3 group do not affect each other. It can be demonstrated by quantum mechanical arguments that equivalent nuclei, that is nuclei with the same chemical shift, will give rise to only one signal whatever the spin coupling constants between them. (See page 115 – 116) (1)

Nuclei with $I > 1/2$ are treated in essentially the same way as those with $I = 1/2$. For example, the NMR studies of boron hydrides have contributed a great deal to the knowledge of the structure of these novel compounds. The NMR of boron is complicated by the fact that boron exists as ^{10}B at 18.83% natural abundance with $I = 3$ and ^{11}B at 81.17% natural abundance with $I = 3/2$. Since the ^{11}B is the isotope in greater abundance, it is possible to subtract out the ^{10}B and concentrate on the ^{11}B spectra. Using the same arguments given above for $I = 1/2$ nuclei acting on $I = 1/2$ nuclei, it can be shown that the ^1H NMR in diborane, $^{11}B_2H_6$, shows a 1 : 1 : 1 : 1 splitting for the terminal hydrogens directly bonded to $I = 3/2$ ^{11}B. The bridge protons, on the other hand, are bonded to two boron atoms and show a seven line pattern with intensities of 1 : 2 : 3 : 4 : 3 : 2 : 1. Discussions on NMR of other boron compounds can be found in (7).

Modern Advances

NMR has made huge strides starting with the early low resolution instruments for protons, then with the high resolution instruments with which spin-spin splitting could be observed. In part, this was due to the abundance of ^1H in nature and also was due to the association of protons in organic compounds with ^{12}C and ^{16}O neither of which has magnetic transitions. This meant that these early experiments could be done with relatively simple instrumentation. With the recent advances in computers and spin physics, a number of nuclei previously considered to be subjects of sophisticated research projects have begun to be routine analytical determinations. Atoms such as ^{29}Si at 4.70% natural abundance can now be studied much more readily than before. This opens up vistas in mineral chemistry as well as in the study of commercially useful silicate chemistry such as clays, zeolites, and related catalysts.

An important instrumental advance has been the pulsed Fourier transform technique. The pulsed NMR differs from the continuous wave NMR in that it measures a response to the disturbance of the system by a pulse of the radio frequency rather than the absorption of energy as the magnetic field is varied while a continuous radiofrequency is imposed on the rotating nuclei. The pulse experiment involves the analysis of the data accumulated as a function of time following the pulse. The time dependent data is collected in a digital computer and processed mathematically using Fourier transformation to generate the analog frequency dependent data as obtained by the continuous wave instrument. A good discussion of the principles involved in pulsed NMR can be found in (3).

With the pulsed instrument, individual spectra are taken rapidly so that signal averaging becomes more feasible than with the old continuous wave instruments. This makes it possible to vastly improve the signal-to-noise ratio. In addition, the pulse instrument makes possible a variety of techniques designed to gain chemical information from NMR data. The two dimensional NMR mentioned below is but one of these.

Another advance has been superconductive magnets that provide high fields. High fields are not necessarily always desirable, but one should keep in mind that chemical shifts are measured in parts per million. If the shift difference is Δ ppm, at higher fields Δ/J is larger and the spectra will be spread out more. The chemical shifts increase linearly with the field, H_o, but the spin-spin coupling constants are independent of H_o. By spreading out the chemical shifts, the chances of the multiplets overlapping with each other is diminished. To achieve a high resolution NMR, the field must be homogenous and stable. These conditions are best met by superconducting magnets with the help of shim coils and by the spinning of the sample.

NMR spectrometers are designated as to field strength by the resonance frequency for the hydrogen nucleus in a particular magnetic field rather than the measure of the magnetic field in Teslas (T). A typical magnetic field for an NMR instrument is 9.4 T. Using equation 2 above, this converts to $v = 4 \times 10^8$ Hz or 400 MHz. One of the highest frequency NMR instruments has a magnet rated at a ^1H frequency of 750 MHz (17.6 T).

One of the techniques made possible by the pulsed techniques is the two-dimensional NMR. In this technique, the sample is pulsed in such a way that the interval between pulses is varied in a systematic manner. If the free induction decay (FID), which is the decay of perturbed nucleus free of the perturbing radio frequency pulse, is Fourier transformed to obtain intensity as a function of frequency, and the frequency data are once again Fourier transformed with respect to the delay interval, a two dimensional spectrum is obtained. The two frequencies selected may be chemical shifts or coupling constants. The chemical data obtained by two dimensional NMR is discussed in Ref. 5 with examples involving compounds of boron, osmium, and tungsten (pp. 76 – 83).

An advance that has excited bioinorganic chemists is that of paramagnetic metal complexes as water proton relaxation agents. Paramagnetic substances can decrease the relaxation times of the proton NMR of water in biological systems. For example, Mn(II) is localized more intensely in normal myocardial tissue than in infarcted tissues. Although Mn(II) is not imaged by NMR, it is able to affect the proton NMR of water surrounding the cardiac tissue. Thus, the damaged tissue of the heart can be determined by proton NMR. Gd(III) has been found to be able to image cerebral tumors by enhancing the proton relaxation in the region of capillary breakdown. The challenge is to find non-toxic paramagnetic substances which will selectively absorb or interact with pathological tissue so that the site can be determined by the effect of the paramagnetic substance on the relaxation of protons in water. A good discussion of this topic can be found in (6).

In the early days, it was generally agreed that solid samples did not provide a lot of useful information via NMR. There were three reasons. One was that the dipole-dipole interactions found in the solid broadened the lines to the extent that the spectra obtained did not convey as much information as solution NMR. A second was that solids exhibit a chemical shift anisotropy. For a single crystal, the chemical shift differs with respect to the orientation of the crystal in the magnetic field. For a powdered sample exhibiting a number of orientations, the peaks are appreciably broadened. A third reason is that the relaxation time, T_1, is very long because of the immobility of the nuclei. With long relaxation times, it is difficult to obtain peaks that are sufficiently narrow for study and, in addition, multi-pulse NMR is difficult and consequently it is difficult to get good signal to noise ratios.

The problem with chemical shift anisotropy can be overcome by a technique known as magic angle spinning. The equation describing the line broadening due to chemical shift anisotropy includes the term $(3\cos^2\theta - 1)$. When θ is 54.7^o this term goes to zero. The line broadening by anisotropic effects can be eliminated by spinning a sample at this magic angle.

If the rotation rate is greater than the linewidth, the dipolar coupling can also be eliminated by magic angle spinning. The effect is equivalent to the molecules tumbling in solution. The rotation rate is limited by mechanical and material strength considerations. Rotation for light elements such as ^{13}C or ^{29}Si can be successfully accomplished. However, it is not possible to get rotations rapid enough for the heavier elements.

With modern instrumentation, it is possible to obtain unique chemical information from solid NMR. For example, single crystal gypsum $CaSO_4.2H_2O$ in a 200 MHz instrument shows two pairs of lines indicating that the two waters have different orientations in the crystal. The ^{13}C NMR at 22.6 MHz of calcium acetate, 2 $Ca(Ac)_2.2\ H_2O$, with proton decoupling and magic angle spinning gives a high resolution spectra indicating different molecules in the unit cell.

Instruments with multinuclear probes are available, but no probe currently available can handle any and all nuclei with I half integral or integral. Probes which will be developed in the future are for those nuclei involved in technologically important materials. Electronic materials, catalysts, or bioinorganic species are some examples.

Helpful Background Literature

1. Pople, J. A.; Schneider, W. G.; Bernstein, H. J. "High-resolution Nuclear Magnetic Resonance"; McGraw Hill: New York, 1959.

2. Drago, R. S. "Physical Methods for Chemists"; 2nd ed.; Saunders College Publishing: Ft. Worth, 1992, pp. 211 – 352.

3. Saunders, J. K. M.; Hunter, B. K. "Modern NMR Spectroscopy"; Oxford Univ. Press: Oxford, 1987.

4. Hore, P. J. "Nuclear Magnetic Resonance"; Oxford Univ. Press: Oxford, 1995.

5. Ebsworth, E. A. V.; Rankin, D. W. H.; Cradock, S. "Structural Methods in Inorganic Chemistry"; 2nd ed.; CRC Press: Boca Raton, 1991, pp. 28 – 114.

6. Lauffer, R. B. *Chem. Rev.* 1997, *87*, 901 – 927.

7. Lipscomb, W. N. "Boron Hydrides"; W. A. Benjamin, Inc.: New York, 1963, pp. 126 – 154.

X-RAY DIFFRACTION

Introduction

X-Ray diffraction methods are some of the most powerful characterization tools known to scientists. In chemistry, and particularly in laboratory experiments concerning solids, the two primary pieces of information most often sought are the structure of the material and its reactivity.

The structures of materials are readily studied by diffraction methods. Structural properties are often responsible for several physical and chemical properties such as electrical conductivity, chemical stability, toughness and others. Presently there is considerable interest in the modification of physical properties by selection of molecular design and by control of structure.

X-Ray diffraction methods can be used to study single crystals, powders and other forms of solids. This chapter will focus on the use of X-ray powder diffraction methods although other methods will be mentioned. The objective of this chapter is to introduce the basic principles of X-ray diffraction, to describe equipment used for such studies and to discuss some practical applications of X-ray diffraction.

Basic Principles

Crystal Systems

There are seven crystal systems including cubic, tetragonal, orthorhombic, hexagonal, monoclinic, triclinic, and rhombohedral. Table I includes the definitions of these seven systems on the basis of the length of the sides of these parallelepipeds and the angles between these sides. Since atoms or molecules in a structure can reside in corner, body center or face type sites or positions of these seven crystal systems, there are more than seven possible structural types. The 14 possible structure types are called Bravais lattices. All crystalline materials basically have one of these 14 structures. These are listed on the following page.

Table I

System	Axes	Angles
cubic	a = b = c	$\alpha = \beta = \gamma = 90^o$
tetragonal	a = b ≠ c	$\alpha = \beta = \gamma = 90^o$
orthorhombic	a ≠ b ≠ c	$\alpha = \beta = \gamma = 90^o$
hexagonal	a = b ≠ c	$\alpha = \beta = 90^o, \quad \gamma = 120^o$
monoclinic	a ≠ b ≠ c	$\alpha = \gamma = 90^o \neq \beta$
triclinic	a ≠ b ≠ c	$\alpha \neq \beta \neq \gamma \neq 90^o$
rhombohedral	a = b = c	$\alpha = \beta = \gamma \neq 90^o$

The seven crystal systems described in Table I above are shown in Fig. 1.

The 14 Bravais space lattices described below are shown pictorially in Fig. 2. The Hermann-Mauguin symbols are given in parentheses following the description of the structure. The 1, 2, 3, etc. in the initial parentheses gives the correspondence of the description and the figure. (In the figures, a_1, a_2, and a_3 are used in place of a, b, and c as in some other sources when there is an equivalence of axes).

The 14 Bravais Space Lattices

(1) primitive (or simple) cubic (P)

(2) body-centered cubic (bcc) (I)

(3) face-centered cubic (fcc) (F)

(4) primitive (or simple) tetragonal (P)

(5) body-centered tetragonal (I)

(6) primitive (or simple) orthorhombic (P)

(7) base-centered orthorhombic (C)

(8) body-centered orthorhombic (I)

(9) face-centered orthorhomic (F)

(10) primitive (or simple) hexagonal (the symbol P is now used exclusively here; C was formerly employed because hexagonal crystals may also be regarded as base-centered orthorhombic)

(11) primitive (or simple) monoclinic (P)

(12) based-centered monoclinic (C)

(13) primitive (or simple) triclinic (P)

(14) primitive (or simple) rhombohedral (R)

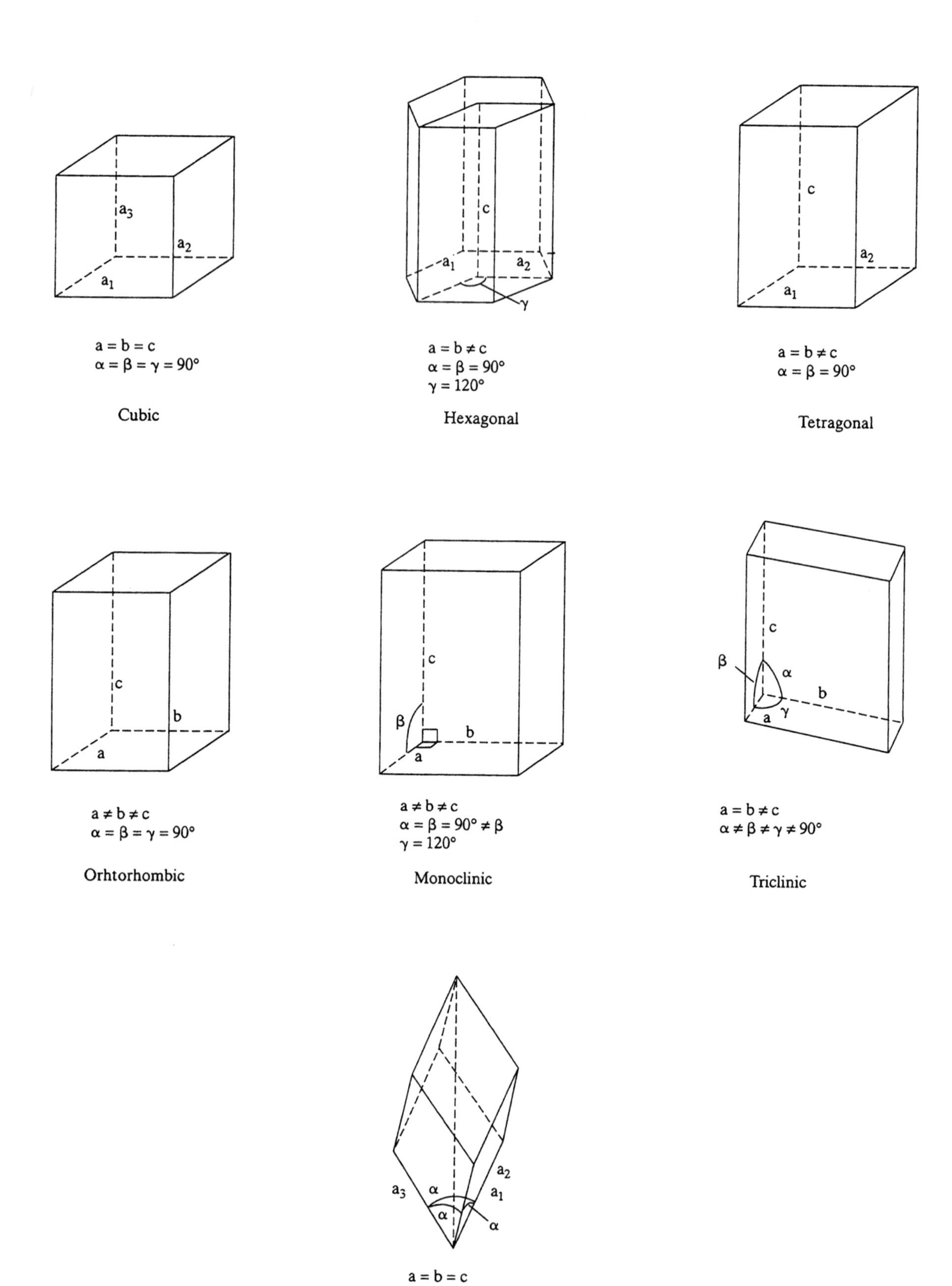

Fig. 1. The seven crystal crystal systems described in Table I.

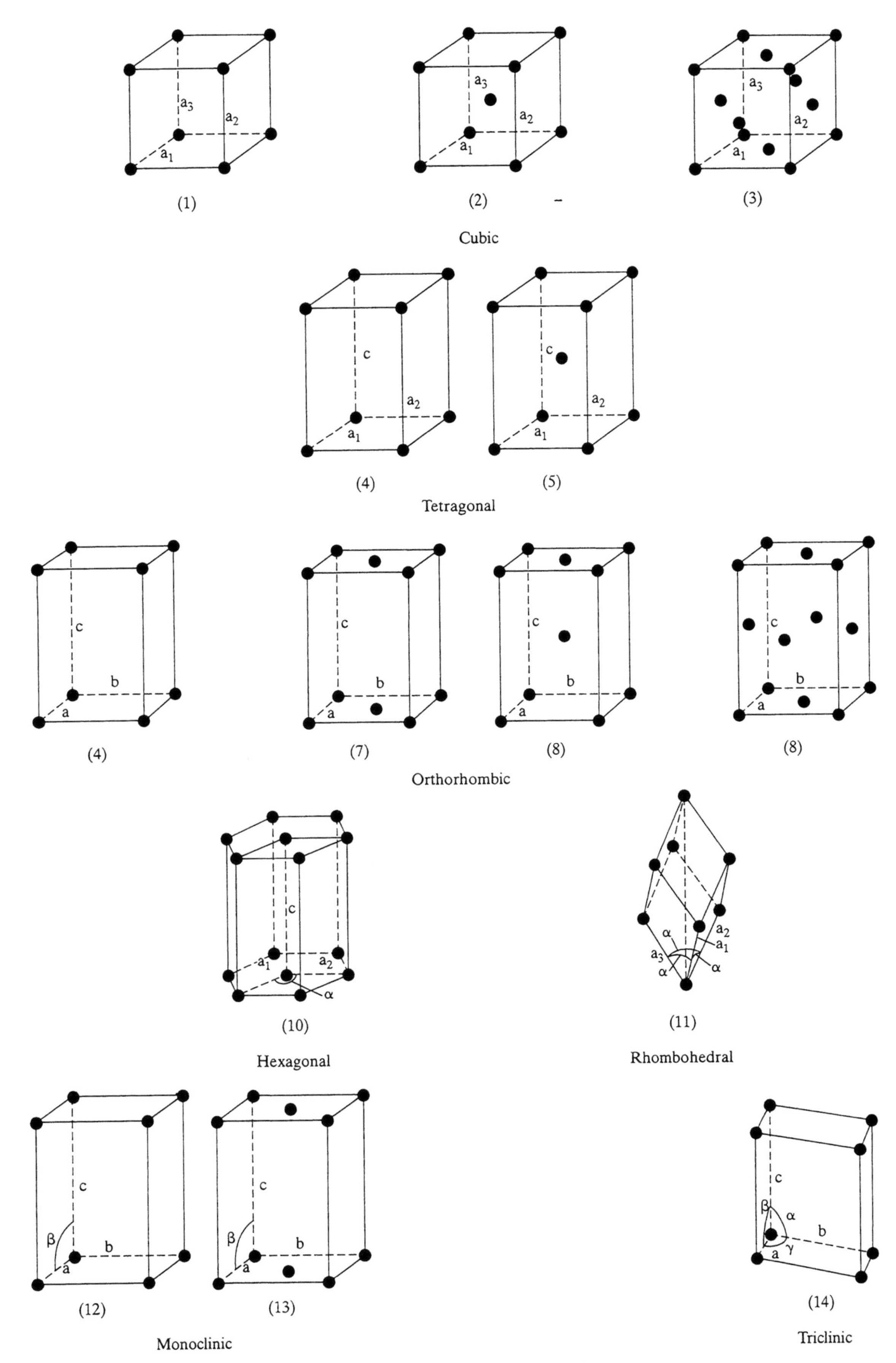

Fig. 2. The 14 Bravais Space Lattices.

Miller Indices and Point Groups

Miller indices are used to designate either directions or a specific plane of solid crystalline materials. Miller indices are given as three different integers, h, k, and l. Examples of use of Miller indices to represent a direction are shown in Fig. 3 for the [100] and [111] directions. Note that brackets are used to indicate a direction whereas a plane is designated using parentheses as shown in Fig. 4.

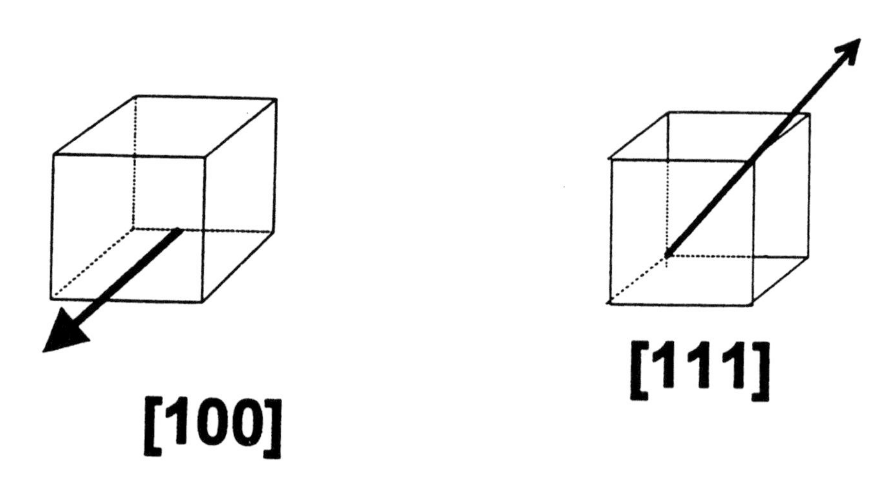

Fig. 3. Examples of Miller Indices to Represent Direction

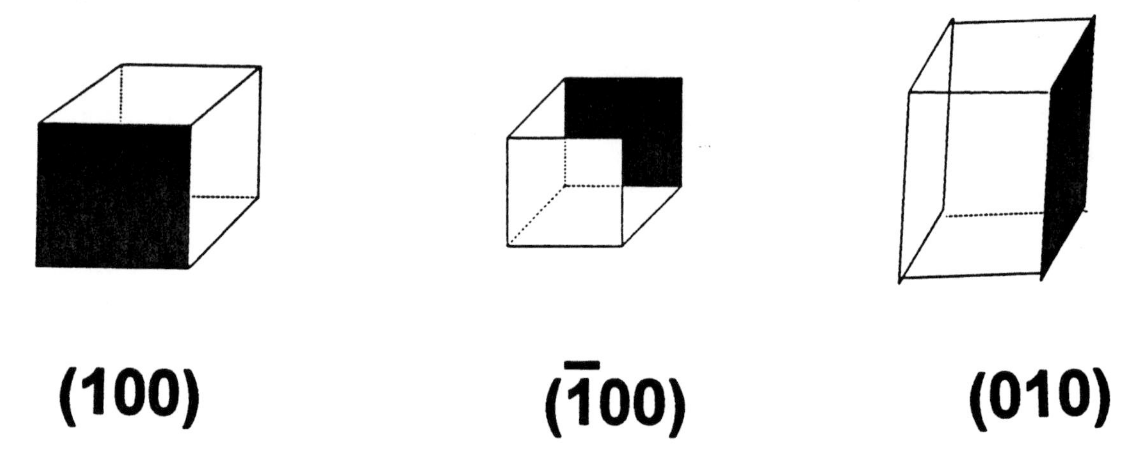

Fig. 4. Examples of Miller Indices to Indicate Crystalline Planes

As stated previously, parentheses are used to identify planes in crystalline materials. In crystallographic studies Hermann-Mauguin symbols are used to represent the symmetry of a particular crystalline system. These Hermann-Mauguin symbols can be arranged according to the crystal system and it turns out that there are 32 different allowable symbols. These are given in Table II along with the corresponding Schoenflies symbols which are more commonly used by spectroscopists to represent the symmetry of different crystalline materials. There are now 32 different point groups rather than 14 because additional symmetry elements such as mirror planes are also needed to describe the symmetry of these groups.

Table II

Hermann-Mauguin and Schoenflies Symbols for the 32 Point Groups or Symmetry Classes

System	Hermann-Mauguin symbol	Schoenflies symbol*
Cubic	23	T
	432 (or 43)	O
	2/m $\bar{3}$ (or m3)	T_h
	$\bar{4}$3m	T_d
	4/m $\bar{3}$ 2/m (or m3m)	O_h
Tetragonal	4	C_4
	$\bar{4}$	S_4
	422 (or 42)	D_4
	4/m	C_{4h}
	4mm	C_{4v}
	$\bar{4}$2m	D_{2d} (or V_d)
	4/m 2/m 2/m (or 4/mmm)	D_{4h}
Orthorhombic	222	D_2 (or V)
	mm2 (or mm)	C_{2v}
	2/m 2/m 2/m (or mmm)	D_{2h} (or V_h)
Hexagonal	6	C_6
	$\bar{6}$	C_{3h}
	$\bar{6}$m2	D_{3h}
	622 (or 62)	D_6
	6/m	C_{6h}
	6mm	C_{6v}

	6/m 2/m 2/m (or 6/mmm)	D_{6h}
Monoclinic	m (or $\bar{2}$)	C_s (or C_{1h})
	2	C_2
	2/m	C_{2h}
Triclinic	1	C_1
	$\bar{1}$	C_1 (or S_2)
Rhombohedral	3	C_3
	$\bar{3}$	C_{3i} (or S_6)
	32	D_3
	3m	C_{3v}
	$\bar{3}$ 2/m (or $\bar{3}$m)	D_{3d}

*Schoenfleis symbols are often used by spectroscopists, but crystallographers now normally employ Hermann-Mauguin notations.

In the Hermann-Mauguin system, the symmetry is described along different directions. For example in the cubic system the first symbol is the symmetry along the [100] direction, the second symbol is the symmetry along the [111] direction and the third symbol is along the [110] direction. In each crystal system, different sets of directions are defined to determine such symmetries.

In the Schoenflies system several questions are asked regarding the symmetry of the crystalline material such as the presence of rotation axes, mirror planes and other symmetry elements and operations. The major point of this discussion is that these two systems describe the symmetry of the crystalline material and, in fact, can be related to each other.

Space Groups

In real materials it turns out that there are actually 230 space groups in which all crystalline materials exist. There are 230 instead of 32 due to the additional symmetry operations of glide planes and screw axes. Examples of glide planes and screw axes are given in Fig. 5.

Fig. 5. Examples for Designation of Glide Planes and Screw Axes

Interplanar Spacings

The relationship in a cubic system between the interplanar distance, d, and the unit cell length, a, and the different Miller indices for different planes, hkl, is the following:

$$\frac{1}{d^2} = \frac{(h^2 + k^2 + l^2)}{a^2}$$

Since other systems have different lengths in their unit cells and different angles between these axes (different than 90° for the cubic system) other relationships between d, hkl and a, b, and c will be more complicated.

In any set of planes, hkl, Bragg showed that X-rays could diffract from different planes only if diffraction occured between planes having a difference of integral numbers of wavelengths. This relationship can readily be obtained from the following diagram:

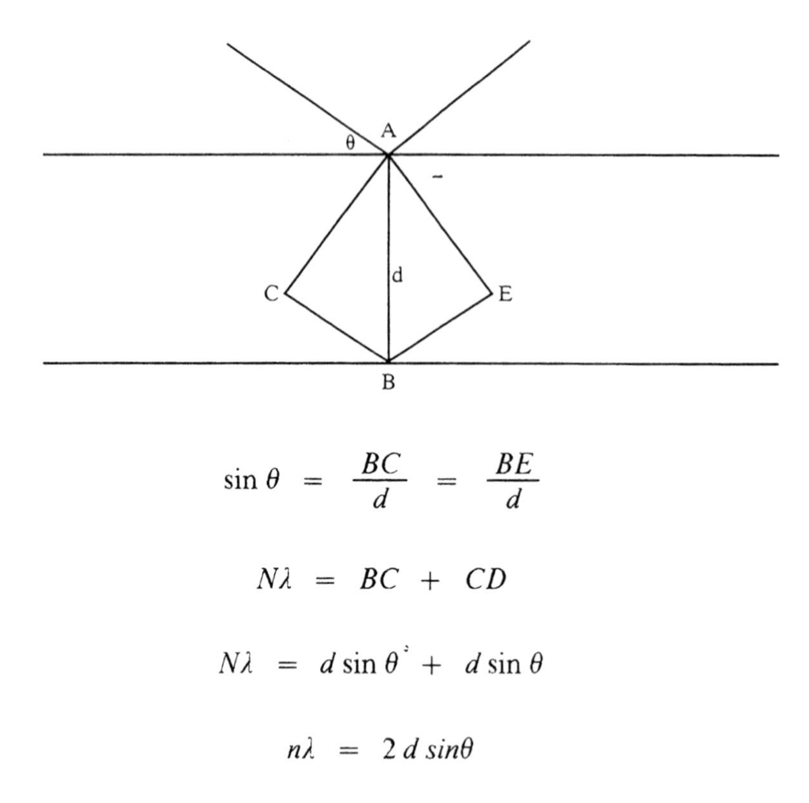

$$\sin\theta = \frac{BC}{d} = \frac{BE}{d}$$

$$N\lambda = BC + CD$$

$$N\lambda = d\sin\theta + d\sin\theta$$

$$n\lambda = 2\,d\,sin\theta$$

Fig. 6. Bragg's Law

In this case wave 1 diffracts at an angle θ and intersects the surface or plane 1 at A. Wave 2 diffracts and intersects plane 2 at point B. By drawing right triangles from point A to wave 2, it is possible to show with trigonometry that the angles between CAB and EAB are both θ. The spacing between planes 1 and 2 is d. With these two triangles it can be shown that:

$$\sin\theta = \frac{CB}{d}$$

and

$$\sin\theta = \frac{BE}{d}$$

Wave 2 travels a distance CB + BE farther than wave 1. Bragg's Law, therefore, states that there must be an integral number of wavelengths for this extra distance:

$$n\lambda = 2\,d\,sin\theta$$

Unit Cells

Unit cells are the simplest representations of the structure of a material. A unit cell must be able to be extended in all directions and form the same exact structure as that of the unit cell. Some very common unit cells of the materials NaCl, CsCl, CaF_2, and TiO_2 are shown in Fig. 7.

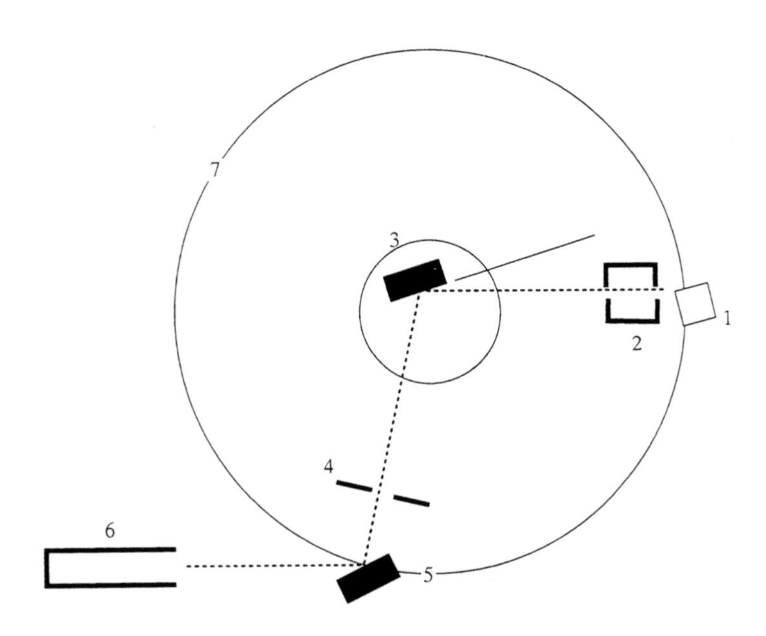

Figure 7. Typical Unit Cells

X-Ray Powder Diffraction

X-Ray powder diffraction methods can be used to study the degree of crystallinity of a material, to determine the basic structure of the material, and to tell whether a material is pure or at times whether it has amorphous impurities. The presence of an amorphous impurity can sometimes be verified by a broad weak background peak centered near 25° 2θ.. A diagram of an X-ray powder diffractometer is given in Fig. 8. X-Rays are produced in the X-ray tube and are collimated onto a sample. In most cases Cu K α radiation is used for studies of powders. The sample is usually moved at angles between

5 degrees 2 θ to 60 degrees 2 θ or larger. A proportional counter is used as a detector and the intensities of the diffraction peaks are recorded on a chart recorder or stored in a computer. Refer to Fig. 8.

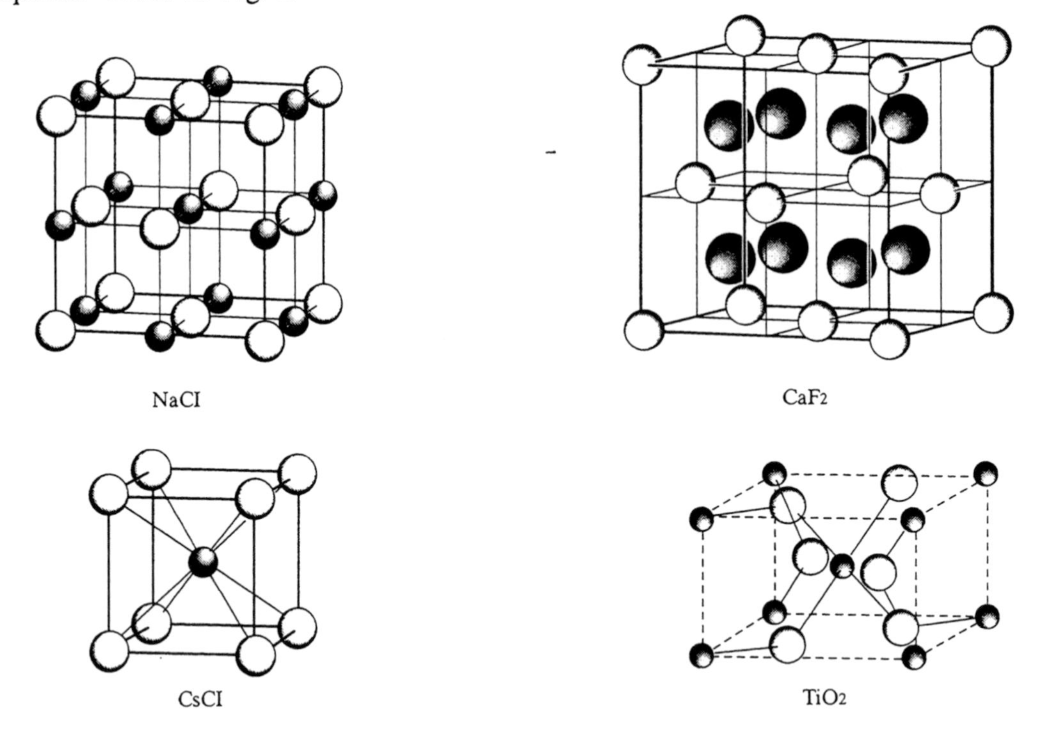

Fig. 8. Diagram of an X-ray Diffractometer. Various Parts: 1 = X-ray Source, 2 = Collimator, 3 = Sample, 4 = Slits, 5 = Monochromator, 6 = Detector, 7 = Focusing Circle.

A random orientation of the powder is needed in order to observe all of the diffraction peaks. Once the 2-theta values are collected they can be converted to theta and by using Bragg's law values of the d spacings can be obtained. If the substance is a common one then the experimental X-ray powder diffraction pattern can be compared to known published patterns such as those found in the ASTM tables. If a powder pattern has never been collected before, analogies to known structural types can be made.

Thin films, wires, fibers and other similar forms of materials can also be studied with the X-ray powder diffraction method. Usually the powder or other sample form is attached to a glass slide by some noncrystalline material such as Vaseline. It is important to not press the sample onto such a slide because a random orientation is necessary in order to observe all of the different planes. During the analysis it is important to compare the peak positions, relative intensities and the general shape of the background signal.

Single Crystal X-Ray Diffraction

With a single crystal the first step is to mount a crystal on a goniometer which is used to rotate the crystal in space with two mutually perpendicular arcs. In addition, the goniometer can be rotated about a spindle axis. In many cases a precession photograph is taken and the crystal is moved in space until the crystallographic axes are aligned with the photographic detector which is behind the X-ray source and the sample. Usually Mo K α radiation is used in single crystal studies.

An example of a precession photograph is given in Fig. 9 on the following page. Note that in this photograph only two of the axes are observed. If the spindel is rotated 90 degrees one of these axes will remain and the third axis previously unobserved will now be seen. Since X-rays cannot be refocused, the pattern observed on the photograph is that of reciprocal space. For a cubic system there is only one unique axis for the cube and the relationship between the reciprocal space axis, a*, and the real axis, a, is a - 1/a*. From similar types of photographs taken at different slices (levels) of the crystal it is possible to assign a space group to the crystal.

Data are then collected on the different diffraction peaks and the intensities are counted in a systematic fashion. The relative intensities of the diffraction reflections are then analyzed with Fourier transform techniques in order to solve for the electron densities of each of the atoms of the structure. Bond lengths and distances are calculated and compared to other materials with similar composition and structure. An example of a refined X-ray single crystal diffraction study is given in Fig. 10 on the following page. The elliptical figures represent the thermal motion of the electrons of the structure.

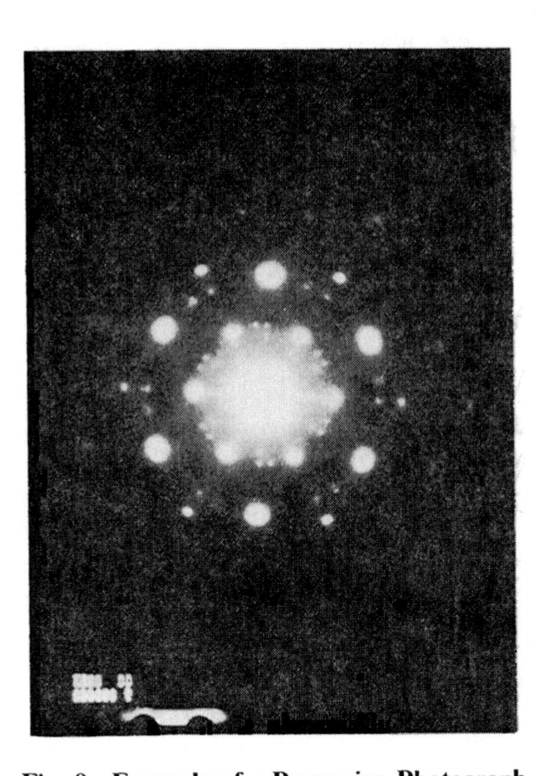

Fig. 9. Example of a Precession Photograph

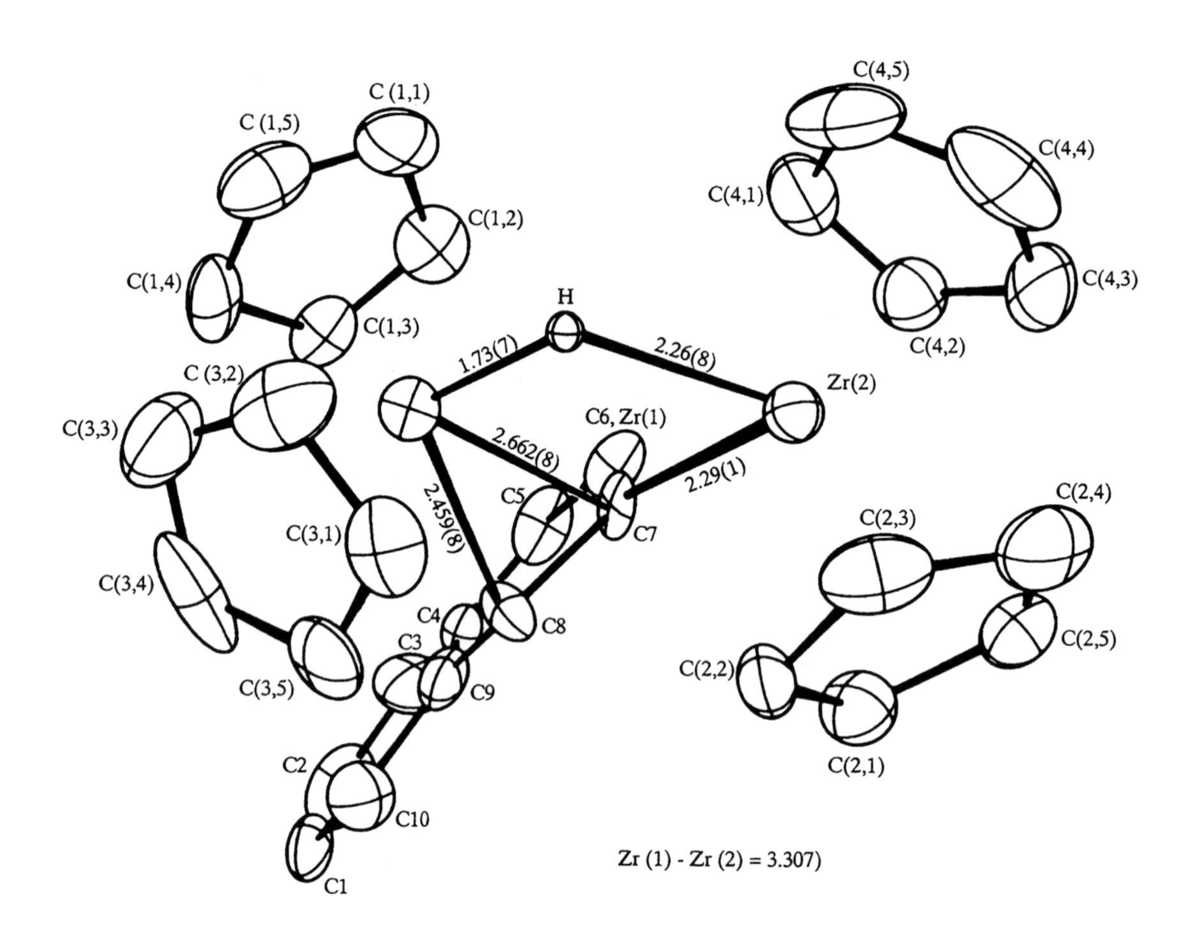

Fig. 10. Structure Obtained from Single Crystal X-Ray Data

Equipment and Prices

An X-ray powder diffraction apparatus that is fully automated can be obtained from several vendors including Philips, Rigaku, Siemens and Nicolet. Typical prices in July 1988 were on the order of $60,000, U.S. dollars. Single crystal diffractometers which these days have automatic determination of space groups are available from Enraf-Nonius, Siemens and Rigaku for prices starting at $200,000.

Practical Applications

X-Ray methods can be used to determine the identity of an unknown species. An X-ray powder diffraction pattern for SiC fibers that have been treated in a mixture of $SiCl_4$ and H_2 is shown in Fig. 11. Analysis of the position of these peaks show that Si has been formed from this chemical vapor deposition. These data suggest that the deposition process is relatively clean and that large amounts of amorphous material are not present. These materials are useful as composites for their potential use at high temperatures in engines.

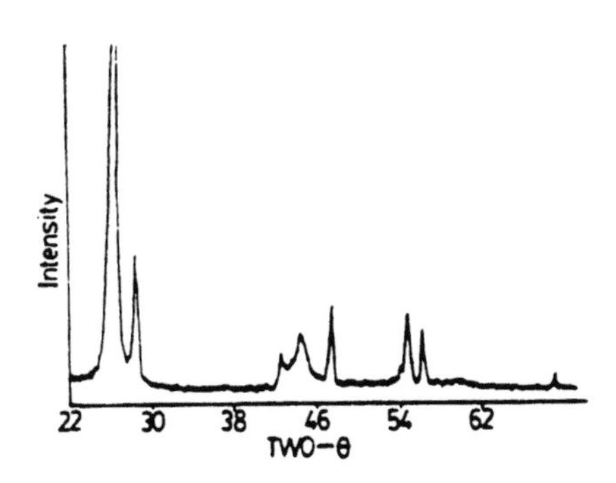

Fig. 11. Silicon Deposited on SiC Fibers

Another example is shown in Fig. 12. The X-ray powder diffraction pattern for a rare earth exchanged zeolite system is shown in curve a. The material is crystalline and of high symmetry. After deposition of vanadium naphthenate and calcination of this material the X-ray diffraction pattern of curve b results. These data suggest that vanadium is detrimental to the zeolite catalyst. On the other hand, if tetraphenyl tin is added to the zeolite catalyst prior to vanadium deposition then the crystallinity of the zeolite is retained. These data suggest that tin can be used to trap the harmful vanadium species and prevent them from destroying the crystallinity of the zeolite. Such materials are used to crack petroleum into gasoline range hydrocarbons.

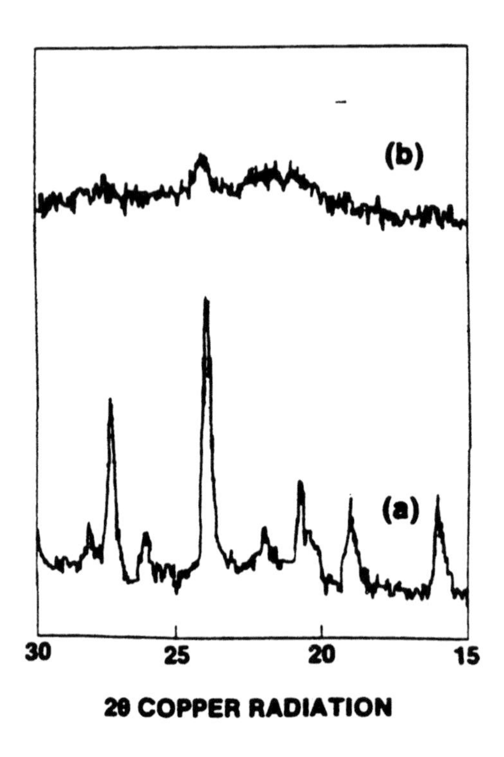

Fig. 12. Rare Earth Exchanged Zeolite Systems

Helpful Background Literature

1. Stout, G. H.; Jensen, L. H. "X-ray Structural Determination", MacMillan Publishing Co., Inc.: New York, New York, 1968.

2. Glusker, J. P.; Trueblood, K. N. "Crystal Structure Analysis", Oxford University Press: New York, New York, 1972.

3. Bragg, W. H.; Bragg, W. L. "X-rays and Crystal Structure", G. Bell and Sons Ltd.: London, 1925.

4. Ebsworth, E. V. A.; Rankin, D. W. H.; Cradock, S. "Structural Methods in Inorganic Chemistry", CRC Press, Boca Raton, 1991, pp. 331 – 377.

ELECTRON PARAMAGNETIC RESONANCE

Introduction

Electron paramagnetic resonance (EPR) is used to study the nature of unpaired electrons in material. Radicals and triplet state species are two of the systems that are readily observed. Since transition metal complexes often exist in several different oxidation states, chances are that a paramagnetic species for most elements will exist. There are several factors that govern whether a material will be EPR active or inactive and some of these will be discussed below.

EPR absorptions occur in the microwave region of the spectrum. The sample is put in a magnetic field on the order of typically 0.34 T which corresponds to an X-band frequency of about 9.5 GHz. Several other frequencies can be used, especially the most common additional Q band frequency of 35 GHz.

Transitions

An equation that can be used to describe an EPR absorption is given below:

$$E = h\nu = gBHM_s \qquad [1]$$

where E is the energy of the microwave absorption, h is Planck's constant, ν is the frequency, g is a proportionality factor, B is the value of the Bohr magneton for a free electron, H is the value of the magnetic field where the transition occurs, and M_s is the electron spin quantum number.

In an EPR experiment, generally a fixed frequency is used and the magnetic field is scanned. The Bohr magneton for a free electron is a constant. The value of the magnetic field (H) is measured from a spectrum. Since h is also a constant, then the g factor can be calculated. The g factor value is one of the most important parameters in an EPR experiment and is specifically related to the number of unpaired electrons and where they reside in a molecule. If for example the unpaired electron is associated with a transition element such as Ti^{3+}, a d^1 system, the specific g value for this system will depend on the local symmetry around the Ti^{3+} ion and the extent of delocalization of the unpaired electron. In other words, the g value may relay information about covalency in a molecule.

If the unpaired electron is centered on an ion that has cubic symmetry, then a single isotropic transition is observed. On the other hand, if the ion is in a low symmetry state such as the case when all axes of a crystal are different (i.e., orthorhombic and lower symmetry systems), then there will be three different g values. An energy level diagram for an isotropic system is given below.

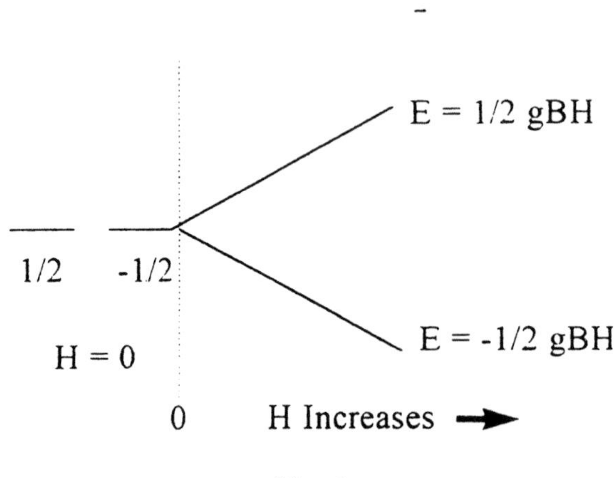

Fig. 1.

In many cases the unpaired electron is shared with neighboring ligand atoms or other atoms such as another metal ion in a dinuclear metal complex. If the other nearby species are high enough in abundance, if they have a nuclear quantum number M_I, and if the isotopic abundance is high, then there can be an interaction between the unpaired electron spin described by M_S and the nuclear spin, described by M_I which gives rise to a hyperfine coupling (a) as described in the following term:

$$a\, M_S M_I \qquad [2]$$

The hyperfine coupling constant will be isotropic for cubic systems and will be split into a variety of terms for lower symmetry complexes as described above for the g factor. The number of lines can generally be determined from the formula

$$2nI + 1 \qquad [3]$$

where n is the number of equivalent nuclei and I is the nuclear spin quantum number. Some nuclear spin quantum numbers are I = 1/2 for H, I = 3/2 for ^{63}Cu and ^{65}Cu, I = 7/2 for V, I = 1/2 for ^{13}C, I = 3/2 for ^{35}Cl, I = 5/2 for Mn, and I = 7/2 for Co. Equation 3 is used for each different type of coupling to determine the total number of peaks in a spectrum. The relative intensities of the peaks can be determined for spin I = 1/2 systems in the same way Pascal's triangle is used in NMR spectroscopy.

Samples can be solids, liquids or gases for EPR experiments. However, there are several caveats. In solution, if there are too many paramagnetic ions, then saturation can result and the signal is broadened so much that it is not observed. The same holds for a solid that has too many paramagnetic centers. Since O_2 is paramagnetic, it can interfere or broaden a signal so it is important to seal off samples under vacuum and use freeze-pump-thaw cycles to remove oxygen. Very often low temperatures are needed to observe EPR spectra. In the case of liquid samples, frozen solutions which typically are prepared from mixtures of solvents are used.

EPR signals are often weak so the spectral data are usually differentiated. Samples are loaded into cells that are composed of high purity quartz in order to avoid paramagnetic contaminants and adsorption of microwaves. A diagram of the basic components of an EPR spectrometer is given below.

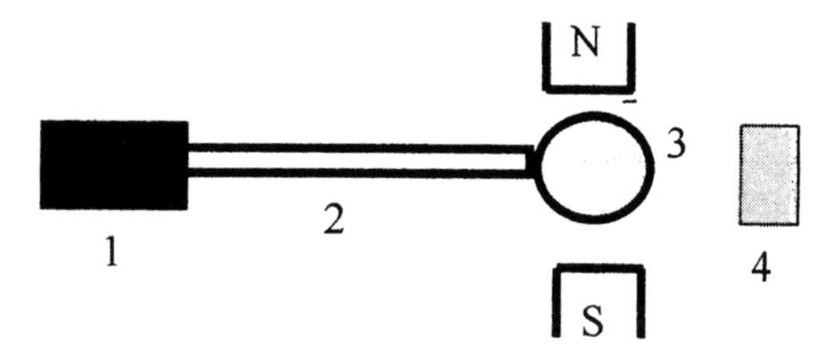

Components of an EPR Spectrometer: 1 = power supply, 2 = Klystron, 3 = cavity, 4 = detector.

Fig. 2.

The sample is placed in the resonant cavity and tuned. The microwave radiation is delivered by the Klystron into the cavity, and a signal is detected as the magnetic field is varied. The magnetic field must be uniform. The spectrum is recorded, differentiated, and then analyzed.

Analysis of transition metal systems can be complicated by a number of factors. Anisotropy can lead to a variety of features in a spectrum. The magnitude of the hyperfine coupling can lead to either resolved or deconvoluted peaks. The number of d electrons is very important with an odd number of electrons being more readily observed especially at room temperature than even electron (unpaired) systems say for example Ni^{2+} systems. The presence of weak or strong field ligands will influence the number of unpaired electrons in a system as will the crystal field. Jahn-Teller effects will lead to a descent in symmetry very often with more peaks than would be observed in an isotropic system. For triplet systems, electron-electron interactions are important and can give rise to splitting of the zero level even in the absence of a magnetic field, a so-called zero field splitting. Quadrupolar interactions can also complicate spectra.

In summary, EPR spectroscopy can be used to study paramagnetic systems, to determine the number of equivalent nuclei that are coupled to the unpaired electron, to determine the degree of covalency in the molecule, and to establish what types of nuclei are in the vicinity of the unpaired electron. There are various excellent sources that describe in great detail the theory and applications of EPR (1-3).

Questions

1. What compounds described in this book can be studied with EPR?

2. For liquid solutions that are saturated due to too many paramagnetic centers, what can be done? What about solid samples?

3. It is sometimes said that EPR and NMR are mutually exclusive. Comment on this statement.

Helpful Background Literature

1. Drago, R. S. "Physical Methods for Chemists"; 2nd ed., Saunders: Fort Worth, 1992, Chapters 9 and 13, 360 – 408, 559 – 603.

2. Ebsworth, E. A. V.; Rankin, D. W. H.; Cradock, S. "Structural Methods in Inorganic Chemistry"; CRC Press: Boca Raton, 1991, Chapter 3, 115 – 141.

3. Wertz, J. E.; Bolton, J. R. "Electron Spin Resonance, Elementary Theory and Practical Applications"; McGraw-Hill: NY, 1972.

COMPUTATIONAL CHEMISTRY

Theoretical Chemistry As Distinct From Computational Chemistry

Where computational chemistry presents a practical face to the chemist, theoretical chemistry appears abstruse and unintelligible to most. Nevertheless, an understanding of theoretical chemistry is essential for using the modern computational tools, so that the boundary of the domain of applicability of any given computational tool can be known beforehand. The term Theoretical Chemistry embodies two broad areas of consideration: quantum mechanics, (the study of the motion i.e., mechanics or ″behavior“ of atomic size elementary particles) and statistical mechanics, or the statistical study of large numbers of quantum mechanical particles.

In quantum mechanics (under normal conditions) the particles under study are electrons, moving in the (spatially) fixed environment of positive charges created by nuclei whose geometry (usually) defines what is meant by the shape or structure of a molecule or atom assembly. Statistical mechanics (and thermodynamics) is that subject which describes the macroscopic properties which emerge from the quantum mechanical behavior of atomic/molecular systems. Thus, a clear distinction between studying a molecule, say water, and the bulk substance (one mole of liquid water), exists. In studying the isolated molecule, one can study its electronic energy levels and associated wave functions for an initial nuclear geometry, (which reflects either the ground state or the excited state geometry of the nuclei) searching in this one electronic state for that special geometry of the molecule whose total energy, electronic energy plus nuclear repulsion energy, is minimum. This sought-after geometry, the nuclear geometry of minimum total molecular energy, is normally called the structure of the molecule or of the excited state.

Continuing, one can study the change in energy as one shifts nuclei about, maintaining the electrons still in a single electronic state. From such a study, the vibrational structure of the molecule should emerge, as well as details of the dissociation of the molecule into that set of substances whose isolated electronic states conform to that required for the original molecule's electronic state (which was being maintained invariant). Further, one can study two electronic states, and the difference in energy between them, which is the precursor to studying visible and ultraviolet spectroscopy.

---

*Written by Prof. Jeffrey Bocarsly and Prof. Carl David of the University of Connecticut.

What is Computational Chemistry?

Computational chemistry is the term used to describe that area of chemistry which uses mathematical models to describe molecular and atomic interactions of all types. The goal of producing an accurate mathematical description of a molecule is to be able to predict all of its properties, including its physical properties (from macroscopic properties such as its melting point to microscopic properties such as molecular structure or spectroscopy) and its chemical properties (such as its site of reaction, its rate of reaction or the structure of its transition state). The ultimate goal of this type of work is to accurately determine through computation all of the desired knowledge about a molecule prior to entering the laboratory for an experimental verification of the computation.

What Problems Are Computational Techniques Currently Applied To?

Almost all current chemical problems are being subjected to computational studies. These include studies which are used to help design new drugs, or those which address how catalysts and enzymes work or how proteins acquire the correct molecular structure after they are synthesized in cells. Computational studies have also been used to understand such diverse areas as how molecules transfer electrons between themselves and the atmospheric chemistry behind the changes in the ozone layer. As the availability of relatively inexpensive computing power has increased (a development of the last ten years), chemists have begun to examine an immensely wide variety of chemical problems from a computational perspective. It is likely that computational chemistry will continue to grow in importance as it becomes a useful tool for all areas of chemistry.

Different Types of Computational Methods for Different Molecular Systems

While there are a variety of ways of categorizing computational methods, generally computational techniques fall into one of three groups: *ab initio,* semiempirical, and empirical. *Ab initio* methods attempt to calculate molecular properties and structure first quantum mechanical principles. That is, they try to create an accurate physical model for the electron density of a molecule, leading to its spectroscopy, structure and reactivity. As such, *ab initio* quantum mechanical computations are the most desirable. These computations in principle contain nothing but atomic constants, i.e., the charges of the electron and proton, the total charges of atoms or molecules, Planck's constant, the masses of the electron and various nuclei, and the velocity of light. The identity of the chemical moiety under study is controlled solely by assigning atomic numbers (hence charges) to the nuclei, and placing them in space (x, y, and z coordinates). It should always be remembered that such computations in their purest form are not possible given today s theoretical understanding, and some assumptions beyond the fundamental physical description of the atom or molecule are usually required (a). This method has the advantage of providing a molecular picture which emerges from the fundamental

(a) A major error in *ab initio* computations is usually in the area of electron-electron correlation energy. This refers to the ways in which electrons in orbitals interact with each other, and the effects this interaction has on the electron distribution in the orbital. The larger the computation, the more likely that a greater fraction of the correlation energy has been accounted for. But, the total correlation energy is never completely obtained, so there is always some error in *ab initio* work.

physics of molecules, however it is computationally expensive. That is, a tremendous amount of computation is required to build up an *ab initio* molecular picture, and currently only small to moderately sized molecules can be treated by this method.

In order to cope with the computational load less rigorous quantum mechanical techniques have been developed. These methods are termed semiempirical because they attempt to simplify quantum mechanics by either ignoring or approximating certain terms (usually integrals) in the *ab initio* computations with empirically determined parameters, thereby rendering such computations tractable, and therefore practical. It is this field which, in large measure, has made possible the flourishing of computational chemistry.

The final type of computation is termed empirical, because it makes no attempt to derive molecular properties from first physical principles. Fully empirical computations discard quantum mechanics completely, and ascribe a classical potential energy model to govern the motion of nuclei. The interactions between atoms in a molecule are modeled by using a sum of relatively simple algebraic expressions from classical mechanics to represent the energy of the physical structure of a molecule. Each of the functions in the sum has constants which are empirically determined, giving this method its name. The goal of this type of modeling is to calculate the energy of different conformations of a molecule, to find which conformation represents an energy minimum. This low energy structure should correspond to the ground state structure of the molecule. The advantage of the method is that it can provide highly accurate structural predictions, even for large molecules, for only a moderate computational price.

How Do These Computational Models Work?

The details of chemical computations can be quite complex, however some simple examples can provide a taste of how computational methods can be used to study molecules. We will examine two computational examples, a semiempirical computation and an empirical calculation, in order to indicate how quantum mechanically-based computations differ from empirical calculations.

Semi-empirical calculations

For our semiempirical calculation, we shall examine a computation named after its developer, E. Hückel , known as the simple Hückel method. The Hückel method was developed to study organic molecules which have π bonds, and one of its basic assumptions is that the π electrons can be mathematically treated while ignoring the σ electrons. We will apply the Hückel method to one of the simplest π systems, the neutral allyl radical, $C_3H_5\bullet$ [1]. The π system of $C_3H_5\bullet$ is made up of the $2p_z$ orbitals on each carbon in the molecule (b).

(b) The σ electrons, which are not treated in the calculation, are assumed to be found in appropriately hybridized carbon orbitals. In the allyl radical case, these would be sp^2 orbitals.

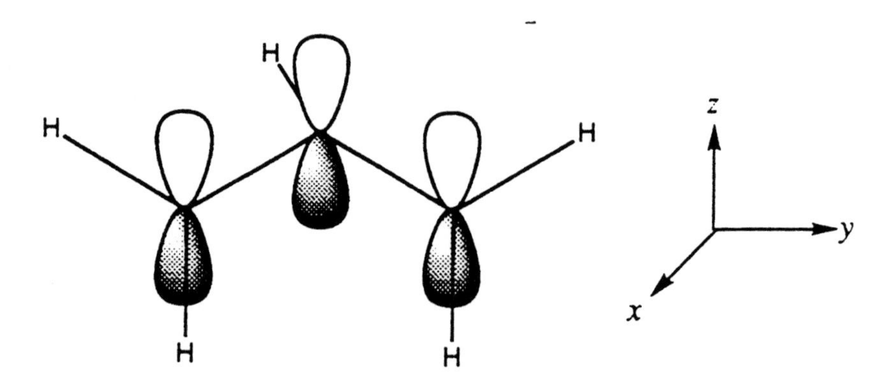

Fig. 1.

The Hückel method utilizes two quantities, α and β in its calculation. α is the average energy of an electron in a $2p_z$ atomic orbital on a carbon. Now, this electron can feel the charge of its own nucleus to a large extent, and the charge of more distant nuclei in the molecule (carbon nuclei) to lesser extents. β, on the other hand, is the energy of an electron which participates in π bonding between carbons, so it experiences the charges of both of the nuclei of the atoms involved in the bond to a significant extent. The electronic energy of the system, E, can be calculated from the following matrix determinant:

$$\begin{bmatrix} \alpha - E & \beta & 0 \\ \beta & \alpha - E & \beta \\ 0 & \beta & \alpha - E \end{bmatrix} = 0 \qquad [1]$$

This Hückel determinant (1) can be evaluated by dividing each element by β, and then by setting $x = \dfrac{\alpha - E}{\beta}$:

$$\begin{bmatrix} \alpha - E & \beta & 0 \\ \beta & \alpha - E & \beta \\ 0 & \beta & \alpha - E \end{bmatrix} = \begin{bmatrix} x & 1 & 0 \\ 1 & x & 1 \\ 0 & 1 & x \end{bmatrix} = 0 \qquad [2]$$

The solution to this determinantal equation can be found by applying the definition of the determinant of a 3 x 3 matrix:

$$\begin{bmatrix} a_{11}a_{21}a_{31} \\ a_{12}a_{22}a_{32} \\ a_{13}a_{23}a_{33} \end{bmatrix} = (a_{11}a_{22}a_{33}) + (a_{21}a_{32}a_{13}) + (a_{12}a_{23}a_{31}) - (a_{31}a_{22}a_{13}) - (a_{12}a_{21}a_{33}) - (a_{23}a_{32}a_{11})$$

In our case, substitution of the Hückel matrix elements in the expression for the determinant gives the algebraic expression:

$$x^3 - 2x = 0 \qquad [3]$$

which can be factored to give $x(x^2 - 2) = 0$. The roots are therefore $x = 0$ and $x = \pm\sqrt{2}$. Substituting for x gives $(\alpha - E)/\beta = 0, \pm\sqrt{2}$. Solving for E leads to $E = \alpha$ and $E = \alpha \pm \beta\sqrt{2}$. These three values for E are the energy levels for the three Hückel molecular orbitals of $C_3H_5\bullet$ which are constructed from the three carbon 2pz atomic orbitals of the radical. The calculated energy levels can be presented graphically:

	________	$\alpha + \sqrt{2}\,\beta \sim +25$ kcal mol^{-1}
E	________	$\alpha \sim 0$ kcal mol^{-1}
	________	$\alpha - \sqrt{2}\,\beta \sim 25$ kcal mol^{-1}

The energy values shown are calculated choosing $\alpha = 0$ kcal/mol (for convenience) and $\beta \sim -18$ kcal/mol. Since the Hückel method is semiempirical, the β value is estimated from experimental data. In a full quantum mechanical *(ab initio)* treatment, α and β would be determined through calculation based on theory. The values shown represent the calculated *electronic* energies only; to calculate the total energy of the system, nuclear repulsion energies must be added in.

The type of computation outlined above can be used to investigate molecular properties beyond the energy levels of molecular orbitals. For example, in the discussion above the values of α and β (which represent the degree to which electrons "feel" the molecule's nuclei) are sensitive to the positions of those nuclei. Thus, by choosing various molecular conformations and the corresponding α and β values, the energies of the orbitals can be found for the different conformations. The conformation with the lowest energy should represent the ground state structure of the molecule. Thus, quantum mechanical computations can be combined in an iterative approach to calculating molecular structure.

Empirical Calculations

This brings us to our second example, an empirical calculation. Empirical calculations sacrifice the theoretical rigor of *ab initio* and (to a lesser extent) semiempirical computations for speed and ability to handle larger molecules. One such type of computation, called a molecular mechanics computation, models interactions between atoms in a molecule using the equations of classical mechanics [2, 3]. For example, the stretching of bonds between atoms are represented by a Hooke's law expression:

$$V_{str} = \Sigma \frac{1}{2} k_{str}(r - r_o)^2 \qquad [4]$$

where V_{str} is the potential energy as a function of bond length, k_{str} is the force constant for the bond, r is the distance between the two atoms involved in the bond and r_o is the equilibrium bond length. The potential energy of bond angle deformation is described by a similar law:

$$V_{bend} = \Sigma \frac{1}{2} k_{bend}(\theta - \theta_o)^2 \qquad [5]$$

In this expression, the bond angle deformation force constant is k_{bend}, the bond angle is θ and the equilibrium bond angle is θ_o.

Rotations about bonds (called dihedral angle rotations) are represented by potential functions of the form:

$$V_{dihedr} = \Sigma k_{dihedr}[1 + \cos(\underline{n}\phi - \gamma)] \quad [6]$$

where k_{dihedr} is a force constant, ϕ is the dihedral angle, and γ is the equilibrium value for a dihedral angle. This equation models all hindered bond rotations, whether they are due to steric hindrance or multiple bonding between atoms. Steric interactions between nonbonded atoms (Van der Waals interactions) in the molecule are modeled by a Lennard-Jones 6,12 potential, which has the form:

$$V_{vdw} = \sum 4\varepsilon_{vdw}\left[\left(\frac{\sigma}{r}\right)^{12} - \left(\frac{\sigma}{r}\right)^{6}\right] \quad [7]$$

Variable r is defined as above, σ is the sum of the two radii, and ε_{vdw} is the depth of the potential well between nonbonded atoms. Finally, electrostatic interactions are included using a Coulomb's law expression:

$$V_{electrostatic} = \frac{q_i q_j}{\varepsilon r} \quad [8]$$

where q_i and q_j are the (partial) charges on the atoms involved, ε is the dielectric constant, and r is the distance between charged groups. Now, each of the potential expressions described above equations [4] – [8] must be summed for each of the relevant pairs or groups of atoms in the molecule:

$$V_{tot} = \sum V_{str} + \sum V_{bend} + \sum V_{dihedr} + \sum V_{vdw} + \sum V_{electrostatic}$$

$$= \sum \frac{1}{2} k_{str}(r - r_o)^2 + \sum \frac{1}{2} k_{bend}(\theta - \theta_o)^2 + \sum k_{dihedr}[1 + \cos(n\phi - \gamma)]$$

$$+ \sum 4\varepsilon_{vdw}\left[\left(\frac{\sigma}{r}\right)^{12} - \left(\frac{\sigma}{r}\right)^{6}\right] + \sum \frac{q_i q_j}{\varepsilon r} \quad [9]$$

This expression for V_{tot} is called a force field, and the constants for each of the terms in the equation (k_{str}, k_{bend}, k_{dihedr}, ε_{vdw}, σ, r_o, θ_o, γ, and ε) are called the force field parameter set. The computational strategy is to use sophisticated search algorithms to find a minimum in the total potential energy for the molecule, V_{tot}. When the values of all of the bond distances, angles, Van der Waals interactions, etc., that lead to a minimum energy are determined, then the resulting molecular structure should be the ground state structure of the molecule.

In order to make the use of molecular mechanics less opaque, let's examine a simple application. Let's consider a hypothetical nonlinear molecule containing three atoms, where the terminal atoms are identical, and the central atom is unique:

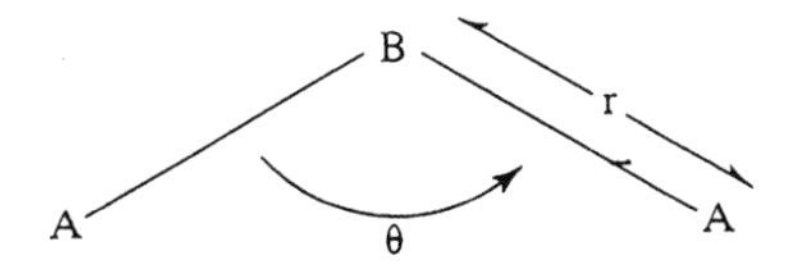

Fig. 2. Nonlinear Three Atom Molecule

Let's construct a simple force field using just bond stretching, bond angle deformation and van der Waals interactions:

$$V_{tot} = \sum V_{str} + \sum V_{bend} + \sum V_{vdw} = \sum \frac{1}{2} k_{str}(r - r_o)^2 + \sum \frac{1}{2} k_{bend}(\theta - \theta_o)^2$$

$$+ \sum 4\varepsilon_{vdw}\left[\left(\frac{\sigma}{r}\right)^{12} - \left(\frac{\sigma}{r}\right)^{6}\right] \qquad [10]$$

In the case of our hypothetical molecule, there are two identical bond stretches (two A – B stretches), one bond angle deformation term (A – B – A) and one van der Waals term, for the interaction of the nonbonded A atoms with each other. Our detailed force field for the A_2B molecule thus becomes the following sum:

$$V_{tot} = \sum V_{str}^{A-B} + V_{str}^{A-B} + V_{bend}^{A-B-A} + V_{vdw}^{AA} \qquad [11]$$

The parameters we will use are as follows (c).

Table I

Parameters used for molecular mechanics minimization of hypothetical A-B-A (d)

parameter	value
k_{str}	553 mdyn/Å
k_{bend}	100 kcal/mol
ε_{vdw}	0.02 kcal/mol
σ	0.6 Å
r_o	0.96 Å
θ_o	104.5°

To gain a visual appreciation of what the potential looks like, let's look at the individual potentials in the force field. In Fig. 3, you can see that the bond stretch potential equation [4] has the algebraic form of a parabola, as does the bond angle deformation equation [5]

(c) H-O-H parameters from the AMBER* parameter set modified for use with MacroModel.
(d) The parameter σ is the sum of the radii of both "A" atoms, where $\sigma_A = 0.3$Å.

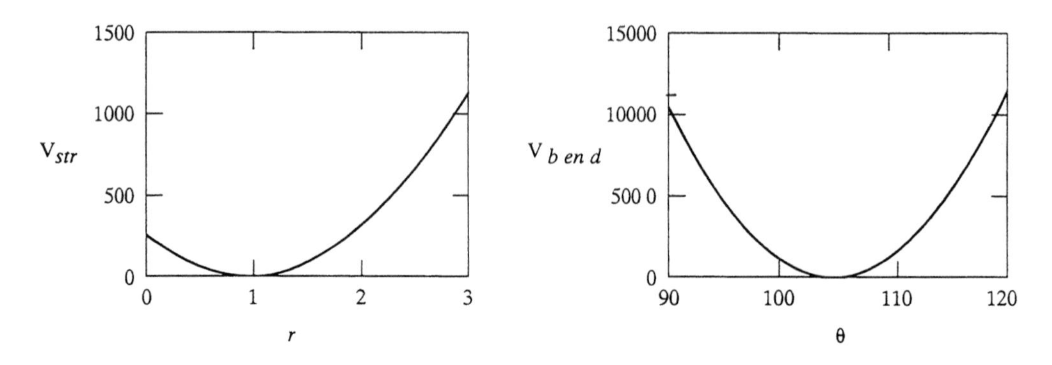

Fig. 3. Graphs of V_{str}(r), equation [4] and $V_{bend}(\theta)$, equation [5].

The van der Waals potential, equation [7], has a different shape; when two atoms get close together, the energy rises sharply as interatomic repulsion grows; at the equilibrium distance (~ 0.65Å), the atoms find themselves in a mildly attractive energy well, and as the atoms separate out to infinity (i.e., no interaction), the energy of interaction tails off to 0 (i.e., no interaction occurs between the atoms) (Fig. 4):

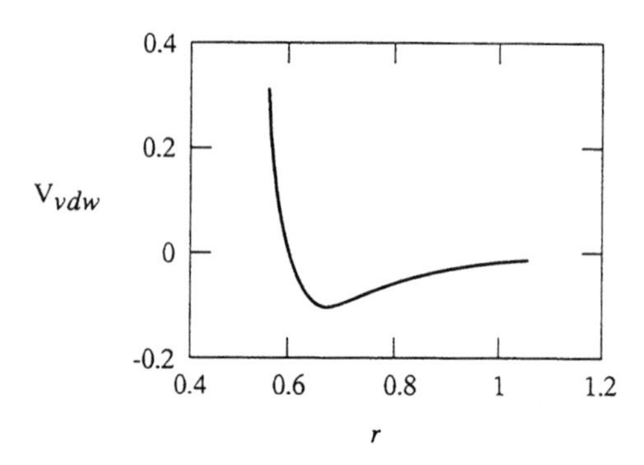

Fig. 4. V_{vdw}(r) equation [7] for A...A interaction in A_2B.

A particularly helpful way of looking at a force field is as a three-dimensional graph. The force field appears as an n-dimensional surface, which is termed a potential surface, since it describes the potential energy of the molecule as a function of its geometry. Now, in our potential function V_{tot} [11], we have three independent, and one dependent, variables: one θ value (the A-B-A angle), two r values (one for each A-B bond) and an additional r value (for the A...A Van der Waals interaction, which is dependent on the θ and r values). To view each of these independently with the total potential would require a five-dimensional graph (one dimension each for the potential function and its four variables), which is not possible. So, let's make two assumptions: first, that both of the A – B bond lengths are always equal, and second, that we can collect all the terms in the potential that depend on interatomic distances (the two bond stretches and the van der Waals interaction) together and view them as a single variable, so that we have a total of three dimensions. The collected terms,

$$\sum V_{str} + \sum V_{vdw} = V_{str}^{A-B} + V_{str}^{A-B} + V_{vdw}^{AA}$$

from equations [4] and [7] appear in Fig. 5.

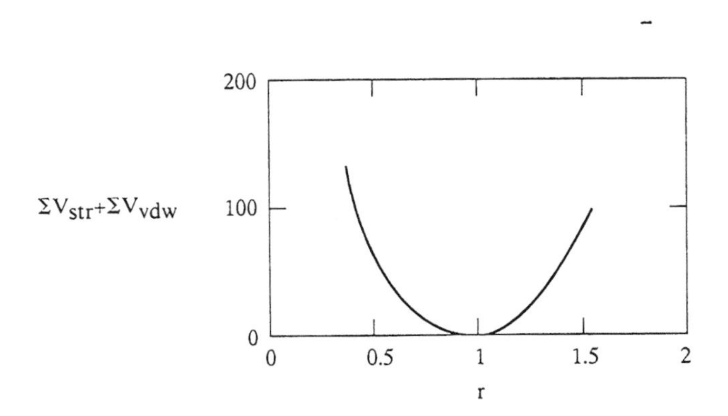

Fig. 5. Combined stretching and Van der Waals potentials [4] and [7].

We are now ready to view our total potential surface for V_{tot} [11]. As shown in Fig. 6, we place our bond angle deformation variable (θ) along the x axis of our graph; along the y axis is the collected atomic distance variable (r), and on the z axis is $V_{tot}(r, \theta)$. In looking at Fig. 6, note that the surface in the r direction roughly has the shape of the combined potential (Fig. 5) while in the θ direction, the shape parallels the angle deformation potential (Fig. 4). We can visually search for the minimum of energy in the surface, and the r and θ parameters which correspond to this minimum should give us the ground state structure of the molecule. A structure based on parameters from this minimum is said to be a molecular mechanics minimized structure. In our case, the minimized r value is $r_{min} \sim 1.05$ Å, while the minimized θ value is $\theta_{min} = 104.5^o$. If you compare these values to our original force field parameters, you can see that θ has not changed, but that r has changed, as the minimization procedure has balanced the competing influences of [4] and [7]

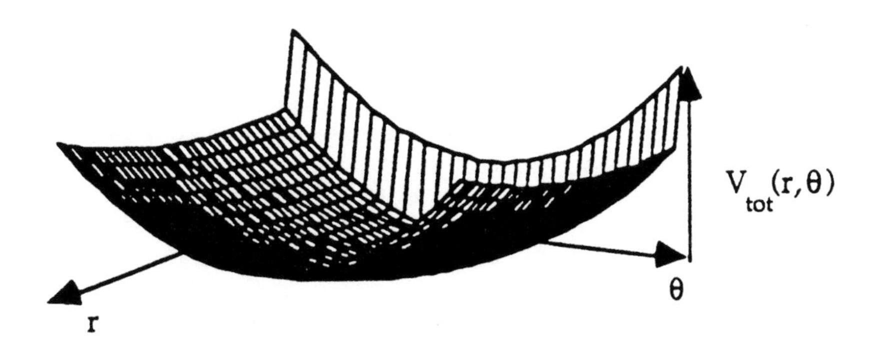

Fig. 6. Potential surface for hypothetical molecule A-B-A [11] (e)

(e) Note that we have simplified our surface by using only one r dimension. Technically, there should be three independent r axes (one for each B-A bond length and one for the A – A distance) in addition to the θ and $V_{tot}(r, \theta)$, however this would require a five-dimensional plot. We have omitted these extra dimensions by collapsing all three r variables into one, for visual simplicity.

Convergence. In the calculation described above, the ground state structure of the molecule is determined when the minimum of the potential surface is located. Locating the minimum of the potential surface is computationally intensive, since a method of searching for this minimum can involve sampling thousands and thousands of the points on the potential surface. Usually, such searches involve calculation of the gradient of the surface, which is a mathematical description of the slope of the surface at a given point. When the slope of the surface is steep, the gradient is large. This typically occurs at points midway between maxima and minima of the surface. At points of high gradient, the calculation descends along the direction of steepest slope to try to find the valley (i.e., the minimum). At the bottom of the valley, the gradient should be quite low, near zero. The calculation therefore is performed in steps, quite analogous to hiking down the side of a mountain into a valley. At each point, after each step, the gradient is calculated and the steepest way down is followed. Remember, each step represents a full set of bond lengths, angles, van der Waals interactions, etc. for the molecule, describing a specific molecular geometry. The process of stepping along the highest gradient path available is continued until the bottom of the valley is found. How is the bottom of the valley identified? In a perfect valley, the gradient at the absolute minimum of the surface should be zero. However, most calculations would have tremendous difficulty finding the point of zero gradient. As a concession to practicality, an acceptably low gradient value is chosen, and when the calculation finds a point which has this low gradient, the calculation terminates. This point is termed the point of convergence of the calculation. The molecular geometry corresponding to this point should represent the ground-state structure of the molecule.

As you can appreciate from the foregoing examples, the molecular mechanics computation is conceptually simpler than the semiempirical Hückel calculation or an *ab initio* calculation, but it sacrifices the theoretical rigor of these latter techniques. That is, molecular mechanics contains no explicit mathematical description of electrons or orbitals. Instead, it folds all the quantum mechanical information that a semi-empirical or *ab initio* computation tries to discover into some fairly simple classical expressions. This strategy works reasonably well (depending on the molecule) because atoms that are bonded together in molecules tend to behave roughly in a classical manner. Of course, this discards any theoretical understanding of the electronic structure of the molecule in an attempt to obtain some geometric information in a computationally inexpensive manner.

A few more points are necessary to complete the picture of molecular mechanics. One of these concerns the nature of the force field parameters. The values in the force field parameter set are not necessarily experimentally observed bond parameters or force constants. Rather, they are thought of as the constants that a particular type of interaction would have in the complete absence of other physical effects. For example, an equilibrium bond length r_o for V_{str} should be the "ideal" value for that bond, without influences from any other interactions, like steric, electronic, hydrogen bonding or electrostatic influences. Such an ideal bond is said to be "unstrained." During the minimization, these equilibrium parameters are distorted as necessary from their "ideal" values to achieve the lowest energy structure. The energy represented by the minimum in V_{tot} is thus the minimum molecular strain relative to an idealized molecular state where no strain exists. Since the energy values calculated for a series of molecular conformations are relative to a hypothetical "strainless" molecule, they are not physically meaningful numbers by themselves, but are only meaningful relative to one another. They allow identification of the lowest energy structure, which should be the ground state structure of the molecule. The calculated energy values are thus termed "strain energy" or "steric energy."

Two other points are worth mentioning. First, we have made no mention of solvent in our discussion. That is because, as we have presented the computations, solvent has been ignored; the molecule is in the isolated gas phase. But solvent can affect molecular structure. Additional refinements in force fields have allowed the introduction of solvents, however this is an improvement we have neglected for the sake of simplicity. Another item which we have ignored is temperature. Molecular structure can surely be affected by temperature. What is the temperature at which we have calculated the structure? The molecular mechanics calculation, in the absence of modification, assumes a temperature of absolute zero (0 K). While this is not a physically realistic temperature for experimental studies of molecules, the differences in structure between 0 K and higher temperatures are often small, so the computational results are not greatly affected by this assumption. Computational chemists, have, however, designed more sophisticated computational techniques to investigate the effect of temperature on molecular structure.

Finally, there is an issue which applies to all computational techniques, whether they are empirical, semiempirical or *ab initio*. This is the question of how one knows whether a search for the minimum of a potential surface is complete. It is always possible (unless one calculates the full potential surface for every conceivable structure, an impossibly large task even for the best technology today) that the minimum that has been found is not the deepest well on the surface. Such a minimum is termed a local minimum; the desired lowest energy minimum is called the global minimum. Fig. 7 compares a global with a local minimum. Computational chemists have developed a number of strategies to try to satisfy themselves that a minimum is global. One of these is to use several different starting geometries and to run the minimization a large number of times. If multiple starting geometries lead to the same final minimum then there is a good likelihood that the minimum is global and not local. For example, a computation which uses a starting geometry at point A maybe trapped in the local minimum L, while a computation starting at B has a good chance of finding the global minimum G (Fig. 7).

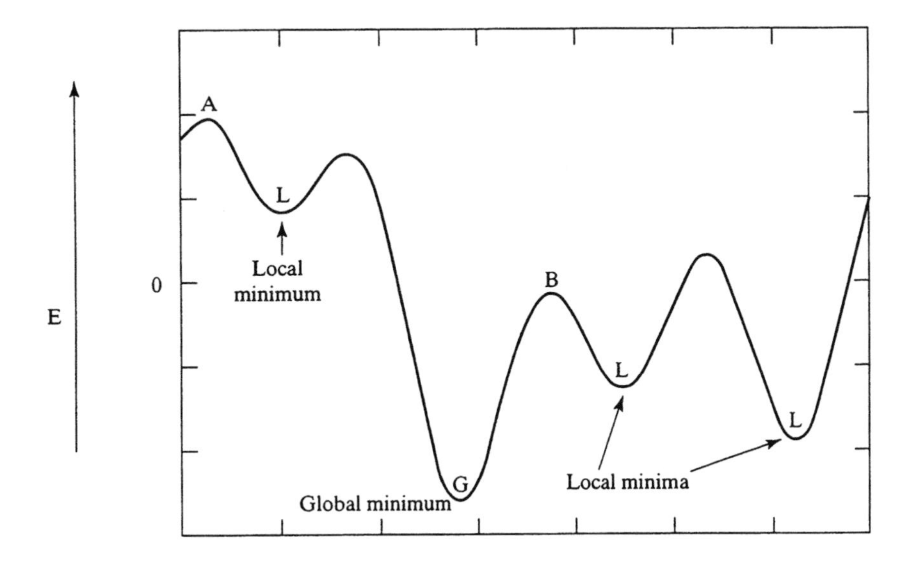

Fig. 7. The problem of local vs. global minima

Helpful Background Literature

1. Lowe, J.P., "Quantum Chemistry"; Academic Press, Inc.: Orlando, 1978, Ch. 8.

2. Allinger, N.L., "Adv. Phys. Org. Chem.", 1976, 13, 1.

3. Boyd, D.B.; Lipkowitz, K.B. "J. Chem. Ed.", 1982, 59, 269.

STRUCTURAL INORGANIC CHEMISTRY

The Importance of the Solid State

Crystalline phases are commonly encountered as reactants and/or products of chemical reactions, but the reactions are usually carried out in a fluid phase (liquid, solution, or gas). The latter fact might, and apparently not infrequently does, give the impression that the atomic structures of the solid phases are of minor importance. Exceptions perhaps are reactions such as the formation of a solid oxide or halide by direct action of oxygen or halogen on the solid metal, where the mechanism of formation of the product can be related to the structures of metal and product. Whether the reaction is considered in the solid state or in the fluid state, the crystal energies must enter into the calculations of the thermochemical cycles.

Despite the neglect of the teaching of the nature of the solid state in most chemistry courses, the importance of this subject is suggested by the following facts:

(i) Some eighty percent of the elements and the majority of inorganic compounds exist as solids under ordinary conditions.

(ii) In contrast to organic compounds, relatively few solid inorganic compounds (or elements) consist of discrete molecules; most consist of arrays of atoms within which the same kind of bonding leads to the formation of one-, two-, or three-dimensional macromolecules. It is clearly illogical to limit our interest to the structures of *finite* groups of atoms.

(iii) Most of the available information about bond lengths, interbond angles, and the nature of chemical bonds has come from studies of the solid state. Its crystal structure represents the ultimate geometrical characterization of a crystalline element or compound.

(iv) Some large and important groups of inorganic compounds exist *only in the solid state,* including hydrates, complex oxides and halides, 'basic' and 'acid' salts, and non-stoichiometric compounds. Intermetallic compounds (alloys) could also be included though these are still conventionally excluded from the teaching of chemistry. The arbitrary exclusion of the extensive structural chemistry of all the above groups of compounds would seem difficult to justify.

(v) Perhaps the most important fact to be emphasized here is that, because the atoms in most inorganic compounds form repeating patterns, their formulas can be understood only in terms of their crystal structures. The formula of a discrete molecule shows the actual numbers of atoms that are bonded together to form a group. However, the same type of chemical formula is used for HI, which consists of discrete molecules H-I in the solid, liquid, and gaseous states, and for

---

Based on unpublished material by A. F. Wells, University of Connecticut

a salt such as NaI, which exists in this form only in the vapor at high temperatures. At ordinary temperatures this salt is a solid in which every ion is surrounded by six of the other kind. Only the first member of the series of compounds:

Table I

AX	HI	AuI	CuI	NaI	CsI
Coordination number of A by X	1	2	4	6	8

exists as finite molecules at ordinary temperatures. In all the other cases there is no direct relation between the chemical formula and the structure of the (solid) compound because the crystal consists of a repeating pattern of atoms in which the coordination number (c.n.) of A by X atoms (the number of X atoms surrounding and bonded to an A atom) is greater than the ratio (unity) of X to A atoms. Fig. 1 shows a number of arrangements, both finite and infinite, of equal numbers of A and X atoms which represent possible structures of a compound AX.

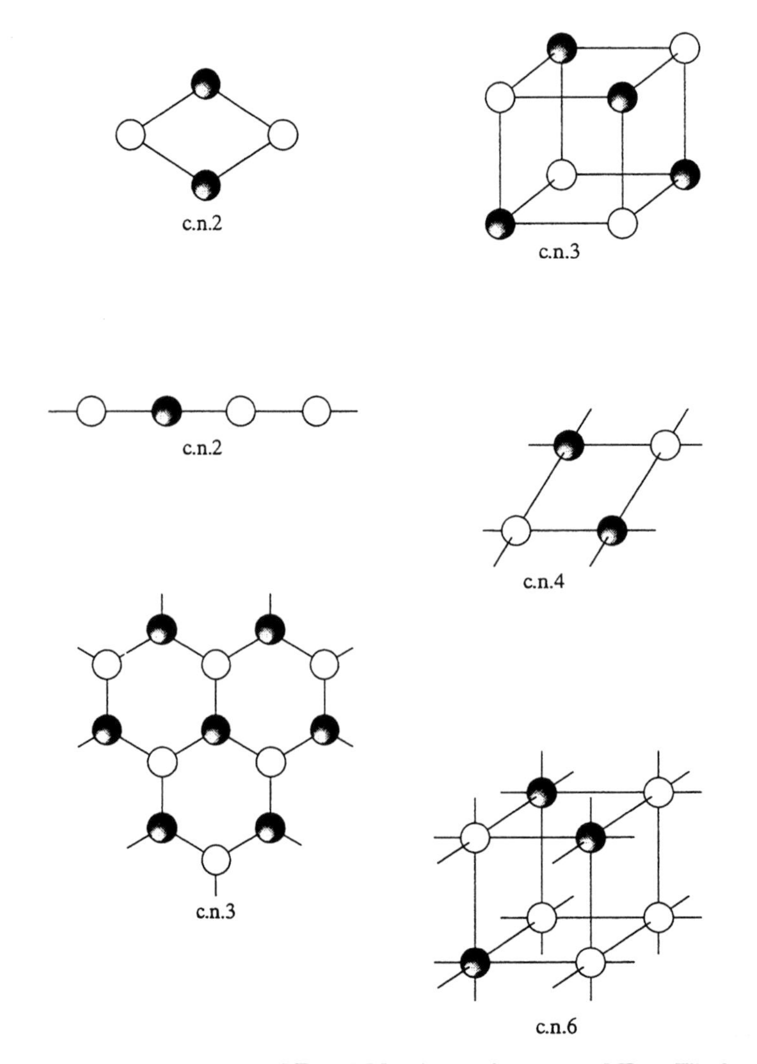

Fig. 1. Arrangements of Equal Numbers of Atoms of Two Kinds

Structural Formulas of Solids

If the bonding between the atoms leads to an atomic arrangement extending indefinitely in one, two or three dimensions, then the structural formula is a portion of the repeating pattern. The repeat unit is a *linear* portion of an infinite chain, an *area* (parallelogram) of a 2D pattern, or a *volume* (parallelepiped) of a 3D structure. The repeat unit (length, area, or volume) must be such that it reproduces the whole structure if it is repeated along the appropriate axis or axes. An illustration or model should show the complete set of bonds formed by at least one atom of each kind. The conventional unit cell diagram (or model) of the crystallographer does not necessarily do this, for it is assumed that the observer understands that the structure repeats along certain axes.

Some Basic Facts and Concepts of Models

There are three topics of a geometrical-topological nature which are basic to an understanding of the structures of crystalline inorganic compounds. They are:

(a) systems of connected points - polyhydra, 1-, 2-, and 3D nets,

(b) sphere packings of high coordination number (c.n.), and

(c) the formation of the structures of binary compounds from tetrahedral and/or octahedral coordination groups.

These three topics correspond to the three types of model that can be used to illustrate the structures of molecules or crystals, namely, the ball-and-stick model, the close-packed (c.p.) sphere model, and models built from tetrahedral or octahedral (or other) coordination groups. These are illustrated in Fig. 2 for the (finite) P_4O_{10} molecules: models of c.p. and polyhedral 3D structures will be made in the model laboratory sessions.

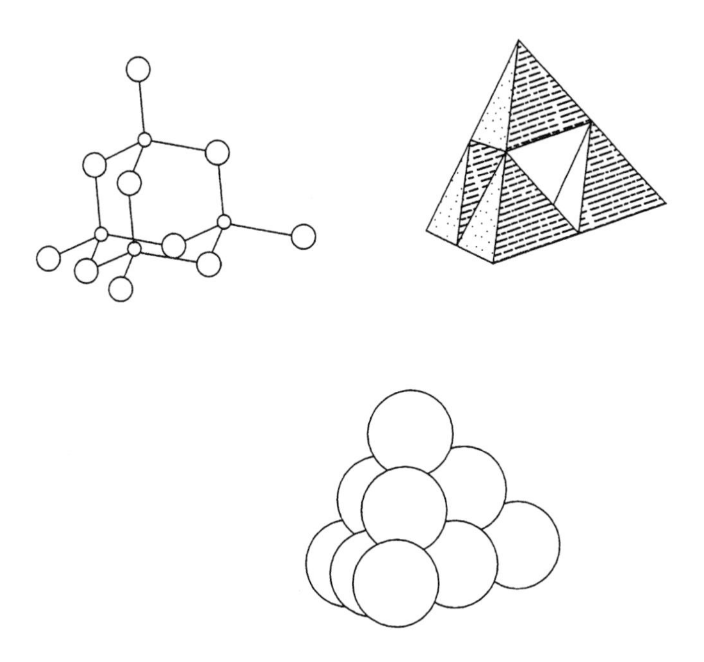

Fig. 2. Three Representations of the P_4O_{10} Molecule

Systems of Connected Points - Polyhedra and Nets

Many structures are built from sub-units of some kind each of which is linked to a number of other similar units. However complex the sub-units that are to be joined together such a structure may be described in terms of a system of connected points, each of which represents either a single atom or a more complex group of atoms. In the simplest system each point is connected to the same number (p) of others:

p = 1 : there is one solution only, a pair of connected points

p = 2 : the only possibilities are closed rings or an infinite chain.

p = 3 : the possible systems now include finite groups (e.g., polyhedra) and arrangements extending indefinitely in one, two, or three dimensions (p-connected nets).

By identifying the underlying p-connected net we emphasize the basic topology of the structure and see how a small number of very simple structural themes are utilized in crystalline compounds of many different chemical types. The description of structures in this way is best suited to those in which p has a small value: the 3- and 4-connected systems are of special importance, in particular the simplest 2D 3-connected and the simplest 3D 3-connected units of Fig. 3 placed at the nodes of the planar hexagon net include:

a. C (graphite), P (black), As, BN, the anion in $Ca(Si_2)$.

b. As_2O_3, As_2S_3.

c. P_2O_5, the anion in $Li_2(Si_2O_5)$.

d. $AlCl_3$, $Al(OH)_3$.

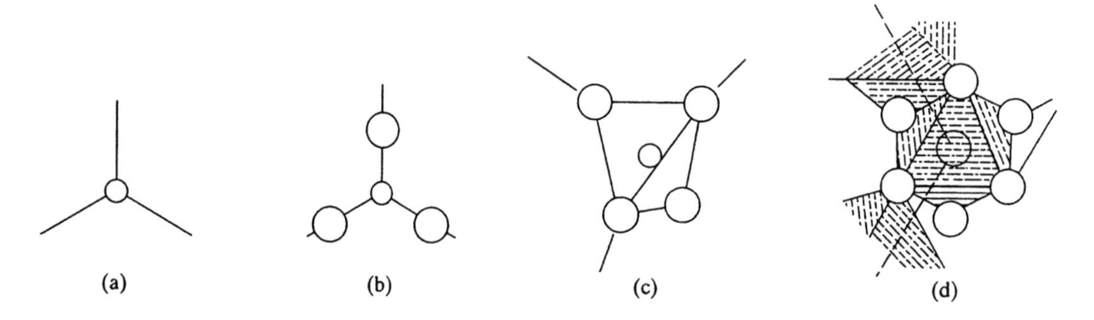

Fig. 3. Some 3-connected units

The numerous structures based on the cubic diamond net include those of elemental C, Si, Co, and gray Sn, compounds such as CuI, ZnS, BN, and SiO (alternate atoms of two kinds), AX_2 structures built of AX_4 groups sharing all four vertices (X atoms) as in cristobalite (SiO_2), and cubic ice, where the links of the net represent hydrogen bonds.

Sphere Packings: the Structure of the Metals Na, Mg, and Al

Of the sixty or so "typical" metals approximately one-half are polymorphic, crystallizing with two or more structures. Three structures have been assigned to these metals, in approximately equal numbers, namely, the body-centered cubic (b.c.c.) structure, the hexagonal close-packed (h.c.p.), and the cubic close-packed (c.c.p.) structures. These are the structures at ordinary temperatures of metallic Na, Mg, and Al respectively.

Closest-packings of equal spheres are of limited interest in the present connection since they represent the structures only of certain metals and noble gases and of a number of molecular crystals in which relatively simple molecules become effectively spherical either by rotation or by being randomly oriented (for example, H_2, HCl, H_2S, and CH_4). There is, however, another aspect of the closest-packing of equal spheres which is of more direct interest in structural chemistry. In many metal oxides and halides, the anions are much larger than the metal ions and their arrangement approximates one of the types of closest-packing, usually h.c.p. or c.c.p. It so happens that the two kinds of larger interstice in a c.p. array of equal spheres are surrounded by a tetrahedral or octahedral group of c.p. spheres and that one or other of the two kinds of interstice can accommodate the metal ions. Many simple structures for compounds A_mX_n in which there is tetrahedral or octahedral coordination of the A atoms may, therefore, be derived by considering the filling of various fractions of the total number of appropriate interstices in a c.p. array of X atoms. The description of the tetrahedral structures in this way is not to be recommended because in actual structures the packing of the X atoms is usually far removed from closest-packing. The bonds have considerable covalent character, and the structures are preferably regarded as resulting from directed bonds and described in terms of 3D 4-connected nets. On the other hand, the c.p. description of many octahedral oxide and halide structures is close to the truth, though interestingly enough this is not so for the simplest of all such structures, the sodium chloride structure, a point to which reference will be made later in this section.

It is not to be supposed that the actual formation of a compound occurs by the insertion or movement of metal ions into interstices between c.p. anions, though as already noted this may be the mechanism of reactions such as the formation of a solid oxide or halide film on a metal surface. The primary consideration in the formation of a crystalline compound A_mX_n is to satisfy the bonding requirements of the A and X atoms (ions), as expressed by their c.n.'s, and such structures are accordingly derived more realistically by joining together, for example, AX_4 or AX_6 coordination groups. Each X atom must be shared between (belong to) some number of such groups, this number being determined by the ratio of X to A atoms in the structure. The X atoms may then rearrange, if this is geometrically possible, to form the most compact arrangement consistent with the mode of linking of the coordination groups; in this way the van der Waals contribution to the crystal energy is increased. This 'coordination group' approach to structures in which there is tetrahedral or octahedral coordination of the metal ions will be illustrated in the following section.

One must be careful in using c.p. spheres to distinguish between the statement that atoms are in "positions of closest packing" and the statement that atoms are "most closely packed." Thus the relative positions of either A or X may be those of c.c.p., but this does not necessarily mean that either A or X atoms are most closely packed in any particular compound with the sodium chloride structure. The two extremes of the structure are: A atoms closest-packed with X in octahedral interstices ($r_A >> r_X$) or X atoms closest-packed with A atoms in octahedral interstices ($r_X >> r_A$). There are an indefinite number of intermediate configurations, and a special case would have A and X equal in size. For NaCl, since $r_{Na^+}/r_{Cl^-} = 0.57$ and for the Cl^- ions to be in contact the radius ratio must be 0.41 or less, the anions are not most closely packed but only in positions related to that arrangement to give a regular octahedron about each

Na^+ ion. See A. F. Wells' article, *Some Simple AX and* AX_2 *Structures*, J. Chem. Educ., 59, 630 – 633, 1982, for more on this topic.

Some Simple Tetrahedral and Octahedral Structures of Binary Compounds

There is a very simple relation between the c.n.'s of the atoms and the chemical formula of any binary compound A_mX_n. In the simplest case let us suppose that the c.n. is the same a for all A atoms and the same x for all X atoms and write these c.n.'s as superscripts - ${}^{a}A_m{}^{x}X_n$. The total number of A-X bonds may be counted as either am or xn, whence am = xn. This relation is of quite general validity: it applies to any system of bonded A and X atoms, whether finite or infinite. In some structures there are A atoms and/or X atoms with different c.n.'s (values of a and/or x). The relation still holds true but we must then use either mean values of a and/or x or write out the chemical formula in an 'extended' form showing the numbers of A (X) atoms with different c.n.'s, when $\sum a_l m_l = \sum x_l n_l$. For example:

Cl Cl Cl
Al Al
Cl Cl Cl

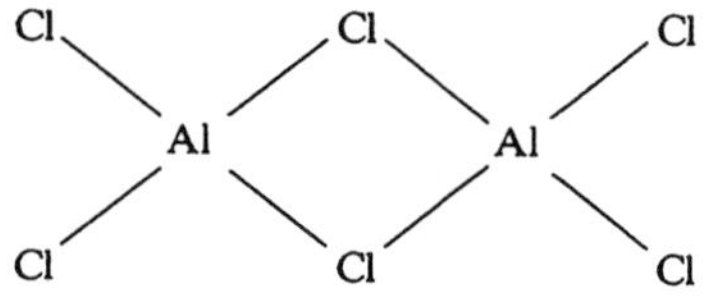

${}^{4}Al^{4/3}Cl_3$ or ${}^{4}A_2{}^{2}Cl_2{}^{1}Cl_4$

Fig. 4.

A similar designation for the P_4O_{10} molecule is: ${}^{4}P_4{}^{8/5}O_{10}$ or ${}^{4}P_4{}^{2}O_6{}^{1}O_4$.

(The mean value of a (x) in such structures is usually, but not necessarily, non-integral. For example, in the double octahedral chain in $K(CuCl_3)$ there are Cl atoms with three different c.n.'s, but the mean value of x is 2 because the formula of the chain is $(CuCl_3)_n{}^{n-}$ or AX_3).

The interest here is in arrangements of atoms that repeat regularly in one, two, or three directions and particularly in those structures in which the c.n. is the same for all A (X) atoms. These structures correspond to integral values of x and are listed in Tables II and III for the two most important c.n.'s of A, namely, 4 (tetrahedral) and 6 (octahedral). (It is a useful exercise for the student to compile the tables for a = 3 and a = 8, and to give examples of simple structures corresponding, for example, to a = x = 1, 2, or 3.)

Table II

Tetrahedral Structures (a = 4)

x	Formula	Structure Type	Examples and Name
1	AX_4	Finite	SiF_4^-
2	AX_2	Chain	$BeCl_2$, SiS_2
		Layer	HgI_2
		3D	SiO_2 (cristobalite)
4	AX	Layer	LiOH, OPb
		3D	ZnS (zincblende, wurtzite)
8	A_2X	3D	Na_2O (antifluorite)

Table II does not include structures with compositions between AX_4 and AX_2 in which fewer than 4 vertices are shared because they have non-equivalent X atoms. In a model-building course it is convenient to deal separately with tetrahedral structures in which only vertices are shared and those in which edges of AX_4 coordination groups are shared. Such a sub-division cuts across the classification according to x values:

Table III

	x = 2	x = 4	x = 8
vertex-sharing	HgI_2 SiO_2	ZnS	
edge-sharing	$BeCl_2$	LiOH	Na_2O

The vertex-sharing AX_2 structures also include 1D and other 2D structures which are illustrated in the laboratory exercises. The ZnS structures are best described as 4-connected nets, since both A and X are tetrahedrally coordinated.

Table IV

Octahedral Structures (a = 6)

x	Formula	Structure Type	Examples and Name
1	AX_6	Finite	SF_6
2	AX_3	Chain	ZrI_3
		Layer	$AlCl_3$
		3D	AlF_3
3	AX_2	Layer	$MgCl_2$ (cadmium chloride)
		3D	MgF_2 (rutile)
4	A_2X_3	3D	Al_2O_3 (corundum)
6	AX	3D	MgO, NaF (sodium chloride)
		3D	FeS (nickel arsenide)

Table IV does not include octahedral structures with compositions AX_5 or AX_4 which contain non-equivalent X atoms. Because they illustrate some of the simplest ways of joining together octahedral coordination groups they should be included in a systematic study of octahedral structures, despite the fact that the examples are not of great importance as regards the structural chemistry of the more common elements:

Table V

Octahedra Sharing:	Structure	Examples
2 opposite vertices	AX_5 chain	BiF_5, UF_5
2 adjacent vertices	AX_5 chain or ring	VF_5, CrF_5, Mo_4F_{20}
4 equatorial vertices	AX_4 layer	SnF_4, PbF_4, NbF_4
2 opposite edges	AX_4 chain	NbI_4

A feature of many fluoride structures is that the octahedral coordination groups share only vertices, the sharing of edges in the rutile structure of MgF_2 has a purely geometrical explanation. In the corresponding chlorides edge-sharing is more usual. $MgCl_2$ and $AlCl_3$ have closely related layer structures formed from octahedral groups sharing respectively 6 or 3 edges. Alternatively these structures are illustrated by models constructed from layers of c.p. spheres.

Some Structures of Binary Compounds of Second Row Elements

It is not feasible to discuss all the geometrical aspects of structural inorganic chemistry in this short essay. Instead an attempt will be made to show how these structural concepts apply to the elements and compounds of the second row elements excluding the non-metal halides. The structures of the latter in the solid state are not known in all cases, but these compounds presumably exist as finite molecules, with the exception of PCl_5 which is $(PCl_4)^+(PCl_6)^-$. The second-row rather than the first-row elements have been selected because the chemistry of certain of the first-row non-metals involves special types of bonds (for example, in elementary boron and its 'electron-deficient' compounds) or is complicated by multiple bonding. Sulfur and phosphorus form the molecules S_2 and P_2 only at high temperatures and at lower temperatures polymerize to single-bonded molecules such as S_8, S_6, or P_4 which persist in certain crystalline forms of the elements. These and other singly-bonded systems in other polymorphs illustrate the formation of $8 - N$ bonds, where N is the ordinal number of the Periodic Group.

The structures of the elements Si, P, S, and Cl

These elements have respectively 4, 5, 6, and 7 electrons in their valence shells and can, therefore, complete these shells by forming $8 - N$ covalent bonds. Accordingly the structures of the elements are:

Table VI

	$p = 8 - N$	
Cl	1	finite molecule Cl_2 in all states of aggregation
S	2	cyclic molecules S_6, S_8, etc. in various crystalline forms and infinite chain in plastic sulfur
P	3	tetrahedral molecule P_4, the layer structure of black P, corrugated configuration of the simplest 3-connected planar net (and a more complex layer in red P)
Si	4	3D 4-connected net (cubic diamond structure under pressure a second 3D net also built of 6-gons

Examples of structures of these types are shown in Fig 5. (The layer of black phosphorous is a more buckled form of the layer shown for As).

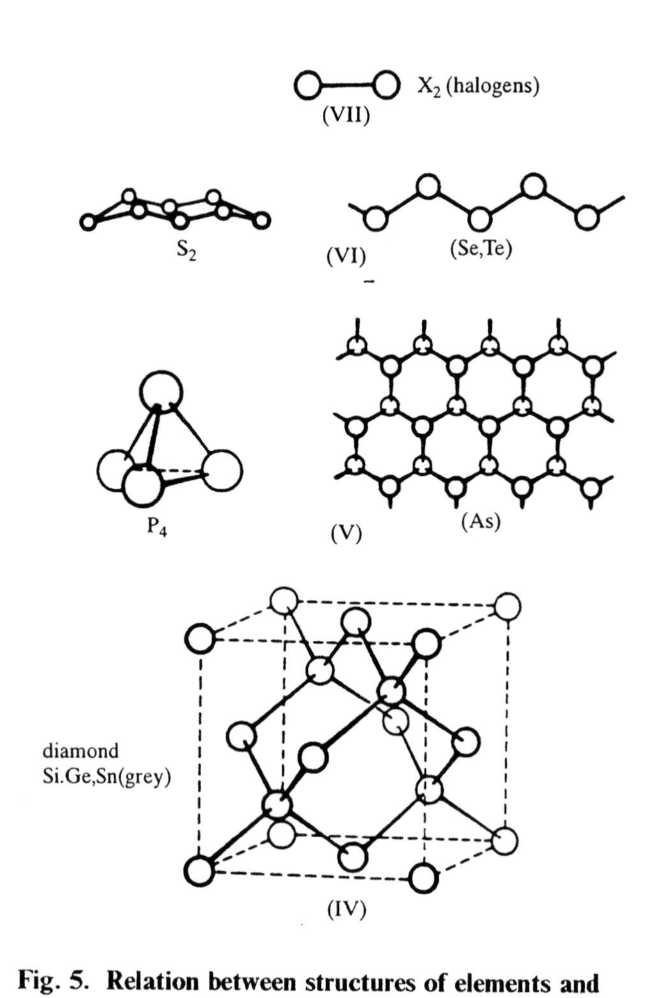

Fig. 5. Relation between structures of elements and position in Periodic Table

Table VII

The Elements of the second short period and their compounds

I	II	III	IV	V	VI	VII
Na	Mg	Al	Si	P	S	Cl
sphere packings			p-connected systems			
Na_2O	MgO	Al_2O_3	SiO_2	P_2O_5	SO_3	Cl_2O_7
tetrahedral edge-sharing structures			tetrahedral vertex-sharing structures			
NaF	MgF_2	AlF_3	SiF_4	PF_5	SF_6	--
$NaCl$	$MgCl_2$	$AlCl_3$	$SiCl_4$	PCl_5	(SCl_4)	
octahedral structures; closest packing of equal spheres						

Table VII indicates that there is a periodic relationship of structural representations. Three of the compounds, MgO, NaF and NaCl have the sodium chloride structure. Comment was made on this in the last paragraph. Some of the oxides, fluorides, and chlorides are discussed in the subsequent sections as is the more complex structure of Al_2O_3.

Table VIII

Structures formed by XO_4 groups sharing vertices

p		Oxy-ion			Oxide molecule
1	pyro	$Si_2O_7^{6-}$	$P_2O_7^{4-}$	$S_2O_7^{2-}$	Cl_2O_7 finite
2	meta	$(SiO_3)_n^{2n-}$	$(PO_3)_n^{n-}$	$(SO_3)_n$	cyclic or linear
3		$(Si_2O_5)_n^{2n-}$	$(P_2O_5)_n$		polyhedral, infinite, 1D, 2D, or 3D
4		$(SiO_2)_n$			infinite 3D

The oxides SiO_2, P_2O_5, SO_3, and Cl_2O_7.

The octet of valence electrons of Si, P, S, or Cl is completed in the tetrahedral oxy-ions SiO_4^{4-}, PO_4^{3-}, SO_4^{2-}, and ClO_4^{-} by acquiring $8 - N$ negative charges. The same result is achieved if these groups share $8 - N$ oxygen atoms, each being shared with one other XO_4 group; the result is the neutral (highest) oxide. The topologically possible structures for these oxides are accordingly of exactly the same kinds as for the elements, namely, for $N = 7$ a finite binuclear molecule, for $N = 6$ rings or chains, and for $N = 5$ or 4 polyhedral groups or infinite 1-, 2-, or 3-dimensional structures. These oxides, in fact, illustrate all the main types of structure built from tetrahedral groups sharing vertices each vertex being common to *two* tetrahedra only:

Cl_2O_7 : finite molecule Cl_2O_7

SO_3 : cyclic trimer S_3O_9 and infinite chain

P_2O_5 : polyhedral P_4O_{10} molecule, infinite layer based on the simplest planar 3-connected net, and a 3D structure based on a 3D 3-connected net

SiO_2 : all polymorphs are 3D frameworks formed from SiO_4 groups sharing all their vertices (with the exception of one polymorph formed under very high pressure which has the rutile structure)

The numbers $(8 - N)$ of O atoms shared by each XO_4 group in these oxides are the maximal numbers that can be shared; the sharing of further O atoms would result in cations, which are not known. However, the sharing of fewer O atoms by each XO_4 group is permissible, giving the anions summarized in Table VIII.

NaF, MgF_2, and AlF_3

All the elements of the second short Period exhibit octahedral coordination by F:

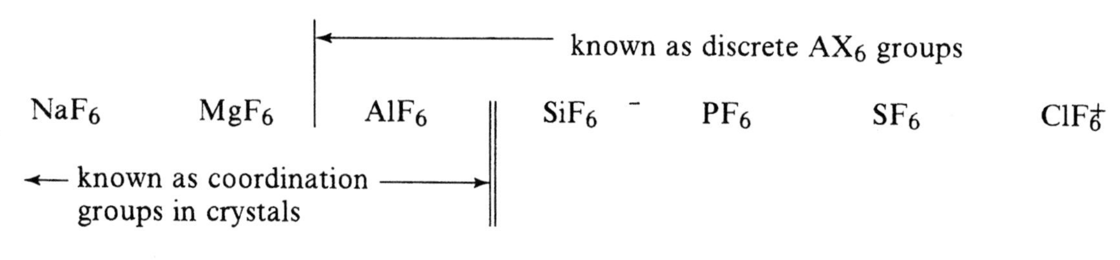

Fig. 6.

The NaF_6 and MgF_6 groups are known *only* as coordination groups in crystals (*not* as discrete ions $NaF_6{}^{5-}$ or $MgF_6{}^{4-}$); the other groups are all known as discrete ions (or molecule in the case of SF_6). Those to the right of the double line are known *only* as discrete groups (ions or molecules); those to the left of the single line are known *only* as coordination groups in structures extending indefinitely in one or more directions as the result of the sharing of F atoms. Aluminum occupies a unique position, for $AlF_6{}^{3-}$ is known as a finite anion in $Na_3(AlF_6)$, the mineral cryolite, and as a component of the infinite $(AlF_5)_n{}^{2n-}$ chain ion in $Tl_2(AlF_5)$ in which each octahedral AlF_6 group shares two opposite vertices, of the $AlF_4)_n{}^{n-}$ layer in $Tl(AlF_4)$ in which similar groups share their four equatorial vertices, and finally of the structure of AlF_3, which is formed of AlF_6 groups sharing all their vertices to form a 3D framework. Further sharing of F atoms between octahedral coordination groups leads to the rutile structure of MgF_2 and ultimately to the sodium chloride structure of NaF. As a group these fluorides, simple and complex, illustrate the formation from octahedral MF_6 groups of structures in which the F : M ratio has all integral values from 6 to 1.

It is instructive to look at the structures of NaF, MgF_2, and AlF_3 in another way. The octahedral coordination of M implies 6-, 3-, and 2-coordination of X respectively. One might expect to find three very simple structures for ionic crystals in which X ions have the most symmetrical possible types of coordination, namely 6 (regular octahedral), 3 (equilateral triangular), and 2 (collinear), with bond angles 90°, 120°, or 180°. The first of these structures is the sodium chloride structure. Examination of the geometry of systems of linked octahedra shows that an AX_2 structure with *regular* octahedral coordination of M and *equilateral* triangular coordination of X is not possible; the interbond angles in crystals with the rutile structure are close to the ideal values, which are 90° (one) and 135° (two). The vertex-sharing AX_3 structure can have various configurations ranging from the least densely packed with A-X-A, 180° (ReO_3 and the anion framework in $K(MgF_3)$), in which the X atoms occupy three-quarters of the c.c.p. positions, to the most dense (A-X-A, 132°) with h.c.p. X atoms (RhF_3, IrF_3, PdF_3). Intermediate configurations are also known (for example, FeF_3, and CoF_3, with A-X-A around 150°); the details of the structure of AlF_3 still appear to be uncertain.

$MgCl_2$ and $AlCl_3$

A feature of many fluoride structures is that the octahedral coordination groups share only vertices; the sharing of edges in the rutile structure of MgF_2 has a purely geometrical explanation (see below). In the corresponding chlorides edge-sharing is more usual. $MgCl_2$ and $AlCl_3$ have closely related layer structures fromed from octahedral groups sharing respectively 6 or 3 edges. Alternatively these structures are illustrated by models constructed from layers of c.p. spheres.

Al_2O_3

In the structure of αAl_2O_3 (corundum, ruby, sapphire) each O^{2-} ion is bonded to 4 Al^{3+} ions, or alternatively, 4 octahedral coordination groups meet at every vertex. The mode of packing of the AlO_6 octahedra is complex, each sharing one face, three edges, and nine vertices (with different octahedra). This unexpected complexity is due to the (geometrical) difficulty of building a 3D ionic structure with a reasonable efficient packing of the anions (here h.c.p.) and at the same time a tetrahedral environment of these ions (see Corundum Structure below).

Two General Points Concerning Crystal Structures

The sections so far are intended to cover simply the facts concerning polyhedra and nets, sphere packings, and the more important tetrahedral and octahedral structures. There are also certain general points that present more difficulty than this purely factual material; two of these are the following.

A crystal structure is not completely described by the unit cell dimensions and the coordinates of the atoms. In a binary compound there are contacts between an A (X) atom and its nearest X (A) neighbors, but we may also be interested in contacts between A atoms or between X atoms. We can decide if these occur only if radii are assigned to the atoms (ions), and this in turn depends on the way in which we divide the distance A-X into n_A and n_X. The description of a structure A_nX_n as a c.p. assembly of X atoms with A atoms in interstices implies contacts between X atoms. However, the X atoms are not necessarily most closely packed even though their centers are in the positions of closest packing, a matter which is best discussed with reference to particular structures.

The second point is that there are limitations on the ways in which atoms can be bonded together to form a periodic 3D structure which are not directly related to the *chemical* properties of the atoms, that is, their electronic structures and bond-forming capacities. These limitations may be divided into topological and geometrical ones, the former being concerned simply with connectedness and the latter involving metrical features (bond lengths and interbond angles). Just as there are only five convex polyhedra (n, p) having p n-gons meeting at each vertex and only three planar nets (n, p), so certain other types of connected systems are not possible. For example, it is not possible to build a simple layer structure for a compound A_2X_3 with 6-coordination of A, or (apparently) *any* 3D AX_2 structure of 10 : 5 or 12 : 6 coordination, that is, with c.n.'s higher than those (8 : 4) in the fluorite structure. In the other category come the purely geometrical limitations on the ways in which *regular* octahedra may be joined together if a lower limit is set to the permissible distance between vertices of different octahedra. This subject has a direct bearing on the rutile and corundum structures, as is explained below. It should also be mentioned here that there is an (apparent) impossibility of building 3D AX or AX_2 structures with square antiprism coordination of A, a point of interest in connection with the cubical coordination of the cations in the cesium chloride and fluorite structures. It might appear that the subject of structural inorganic chemistry is being unnecessarily complicated by introducing these topological and geometrical matters. However, since the formation of 3-dimensional structures is being considered, it is only to be expected that there will be restrictions on the ways in which atoms or groups of atoms can be joined together, these restrictions being of a quite different nature from those arising from the electronic structures of the atoms.

Geometrical Limitations in Octahedral Structures

If two regular octahedra AX_6 have a face in common, the angle $A - X - A$ has the value $70^o\ 32'$. If an edge is shared, rotation about the shared edge is possible; but, if it is assumed that the minimal distance (d) between X atoms of different octahedra is not less than the edge length (ℓ) of an octahedron, then the system is invariant with the angle $A - X - A$ equal to 90°. For a pair of vertex-sharing octahedra with the same restriction ($d \geq 1$) the angle $A - X - A$ can range from 180° in the fully extended case to $131^o\ 48'$ at the lower limit (Fig. 7) when $d = 1$. (For simplicity the approximate value 132° will be used for this angle, which is $2 \sin^{-1}\sqrt{5/6}$. These angles have a direct bearing on the structures of many molecules and crystals built from octahedral coordination groups, including compounds with the rutile and corundum structures.

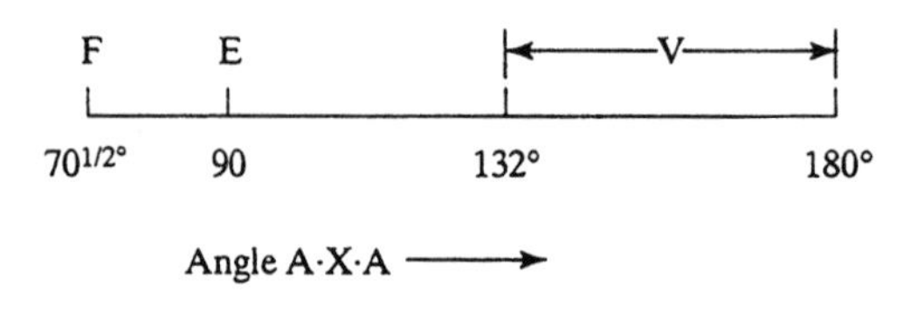

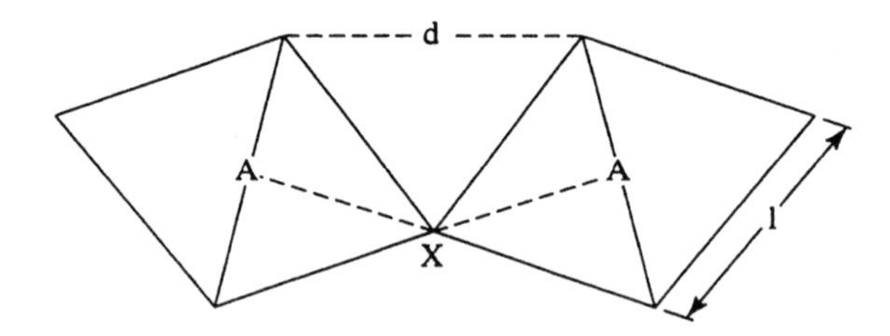

Fig. 7. The possible values of the angle $A - X - A$ for regular octahedra AX_6 sharing a face (F), an edge (E), or a vertex (V) subject to the restriction that $d \geq 1$

Rutile structure. From the permissible range (132° - 180°) of $A - X - A$ angles it follows that if more than two regular octahedra meet at a point (that is, share a common vertex (X atom)) they must share one or more edges and/or faces. In the rutile structure *three* octahedra meet at each X atom. The maximal value of $A - X - A$ would be attained when the three A atoms at the centers of the octahedra are coplanar with X, when the angle would be 120°. This value falls in the *forbidden* range, between 90° and 132°, and is therefore not possible for regular octahedral coordination groups. One edge is, therefore, shared at each X atom (Fig. 8), and in fact the 3D rutile structure follows directly from the geometry of the octahedron assuming only that the difference between the electrogenativities of A and X is sufficient to insure ionization and that the relative sizes of the ions are consistent with octahedral coordination of the cations.

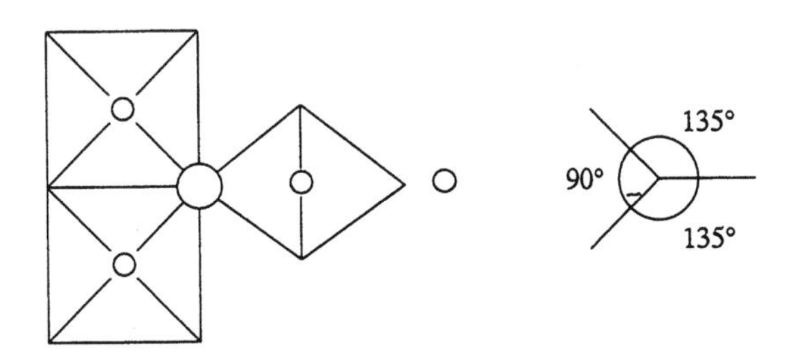

Fig. 8. Three octahedra meeting at a common vertex in the rutile structure.

Corundum structure. In an octahedral A_2X_3 structure *four* octahedra meet at each vertex (X atom), assuming all X atoms to be equivalent. Since the value 109.5° falls within the *forbidden* range of A – X – A values, a structure with *regular octahedral* coordination of A and *regular tetrahedral* coordination of X is not possible; there must also be sharing of edges and/or faces. The actual Al – O – Al angles in Al_2O_3 are: 85° (one), 94° (two), 120° (one), and 132° (two), rather than the six of 109.5° characteristic of regular tetrahedral coordination.

The topics discussed in this essay should point out the importance of geometrical considerations in the study of solids. Preferred bond angles based on maximum orbital overlap considerations that are valid for molecular compounds are modified by geometrical considerations in the solid state. Many of the solid state structures can only be understood in the context of geometrical limitations. This introductory essay is intended, therefore, to expose the student to some of these concepts.

Helpful Background Literature

1. Wells, A. F. "Structural Inorganic Chemistry", 5th ed.; Clarendon Press: Oxford, 1984.

2. Wells, A. F. "Models in Structural Inorganic Chemistry", Clarendon Press: Oxford, 1970.

3. Wells, A. F. "The Third Dimension in Chemistry", Clarendon Press: Oxford, 1962.

Part II LABORATORY EXERCISES

CRYSTAL GROWTH IN GELS

Equipment Needed: 2 test tubes, 25 mm x 150 mm
1 U-tube 50 mL
Temperature bath, 35^oC to 55^oC (optional)

Chemicals Needed: 30 mL of 1.06 density sodium silicate
30 mL of 1.0 M acetic acid
1 mL of 1.0 M lead(II) acetate
3 g potassium iodide
0.2 g $CuSO_4 . 5 H_2O$
3 mL of 1% hydroxylamine hydrochloride
1 g of mercury(II) chloride
3 g calcium chloride
3 g sodium carbonate

Time Needed: 8 hours initially for gel to set
2 hours or less to prepare gel
If lab only meets one time per week solutions can be added to the set gel after one week.

Safety Notes:

1. Glacial acetic acid is a weak acid but can cause burns to skin if handled improperly. Gloves should be worn during dilution of the acid.

2. Hydroxylamine hydrochloride should be handled with gloves. It is used as a reducing agent in photography. Keep the solution tightly stoppered.

3. Mercury(II) chloride is a highly toxic chemical. Gloves should be worn while using this reagent. It is used in photography, embalming, leather tanning, and is an important reagent in analytical chemistry.

4. Lead(II) acetate is used in the dyeing industry. Lead(II) iodide is photoactive, poisonous if ingested, and is used in printing and bronzing applications.

Purpose

Crystal growing is an important part of solid state chemistry. The development of most aspects of solid state chemistry – photovoltaics, solid electrolytes, catalysts, integrated circuits, rectifiers, etc. – all depend on studying single crystals. A technology such as the electronics on a silicon chip depends on the ability to grow large single crystals. Production of large single crystals of silicon involves high temperature melt pulling which uses specialized equipment. On the other hand, the oldest and most familiar laboratory technique for growing single crystals is that of slow evaporation or cooling of aqueous solutions. Other solution growth methods include (1) flux growth, (2) electrolytic methods, (3) hydrothermal synthesis, and (4) gel growth. This latter method will be illustrated in this experiment.

Introduction

A gel is a colloidal system in which the dispersed phase forms a three dimensional network in the host liquid such that the material acquires a semi-solid or gel consistency. Gels can be formed from water by the addition of silicates, oleates, gelatin, polyvinyl alcohol or agar. Gels formed from water as opposed to other liquids are called hydrogels. Each of the types of hydrogels mentioned above have been used in various ways for crystal growth, but the silica type materials derived from water glass or sodium metasilicate have been the most widely used.

When a soluble silicate solution is acidified, the silicate ion reacts to form monosilicic acid which can react with itself to to produce water and Si-O-Si bonds. These siloxane bonds are the basis for the formation of the three dimensional network responsible for gel formation. An example of the formation of the gel is given below.

```
                          --O--Si--O--Si--
                            |   |   |   |
SiOH  +  HOSi   →         --Si--O--Si--O--   +   HOH
                            |   |   |   |
                          --O--Si--O--Si--
```

After the initial gel formation, siloxane bonds are formed and the expelled water which is now on the top of the gel evaporates. This process is known as syneresis. Because the resultant gel is still 90% to 97% water, it is not surprising that the three dimensional siloxane network includes open channels and pockets. Ions diffuse down the channels and when two different ions meet crystallization occurs.

Crystal growth can be accomplished by dissolving one reactant in the gel before the gel is allowed to set. A solution of the other reactant is added to the top of the gel. In solution, cooling allows nucleation to occur because saturation is approached slowly. In a gel, nucleation is slow because the matrix hinders the diffusion of ions. Slow crystallization often leads to well formed single crystals. Because the ions in solution only diffuse slowly through the open channels and pockets in the gel, the crystal growth rate is diminished with respect to pure solution growth.

Often another phenomenon is noticed when the two ions which interact on diffusion through the gel form an insoluble precipitate. Instead of the formation of a uniform mass of precipitate or of interdispersed crystals, a ring or rings of precipitate are formed with interdispersed layers of gel. These rings are known as Liesegang rings and have been studied for at least a century. Natural formations such as growth rings in trees and rings in minerals like agates are believed to be the result of Liesegang ring formation.

The most important variables in the growth of crystals in gels are given below.

1. The pH of the gel formation process is especially important in forming silicate gels. If the pH is too low a gel may never form. If it is too high the gel may form instantaneously incorporating a large number of bubbles. Usually a setting time of 8 to 10 hours is desirable to form a gel which is suitable for forming large single crystals. The changes in pH during the formation of the gel can be monitored with a pH meter.

2. The concentrations of the reagents depend ultimately on the nature of the solubility product of the compound forming the single crystal and the diffusion rates of the ions forming the crystal. There are no useful theoretical guidelines for determining the concentrations of reagents for a particular experiment. The optimum concentration is arrived at by experimental methods.

3. A choice must be made as to which reagent must be incorporated in the gel fraction and which is to be in the solution above the gel. Preliminary experimentation is needed to determine which reagent is to be located where.

4. In formation of a silicate gel phase, it is usually best to add the selected reactant to the unset gel solution and then to add this solution to the acid. The reverse procedure of adding acid to the gel solution often results in instantaneous gel formation which is not desirable.

5. The nature of the crystals being formed will determine to some extent whether test tubes or U-tubes are used. If both reagents are to be diffused into the gel, then a U-tube must be selected as the container. This method will often produce very large single crystals.

If the original reactant concentrations in the U-tube experiment are fairly low and if the solutions are periodically replaced with successively higher concentrations of reactants then exceedingly large crystals can be formed. This method is known as concentration programming and relies on the phenomenon of Ostwald ripening. Originally only a few crystals are formed. After the solutions are changed and the concentrations are higher, the original crystals will become larger rather than new crystals being formed.

6. The temperature during gel formation is usually room temperature but, during crystallization (after the addition of a reagent to the top of the gel) the temperature can be an important factor with respect to induction effects. Other variables in the preparation of the silicate gel phase are the type of silicate selected and the concentration of the silicate prepared for the experiment. Electrodes can also be incorporated into the gel in order to induce crystallization.

Experimental Section

Each student should do two of the following experiments. The preparations are to be labelled and allowed to stand in the laboratory for observation during the semester.

First, prepare two test tubes, ca. 25 x 150 mm. These will be used to hold the gel. Prepare 30 mL of sodium silicate solution of density 1.06 g/mL by dilution of the commercial solution. The volume of commercial silicate solution to make 30.0 mL of density 1.06 may be calculated from the following relationship:

$$V = \frac{30(1.06 - 1.00)}{(D - 1.00)}$$

where D = density of commercial solution
V = volume of commercial solution to be diluted to 30 mL with a density 1.06

For example, sodium silicate solution from Fischer Scientific Company has a density of 1.37 g/mL. 15 mL of the silicate solution will be used in each experiment. There are a number of variables that will influence the growth of the crystals. Be sure to vary at least one of the parameters in your synthesis. Keep a record of the growth of your crystals during the semester. In particular note the color, size and shape of these crystals. The variables that are important in crystal growth in gels are:

1. Temperature.
2. Concentration of the reactants.
3. Type of tube.
4. Whether the acid is added to the gel or vice versa.
5. How the reactants are dissolved and dispersed (i.e., in the gel or in the layer above the set gel.)

The time for the gel to harden is usually 8 hours. Plan to let the gel set from one lab period to the next. Then add solutions to the top of the set gels during the period following the preparation of the gel.

Occasionally, the gel sets up too fast and forms before it can be poured out of the beaker. If this happens, start over again.

Lead(II) Iodide Crystals

Add 1 mL of 1 M lead(II) acetate solution to 15 mL of 1 M acetic acid solution. Mix the resulting solution with 15 mL of the diluted sodium silicate solution. Pour the mixture into one of your test tubes (to about 2/3 full) and allow to set. When the gel is firm, add a solution of 1.0 g KI in 5 mL water and cork. At first a solid layer of lead(II) iodide is formed at the surface, but soon frond-like crystals grow down into the gel. In the bulk of the gel, single hexagonal crystals of the salt gradually appear and grow to the size of several mm in the course of a few weeks. On prolonged standing in light the crystals darken somewhat. This will not happen if the experiment is kept in your locker.

Copper Crystals

Mix 15 mL of the diluted sodium silicate solution, 15 mL of 1 M acetic acid and 0.2 g copper(II) sulfate. Fill one of your test tubes 2/3 full and allow it to set. Then add a solution of 1% hydroxylamine hydrochloride. Add enough hydroxylamine hydrochloride solution so that a white ring forms at the interface. (It will take about 2.5 mL.) In the course of a few weeks fine tetrahedral crystals of copper appear.

Banded Mercury(II) Iodide

Mix 15 mL of the diluted sodium silicate solution and 15 mL of 1 M acetic acid containing KI. Use between 0.2 g and 2 g of KI; record the amount on the label. Fill one of your test tubes 2/3 full of this solution and when set add a solution of mercury(II) chloride (1.0 g in 15 mL of water). In a few days bright red bands of HgI_2 will have formed. The form and color of the bands depends greatly on the concentration of KI. These red bands are known as Liesegang rings and are similar in appearance to growth rings found in trees.

Calcite Crystals

Mix 15 mL of the diluted sodium silicate solution with 15 mL of 1 M acetic acid. Fill a U-tube with the gel-acid mixture and allow to set. Dissolve 3 g. of $CaCl_2$ in about 10 mL of water and add to one side of the U-tube. Dissolve 3g. of Na_2CO_3 in about 10 mL of water and add to the other side. The reagents will diffuse from both sides and form calcite (calcium carbonate) where they meet.

Auxiliary Experiments

Crystals can be extracted from the gel with spatulas. If the crystals are insoluble in dilute sodium hydroxide, the gel can be washed off with 1 M NaOH. Before extraction, the crystalline habits can be observed under a microscope. If a lens adapter is available photomicrographs can be taken. The electrical conductivity of the copper crystals can be tested by touching the probes of a voltohm meter to the faces of the copper crystals. Reseeding and concentration programming experiments can be carried out to obtain even larger crystals.

Questions to be Considered

1. Freshly formed barium sulfate crystals are often too fine to be filtered. How is Ostwald ripening used to convert the fine precipitate to a filterable precipitate?

2. Explain the mechanism of Ostwald ripening.

3. Write the reaction for the formation of copper crystals.

4. Write the reaction for the formation of calcite ($CaCO_3$).

5. Where in nature are calcite crystals observed?

6. Suggest a crystal which might be grown in a gel analogously to the above experiments.

Helpful Background Literature

1. Timofeeva, V. A.; Belyaev, L. M.; Ursulyak, N. O.; Belitskii, A.V.; Bykov, A. B.; Prilepo, V. M. "Use of Recrystallized $Y_3Fe_5O_{12}$ Seeds for Growing Yttrium Iron Garnets from Flux", *J. Crystal Growth*, 1981, 52, 633 – 638.

2. George, M. T.; Vaidyan, V. K. "An Electrolytic Method to Grow Copper Dendrites and Single Crystals in Gels", *Krist. Tech.*, 1980, *15*, 653 – 659.

3. Robson, H. "Synthesizing Zeolites", *Chem. Tech.*, 1978, *8*, 176 – 180.

4. Henisch, H. K. "Crystal Growth in Gels"; The Pennsylvania State University Press: University Park, PA, 1973.

5. Suib, S. L.; Weller, P. F. *J. Crystal Growth*, 1980, *48*, 155 – 160.

6. Suib, S. L. *J. Chem. Ed.*, 1985, *62*, 81 – 82.

STRUCTURAL INORGANIC CHEMISTRY VIA MODELS - 1

Session One of Three

Equipment Needed:

10 trigonal balls (dark blue)
6 trigonal balls (silver)
20 tetrahedral balls (black)
3 distorted tetrahedral balls (red)
7 octahedral balls (light blue)
14 octahedral balls (yellow)
50 bonds for above balls
4 clear plastic spheres
1 cemented triangle of 6 spheres
or one wooden triangular tray
4 plastic octahedra
6 plastic tetrahedra
14 tetrahedra (folded paper)
1 tin of marbles (Td and Oh size)
1 hot glue gun (per 3 students)

Chemicals Needed: None

Time Needed: 3 hours

Safety Notes: None

Purpose

Inorganic structures are best understood by the use of models. Three different structural representations commonly used to represent inorganic structures are demonstrated. Exercises are provided for the transposition of these three-dimensional structures to a two-dimensional drawing on paper.

Introduction

Solid state structures are varied and complex. In order to be able to think about and visualize the relative atomic arrangements, models are needed. This laboratory introduces models which can be used to visualize solid state structures.

It is important to understand solid state structures since solid state chemistry plays such an important part in modern technology. Solid state circuits are used in radios, television, calculators, computers, radars, automobile ignitions, etc., and this technology would not be possible without the knowledge of solid state chemistry.

Solid state chemistry is not only an often neglected aspect of undergraduate education, but it is as often misunderstood. Most of undergraduate chemistry deals with molecular compounds or solution chemistry. Because the concept of infinite repeating patterns is not emphasized, recognition of the importance of these concepts is lacking. In addition, solid state facts are often confused with the concepts of molecular chemistry. For example, sodium chloride has an infinite three dimensional repeating pattern. Yet most chemists think of sodium chloride as sodium ions and chloride ions in solution. In dealing with moles of sodium chloride, the subconscious tends to relate a mole of sodium chloride to a mole of ethyl alcohol. One needs to be reminded that there is no such thing as a molecule of sodium chloride!

There are, in fact, molecular inorganic compounds. Sulfur(VI) fluoride (sulfur hexafluoride), carbon monoxide, carbon dioxide, dinitrogen tetroxide, diborane, and uranium(VI) fluoride are but some. Many more inorganic compounds are infinite structures. Some of these are one-dimensional, i.e., chains. Some are two-dimensional, i.e., sheets. Some are infinite three-dimensional lattices. The structures reported for many of these compounds in the literatures are complex enough so that few can read the descriptions and look at the two-dimensional picture and later reproduce the structure. Models are designed to bring some order to the seeming chaos. They are a crutch in correlating and remembering structural relationships.

The three types of models to be illustrated in this experiment can be described as:

(1) ball-and-stick or connector-and-stick,

(2) connected polyhedra, especially tetrahedra and octahedra,

(3) packed spheres.

Experimental Procedure

Prelab Assignment

Put together 14 paper tetrahedra from the sheets provided before coming to the laboratory. Crease all folds inward including the tabs. There should be no tabs visible when the tetrahedron is completed. Bring the tetrahedra to the lab in a grocery bag. This same bag can be used to take the completed models back home.

Exercise #1

One crystalline form of diphosphorus pentoxide (phosphorous pentoxide) P_2O_5, is in fact a molecular compound consisting of P_4O_{10} units. Four of the ten oxygen atoms are terminal atoms, bonded to only a single phosphorus atom. Six of the oxygen atoms act as a bridge between two phosphorus atoms.

Using four tetrahedral balls (black) as the phosphorus atoms, ten trigonal balls (dark blue) for oxygen atoms (this is a reasonable approximation to the actual 123.5° bond angle), and 16 sticks, construct a model which corresponds to the structure illustrated in Fig. 1.

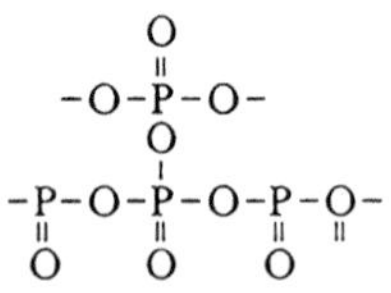

```
         O
         ||
      -O-P-O-
         O
         |
   -P-O-P-O-P-O-
    ||   ||  || ||
    O    O   O
```

Fig. 1

To aid in construction, leave off the terminal oxygens (those that are double bonded) until the end. At first, you will have a somewhat planar structure. However, recognizing that all of the oxygens on the structure at this point are bridging oxygens, twist about the bonds until the four phosphorus atoms are connected to each other via three bridging oxygen atoms. Note that the four phosphorus atoms are at the corners of a tetrahedra with the oxygen atoms located along the edges of the tetrahedra. Now add the terminal oxygen atoms to the model.

Inspect the model which you have constructed and answer the following questions.

1. Atomic composition (formula):

2. Number of bridging oxygen atoms:

3. Size of closed rings:

4. Conformation of the rings:

5. Number of rings in the structure:

6. The spatial arrangement about the phosphorus atoms:

7. Sketch the structure of your model below, making sure that all of the atoms are visible.

8. What structure would you predict for the molecular form of P_4O_6?

Exercise #2

The molecular P_4O_{10} unit might also be represented by an assemblage of four tetrahedra. A phosphorus atom is visualized to be at the center of the tetrahedron. Oxygen atoms are located by the vertices of the tetrahedron.

Fold up four tetrahedra. Make sure that the creases are made sharply. Make sure that the tabs are folded in toward the body of the tetrahedron and that they are not exposed when the tetrahedron is finished.

Arrange three of the paper tetrahedra so that they are face down on the table with adjacent vertices on the plane of the table top touching each other. Carefully hold a third tetrahedron so that three of its vertices are touching the three vertices of the tetrahedra on the table not touching the surface of the table. Glue the corners which touch using the hot glue gun. This is a polyhedral representation of P_4O_{10}.

Look at the ball and stick representation of P_4O_{10} and the tetrahedral representation of P_4O_{10} and note that they represent the same molecule.

Exercise #3

A short diversion is needed at this point to consider regular polyhedra. A regular polyhedron is defined as one with the same regular polygon as faces. A regular polygon is defined as a polygon with equal sides and equal angles. Thus a regular triangle is an equilateral triangle, a regular tetragon is a square, etc.

Three regular polyhedra can be constructed using equilateral triangles as faces. If three triangles meet at each vertex of the polyhedron, a tetrahedron results. If four triangles meet at each vertex of the polyhedron, an octahedron results. If five triangles meet at each vertex of the polyhedron, an icosahedron results.

Examine each of these regular polyhedra with triangular faces.

Draw the tetrahedron face down and edge down.

Draw the octahedron point down, edge down, and face down.

Place the octahedron face down on the table and note that it is a special case of a trigonal antiprism.

Place the icosahedron point down on the table and note that it is a special case of a bicapped pentagonal antiprism.

Recognizing that the icosahedron is a bicapped pentagonal antiprism, deduce the number of faces in an icosahedron.

One regular polyhedron can be formed using square faces. The polyhedron thus formed is a cube.

Draw the cube face down and edge down.

One regular polyhedron can be formed using pentagonal faces. The polyhedron thus formed is a pentagonal dodecahedron.

How many faces and how many vertices does a pentagonal dodecahedron have? How many faces and how many vertices does an icosahedron have? These polyhedra are reciprocal to one another. What is the definition of reciprocal polyhedra? Show that the cube and octahedra are also reciprocal.

How many regular polyhedra can be formed?

Exercise #4

The P_4O_{10} molecule can also be represented as a close-packed structure of oxygen atoms or ions with phosphorus in some of the tetrahedral holes. Using the triangular array of six plastic spheres, pile four more spheres of the same size on this array in order to represent the positions of the oxygen atoms. Look at the ball and stick model and the tetrahedral model and note the equivalence of the relative positions of the oxygens. Locate a tetrahedral hole where the phosphorus is located by looking at the tetrahedral structure. Select four small beads which will fit into the tetrahedral hole and place in the appropriate places to correctly represent the location of the phosphorus atom. Compare the three structures which represent the same P_4O_{10} molecule.

Note that in the center of the close-packed structure between the glued array of six spheres and the second layer, there is an octahedral hole. Select a marble which fits into this hole. By noting the relative size of the spheres which fit into a tetrahedral hole and into an octahedral hole, can you make a statement as to the relative size of these two holes? Keep in mind that a close-packed structure represents the maximum sphere packing into a given volume.

Exercise #5

Compounds with the same formula, AX, can have different structures depending on the coordination numbers for A and X. If each has a coordination number of 1, only a diatomic molecule is possible. If each has a coordination number of 2, only chains or rings are possible. With coordination numbers of 3 or more, two- and three-dimensional nets are possible.

Table I

	AX COMPOUND				
	HI	AuI	CuI	NaI	CsI
Coordination No. of A by X:	1	2	4	6	8
Structure Type:	Diatomic molecule	1D chain	3D zinc blende	3D NaCl structure	3D CsCl structure

There is no connection between the simplest formula of a compound and its structure or between the formula and the coordination number of atoms. It is, however, possible to derive the simplest formula from repeat units of structures.

1. Using two different colored octahedral balls, prepare a ball-and-stick model representative of AuI. If, for example, the balls are pale blue and yellow, use the 180° bond angles of the yellow balls and the 90° bond angles of the pale blue balls. A zig-zag chain structure should result. If the yellow ball represents a gold atom, then the pale blue ball represents an iodine atom.

   Draw the structure for AuI.

   Indicate the repeat pattern. What is the formula for each repeat unit?

2. Boron nitride, BN, is an example of an AX formula with a 2D layer structure analogous to that of graphite. Using the ball-and-stick representation prepare a model of a BN layer with each atom coordinated only to the other type of atom.

Draw the structure for boron nitride.

What is the bond angle in the BN layer?

Indicate the repeat pattern for the BN layer. What is the formula for each repeat unit?

3. Red mercury(II) iodide, HgI_2, is an example of a 4-connected net. The structure of red mercury(II) iodide can be represented as follows:

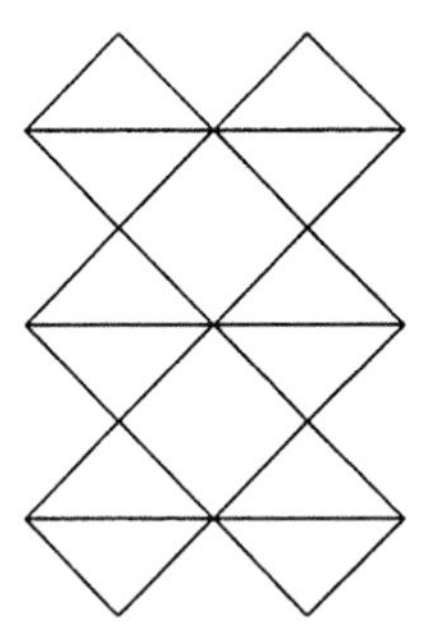

Fig. 2

Make six paper tetrahedra as previously directed. Using the drawing as a guide, glue the corners which touch to make a polyhedral representation of red mercury(II) iodide.

If mercury is located at the center of the tetrahedra and the iodines at the vertices of the tetrahedra, draw a structural representation using Hg and I atoms and bonds. Show why this is a 4-connected net.

Look at the tetrahedral model and identify a repeating unit. Sketch this repeating unit on the drawing of the tetrahedral representation of red mercuric iodide. (You may wish to extend the structure shown in Fig. 2.)

What is the formula for each repeat unit?

4. The ZnO_2 in $SrZnO_2$ also forms a 4-connected net. However, unlike HgI_2, the ZnO_2 layer must accomodate the accompanying strontium ion. In order to do this, the layer is found to be puckered. By being puckered, or corrugated, the strontium ions are able to be accomodated between the ZnO_2 layers. Look at the polyhedral representation

drawn below and explain how the model would differ from the red mercury(II) iodide model made in part 3.

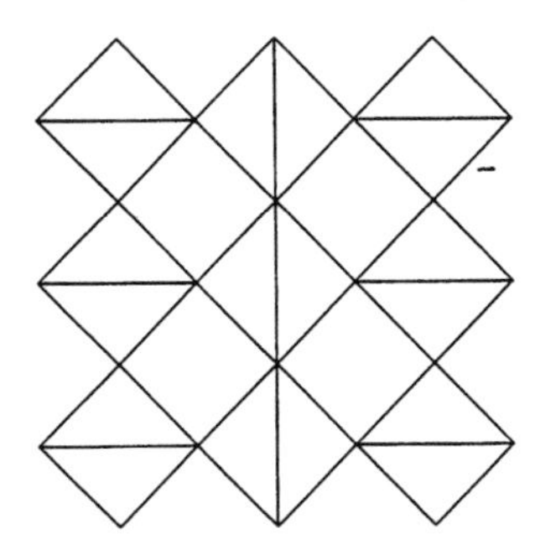

Fig. 3

5. Although mercury is 4-connected (tetrahedrally coordinated), it forms a two-dimensional net. The sulfur around Pd in PdS is 4-connected (square planar). However, PdS forms a three-dimensional net. Study the PdS model made in the CuO experiment.

6. $PdCl_2$, like PdS, is square planar. Whereas sulfur in PdS is 4-coordinated, the chlorines in $PdCl_2$ are 2-coordinated. $PdCl_2$ forms chains. Using octahedral balls (yellow) for palladium and octahedral balls (pale blue) for chlorine, make a partial structure of $PdCl_2$.

Write the structure for $PdCl_2$.

Indicate the repeating pattern. What is the formula for each repeat unit?

7. $BeCl_2$ consists of tetrahedrally coordinated beryllium and 2-coordinated chlorines. Using pairs of 90° bonds in the distorted tetrahedral balls (red) and the 90° bonds in the octahedral balls (yellow), make a structure of $BeCl_2$.

The $BeCl_2$ structure may also be represented by a model with tetrahedra which are sharing opposite edges. Make four paper tetrahedra and glue them together to represent $BeCl_2$.

Locate the positions of the atoms in the tetrahedral structure.

Draw a structure of $BeCl_2$ showing Be and Cl atoms.

Draw a structure of the tetrahedral representation of $BeCl_2$.

-

Indicate the repeat unit. What is the formula for each repeat unit?

8. Selenium(IV) oxide, $(SeO_2)_n$ consists of a chain wherein tetrahedra are sharing two vertices. One of the unshared vertices of each tetrahedra represents the location of an electron pair. The other unshared vertex locates a terminal oxygen atom. Arrange some tetrahedra to represent the SeO_2 structure and sketch two representations of the structure, one showing the tetrahedra and the other showing the selenium and oxygen atoms.

Indicate the repeat unit for $(SeO_2)_n$ and give the formula for the repeat unit.

9. Sulfur(VI) oxide (sulfur trioxide), $(SO_3)_n$, also consists of tetrahedra sharing vertices. Each of the vertices represent positions of oxygen atoms. Sketch the structure of the SO_3 chain showing the sulfur and oxygen atoms.

Indicate the repeat unit for $(SO_3)_n$ and give the formula for the repeat unit.

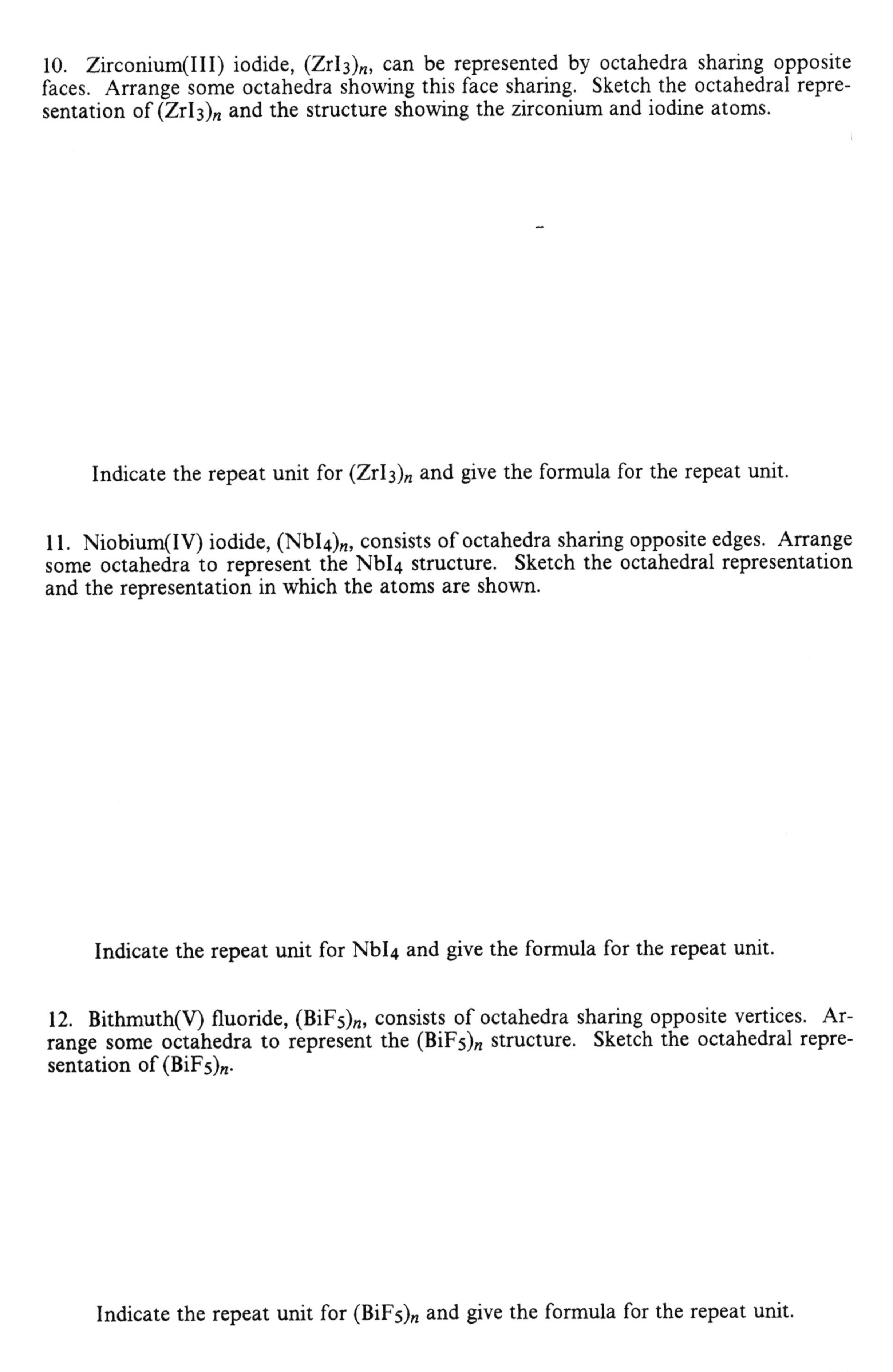

10. Zirconium(III) iodide, $(ZrI_3)_n$, can be represented by octahedra sharing opposite faces. Arrange some octahedra showing this face sharing. Sketch the octahedral representation of $(ZrI_3)_n$ and the structure showing the zirconium and iodine atoms.

Indicate the repeat unit for $(ZrI_3)_n$ and give the formula for the repeat unit.

11. Niobium(IV) iodide, $(NbI_4)_n$, consists of octahedra sharing opposite edges. Arrange some octahedra to represent the NbI_4 structure. Sketch the octahedral representation and the representation in which the atoms are shown.

Indicate the repeat unit for NbI_4 and give the formula for the repeat unit.

12. Bithmuth(V) fluoride, $(BiF_5)_n$, consists of octahedra sharing opposite vertices. Arrange some octahedra to represent the $(BiF_5)_n$ structure. Sketch the octahedral representation of $(BiF_5)_n$.

Indicate the repeat unit for $(BiF_5)_n$ and give the formula for the repeat unit.

Exercise #6

The diamond structure is an example of a 3D 4-connected net. Using the sticks and tetrahedral balls (black), prepare the cubic diamond structure. In the diamond structure, every tetrahedral ball is connected to four other tetrahedral balls. Any two tetrahedral balls connected to each other must be in the staggered ″ethane″ configuration. All the rings formed will be six membered rings and will be_of the chair configuration.

There is a hexagonal analogue to the cubic diamond structure. If the hexagonal structure is available, answer the following questions.

> How does the configuration of the rings differ in the cubic and the hexagonal structures?
>
> All of the ″ethane″ configurations in the cubic structure were staggered. What can be said about the ″ethane″ configurations in the hexagonal structure?

Your instructor will explain how zinc blende and wurtzite structures relate to the cubic and hexagonal structures being discussed. Some forms of ice and SiO_2 also relate to these two structures. Make notes on how these structures relate to the cubic diamond structure and to the analogous hexagonal structure.

Preparation for the Next Class

Take home 18 paper octahedra. Assemble these octahedra and bring them back to the next class in a grocery bag. The same bag can be used to take the finished models home. To assemble the octahedra, fold each crease sharply and all of the tabs in toward the body of the octahedron. None of the tabs are to extend outside the body of the finished octahedron.

Auxiliary Experiment

Make the hexagonal structure related to the cubic diamond structure. In this structure, all of the ″ethanes″ in a layer will be staggered, and all the ″ethanes″ perpendicular to these layers will be eclipsed.

Questions to be Considered

1. What structural features are emphasized in each of the structural representations?

2. Bonding in the solid state is very much affected by geometry. Consider that the bond angle of H-O-H in water is 104.45^o, in OF_2 the F-O-F angle is 103°, in dimethyl ether the C-O-C angle is 111°, and in α-quartz the Si-O-Si angle is 144°. The Si-O-Si bond angle varies from 125° to 160° in various felspars. Comment on these angles in comparison to what might be expected on the basis of the hybridization of the oxygen orbitals.

Helpful Background Literature

1. Wells, A. F. "Structural Inorganic Chemistry", 5th ed.; Clarendon Press: Oxford, 1984.

STRUCTURAL INORGANIC CHEMISTRY VIA MODELS - 2

Session Two of Three

Equipment Needed:	6 trigonal balls (dark blue) 14 octahedral balls (yellow) 1 board with nine rod array 4 close-packed layers 2 layers with 6 octahedra 1 layer with 7 octahedra 3 close-packed layers with some blue spheres 18 octahedra (folded paper) 8 2-connecters 12 4-connecters 1 plastic blue sphere 1 tin of marbles 5 hot melt glue guns (1 per 3 students) bits of polycrystalline wax wooden shish kebab skewers
Chemicals Needed:	None
Time Needed:	3 hours
Safety Notes:	None

Introduction

In *Session One,* inorganic structures were represented by ball and stick models and by polyhedral models. In *Session Two,* the use of polyhedra to represent models is continued. The concept of nets is introduced. In addition, the third type of structural representation, that of packed spheres, is introduced.

Experimental Procedure

Prelab assignment

Before coming to lab, construct 18 folded paper octahedra. Bring them to lab in a grocery bag. The same grocery bag can be used to carry the finished models back home. To assemble the octahedra, fold each crease sharply and all of the tabs in toward the body of the octahedron. None of the tabs are to extend outside the body of the finished octahedron.

Exercise #1

In *Session One,* examples of structures in which tetrahedra are sharing vertices and edges were illustrated. In addition, octahedra were shown to share vertices, edges, and faces. As a review, sketch each of the following:

1. In α-UF_5, octahedral UF_6 groups share opposite vertices to form chains.

   Draw the octahedral representation.

   Find the repeat unit and confirm the formula.

2. Niobium(IV) iodide was discussed in *Session One,* Exercise #5, Part 11. Octahedral NbI_6 groups share opposite edges. Without looking at *Session One*, sketch the niobium(IV) iodide chain and confirm the formula.

3. In BiI_3, SbI_3, AsI_3, $FeCl_3$, and $CrBr_3$ octahedral MX_6 groups share three alternate edges to form sheets or a two-dimensional repeating net. Arrange 12 octahedra on the table top to illustrate alternate edge sharing. Note the rings of six which are formed.

Draw the octahedral structure of BiI_3.

Pick out a repeating unit and confirm the stoichiometry.

4. In the Mo_4F_{20} structure, two adjacent corners of the octahedra are shared. This leads to four octahedra forming a ring. This is in contrast to the chain structure α-UF_5 and BiF_5 which results from two opposite corners of the octahedra being shared. Make a structure of Mo_4F_{20} using the hot melt glue gun and the paper octahedra. First glue two octahedra together as shown following the directions below the sketch.

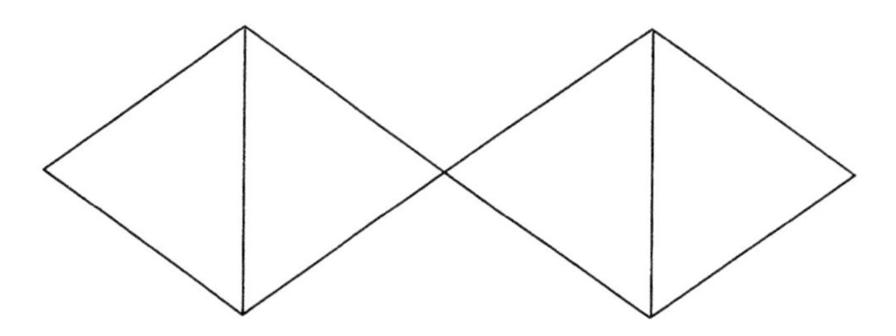

Fig. 1.

In order to get an accurate model, it is important that the distance between the edges (1 to 2) be the square root of twice the square of the length of the edge of the octahedra. Why?

In order to get accuracy, stand the two octahedra being glued on an edge along the lines of the template. Glue the corners touching with the hot melt glue. Allow to stand until the glue hardens. Join four pairs of octahedra in the identical manner.

When the glue has hardened on the pairs of octahedra, join the two pairs into the Mo_4F_{20} ring system.

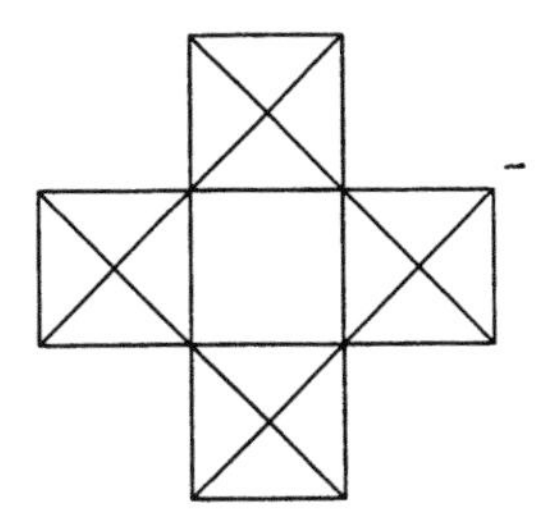

Fig. 2

Make sure the hole in the center is square. The four corners of the octahedra touching the table top should touch the template at the points marked A, B, C, and D.

When the glue has hardened sufficiently to remove the Mo_4F_{20} from the template, make a second Mo_4F_{20} model.

Exercise #2

1. Take two of the Mo_4F_{20} structures and arrange them so that the structure can be shown to continue in an infinite two-dimensional array. This is a representation of the SnF_4 structure. One way of describing the SnF_4 structure is to say that all of the equatorial vertices are shared to form an infinite two-dimensional array.

   Sketch a partial structure of the SnF_4 structure.

2. Pile one of the Mo_4F_{20} structures on top of the second Mo_4F_{20} structure. Put glue at the corners to make the ReO_3 structure. The ReO_3 structure is an infinite three-dimensional array in which all six vertices of each octahedra are being shared. Further characteristics of this model will be described later in this experiment.

Exercise #3

At this point, it is instructive to consider the concept of nets.

If each unit, whether it be an atom or a polyhedron is connected by only one connection (coordination number of 1), only discrete molecules can be formed. Note the HI structure at the beginning of Exercise #5 of *Session One*. If the atom or polyhedron is connected by two connections (connected by single or multiple connections to two others), either chains or rings are formed.

Confirm this for the Mo_4F_{20}, AuI, SeO_2, $BeCl_2$, and NbI_4 structures. The chains from amongst this set can be considered to be one dimensional nets.

If each atom or polyhedra is 3-connected, either two-dimensional or three-dimensional nets are possible.

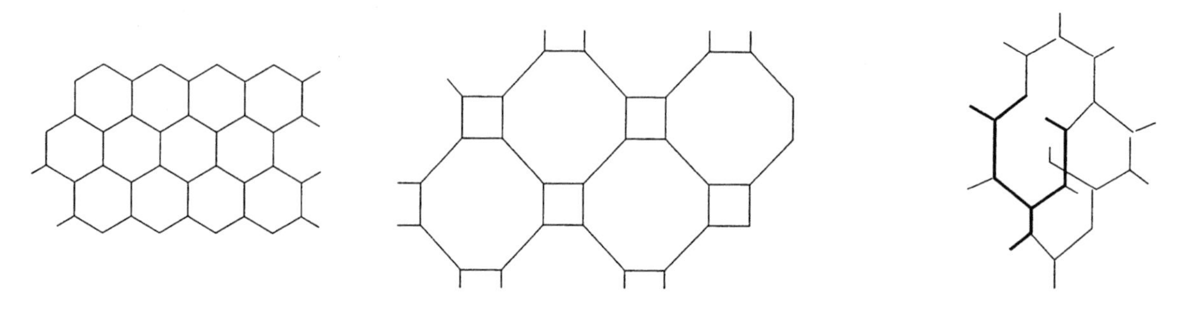

Fig. 3.

If each atom or polyhedron is 4-connected, two-dimensional or three-dimensional nets are possible. The simplest two-dimensional 4-connected net is shown.

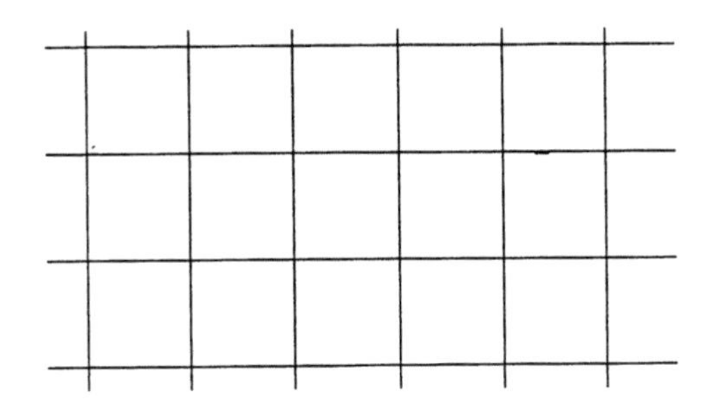

Fig. 4.

Recall that the red mercury(II) iodide structure discussed in *Session One*, Exercise #5, part 3 was an example of a 4-connected net of the type shown above. Each tetrahedra shared all four vertices.

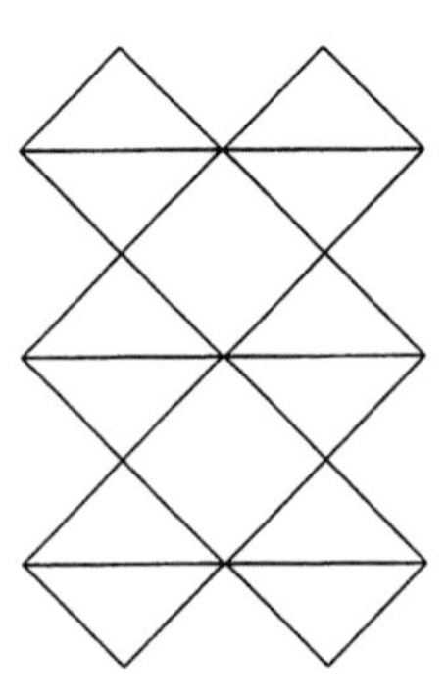

Fig. 5.

What atom is at the interconnection of the two-dimensional 4-connected net?

Do atoms lie on the lines of the net?

Are all atoms of the net on the same plane?

In classifying actual chemical structures, the mathematical concepts such as points, lines, polygons, etc. are applied to real atoms occupying space. Atoms which are located at the intersections of a four-connected net are four-coordinated. For some structures, atoms lie along the lines of the net. The mathematical concepts tell us the range of possibilities and help us bring more order to the structures which we observe.

An example of a three-dimensional 4-connected net is the diamond structure made in *Session One*.

Review question. What was the conformation of the "ethane structure" in the diamond structure?

Review question. What was the conformation of the six membered rings in the diamond structure?

As previously stated, the concept of nets is a part of mathematics. For example, calculations can be made for the types of polygons possible for 3-connected nets. Of the two examples shown for two-dimensional nets, the first had polygons which were all hexagons. The second had polygons half of which were squares and half were octagons.

There are polymorphs of SiO_2 which are representatives of the two-dimensional 4-8 net. These are $BaFeSi_4O_{10}$ (gillespite), $CaCuSi_4O_{10}$ (Egyptian blue), and $Ca_4Si_8O_{20}.KF.8H_2O$ (apophyllite).

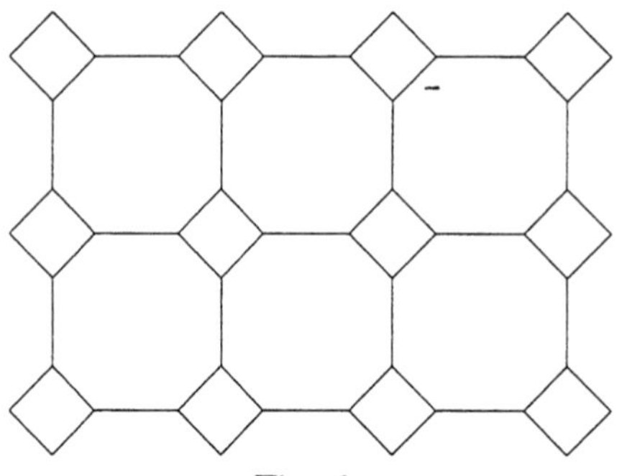

Fig. 6.

On the net shown, locate the silicon atoms and the oxygen atoms. As with the P_4O_{10} structure, there are two different kinds of oxygen atoms. Where are these atoms and how do they vary? The layers of composition Si_2O_5 are held together by the cations. What gives rise to the negative charges on the layers?

In a similar way, the HgI_2 structure is related to the simple 4-connected two-dimensional net. The mercury and iodine atoms are not planar, yet topologically, the HgI_2 structure is identified with the 4-connected 2D net.

What kind of net is associated with a single layer of the BiI_3 structure?

The following table shows the relation between the number of connections, the polyhedra which can be related to this connectedness, and the nets which are associated with each connectedness.

Table I

no. of connections			
2	rings or infinite chains only		
	polyhedra	2D	3D
3 4 5	x x one only*	x x x	x x x
6 7 8		one only	x x x

* only true for simple systems with all faces of the same kind.

Exercise #4

1. Make ten more paper octahedra. Arrange the ten paper octahedra to demonstrate the BiI_3 structure. Show the alternate edge sharing by forming one 6-membered ring with

the remaining octahedra placed in such a way that at least one octahedron in the incomplete structure will have the three alternate edges shared.

2. Place an octahedron into the holes of the six membered ring in the BiI_3 structure. This produces the partial CdI_2, $CdCl_2$, or $MgCl_2$ structure. The complete structures will be considered in Exercise #7.

How many edges are shared by each octahedra in a single layer?

Sketch the octahedral representation of the CdI_2 layer.

Exercise #5

If the maximum number of spheres are to be packed into a given volume, the spheres must be close-packed. The two common types of close-packing observed in solid structures are cubic close-packing (ccp) and hexagonal close-packing (hcp). The layers for either close-packed arrays are the same.

1. Examine the close-packed array provided and note how the second row of spheres nests in the first row of spheres, i.e., the spheres in the second row are positioned in between two spheres in the first row.

Sketch a close-packed array.

How many spheres surround each sphere in this close-packed layer?

2. Place a second close-packed array on the top of the first close-packed array so that each sphere of the second layer nests in a notch of the first layer. Place the third close-packed array on top of the second.

Note that there are two choices in placing the third layer. In one, each of the spheres is located directly above the spheres of the first layer. In the second, the spheres of the third layer are not directly above either a sphere in the first layer or a sphere in the second layer. In hexagonal close-packing the atoms in the layers are arranged to form AB, AB, AB layers. A cubic close-packing of spheres results when the close-packed layers form ABC, ABC, ABC layers.

3. Arrange the close-packed layers provided into a hexagonal close-packed arrangement. Using a thin stick of 2 or 3 mm in diameter, show that there are certain places in the stack where the stick will pass through.

What is the three dimensional unit cell for hexagonal close-packing? Show that there are two spheres per unit cell.

4. Arrange the close-packed layers provided into a cubic close-packed arrangement. Using the thin stick of 2 or 3 mm diameter, probe to see if there are places where the stick will pass through. Are there any?

5. A unit cell for cubic close-packing is a face centered cube. Take the board with nine brass rods. Using yellow balls, slip a ball on each of the four corner rods and one on the center rod. Next place four balls on each of the rods between the corner rods. Finally, slip a ball on each corner rod and the center rod. The array is the fcc unit cell.

Sketch the fcc unit cell.

How many spheres are in the cubic unit cell? Confirm that corner atoms are shared with 8 unit cells, edge atoms with 4 unit cells, and face atoms with 2 unit cells.

6. The unit cell in a ccp array can best be located by piling loose wooden or plastic balls in a cubic close-packed arrangement. This can best be done by using a piece of fibrous bulletin board material about 12 inches by 12 inches. Strips of the same board material can then be attached to the larger board by using straight pins toed into the strips and into the large board. These strips, when placed in the appropriate positions will prevent the loose balls from rolling about. Make a four layer ccp structure using the loose balls. By using the unit cell made by slipping balls on the rods as a guide, remove or add loose balls to the ccp structure to show a face of the unit cell.

Metals such as Mg, Co, Ti and Zn have the hcp structure, while others including Al, Ca, Cu, Ag and Au have the ccp arrangement. These two types of packing are used in the structures of about 70% of all the metals. The other type of packing of interest for metal structures is the body-centered structure. This structure which is not closest packed is found in metals like Na, K, Ba and Cr.

Exercise #6

The spheres in a close-packed array occupy only 74% of the space. The voids are characterized as tetrahedral holes and octahedral holes. The tetrahedral hole is surrounded by four spheres. The octahedral hole is surrounded by six spheres. In halides and oxides, the large negative ions form close-packed arrays. The smaller positive ions are located in the octahedral or tetrahedral holes.

1. Place an appropriate plastic ball into an octahedral hole and a tetrahedral hole.

   The two holes differ in size. Which is larger?

   How many octahedral holes are there per close-packed sphere?

   How many tetrahedral holes are there per close-packed sphere?

Exercise #7

If the halide ion in CdI_2 or $CdCl_2$ is recognized to be much larger than the dipositive cadmium ion, then the structure can be visualized to be represented by close-packed spheres representing the halide ion with the small dipositive cadmium ion fitting into selected holes formed by the close-packed spheres.

1. In CdI_2 all the octahedral holes are occupied between alternate layers in a hcp array. Put the appropriate sized "marbles" in the octahedral holes of the close-packed array to simulate the CdI_2 structure.

2. Arrange the cardboard octahedra to simulate the CdI_2 structure. Indicate which octahedra contain cadmium ions and which octahedra are empty. Many compounds are found with the CdI_2 structure. Some of these are CaI_2, MgI_2, ZnI_2, PbI_2, MnI_2, FeI_2, CoI_2, $FeBr_2$, $NiBr_2$, etc.

3. Arrange the close-packed spheres in a ccp array with all the octahedral holes in alternate layers filled. This is the $CdCl_2$ structure. Other compounds exhibiting this structure are $FeCl_2$, $CoCl_2$, $NiCl_2$, $MgCl_2$, $ZnBr_2$, and $MnCl_2$.

4. Arrange the cardboard octahedra to simulate the $CdCl_2$ structure. Be able to indicate to the instructor the correspondence between the close-packed model and the octahedral model.

5. Refer to the BiI_3 structure made with octahedra in Exercise #4. Arrange the octahedral hole filling to make the close packed analog to the BiI_3 structure. The BiI_3 structure is hcp with alternate octahedral layers 2/3 filled. The question then arises how the succeeding layers should be filled relative to the first. The elevation can be shown as follows:

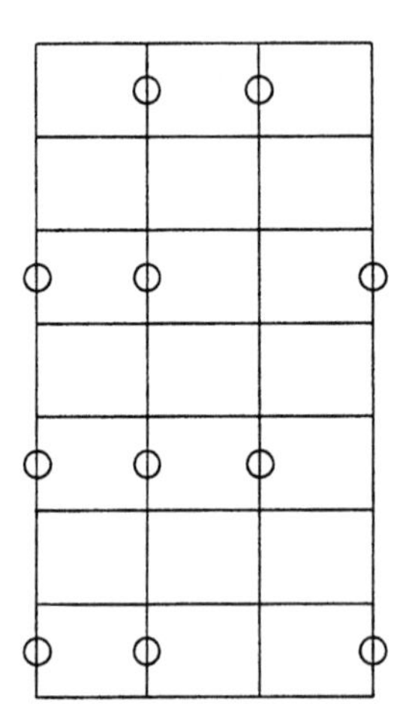

Fig. 7.

Explain why the stoichiometry is 1 : 3.

The $CrCl_3$ structure is related to the BiI_3 structure except that the array is ccp rather than hcp. There are not as many examples of the $CrCl_3$ structure as there are for the BiI_3 structure. This latter structure is exhibited by SbI_3, AsI_3, $FeCl_3$, $CrBr_3$, $ScCl_3$, $TiCl_3$, VCl_3, $FeBr_3$, and $TiBr_3$.

The family of close-packed octahedral structures is shown in Table II on the following page.

Table II

Close-packed Octahedral Structures

Fraction of Octahedral holes occupied	Sequence of c.p. layers		Formula	C.N.s of M and X
	ABAB. . .	ABCABC. . .		
All	NiAs	NaCl	MX	6:6
2/3	α-Al_2O_3 $LiSbO_3$		M_2X_3	6:4
1/2	Layer structures CdI_2	$CdCl_2$	MX_2	6:3
	Framework structures $CaCl_2$ Rutile (TiO_2) $NiWO_4$ α-PbO_2 α-AlOOH	Atacamite ($Cu_2(OH)_3Cl$) Anatase (TiO_2)		
1/3	Chain structures ZrI_3		MX_3	6:2
	Layer structures BiI_3	YCl_3		
	Framework Structures RhF_3			
1/4	α-NbI-4	NbF_4	MX_4	6:1&2
1/5	Nb_2Cl_{10}	UF_5	MX_5	6:1&2
1/6	α-WCl_6		MX_6	6:1

Exercise #8

The ReO_3 structure constructed with eight octahedra is a close-packed oxygen structure with every fourth oxygen position vacant. The location of the Re atom is at the center of each of the octahedra.

1. Tip the ReO_3 structure on a face at one of the corner octahedra. Locate the close-packed layers. The position of the missing oxygen is at the center of the space created by the eight octahedra. This hole is called a cuboctahedral hole. What is a cuboctahedron?

2. The perovskite structure is related to the ReO_3 structure in that it has the same framework structure with the addition of a positive ions larger than Re which occupy the position of the missing oxygen. An example of a perovskite structure is the mineral perovskite, $CaTiO_3$. A large number of complex oxides exhibit this structure as well as some complex halides such as $RbCaF_3$. In addition to the structures which exhibit the

ideal cubic symnmetry such as $SrTiO_3$, $BaZrO_3$, and $EuTiO_3$, there are a number of compounds which are not cubic but which are slightly distorted perovskite structures. Perovskite itself is not perfectly cubic, but belongs to the distorted group of compounds.

> Look at the ReO_3 structure and visualize a perovskite $SrTiO_3$. On the basis of the structure, justify the stoichiometry, i.e., one strontium and one titanium to three oxygens.

Exercise #9

Table II tabulates the variety of close-packed octahedral structures. A similar table can be constructed for tetrahedral structures.

> The CdI_2 and $CdCl_2$ structures illustrate one way to fill 1/2 of the octahedral holes. Suggest other patterns of 1/2 hole filling.

> Zinc blende, ZnS, is a cubic close-packed array of sulfur atoms with the small dipositive zinc ions occupying half of the tetrahedral holes between each layer of the ccp sulfide ions. Wurtzite is ZnS wherein the dipositive zinc ions are occupying half of the tetrahedral holes between each layer of the hcp sulfide ions. Show the relation between the percent hole filling and stoichiometry of ZnS and cadmium halides.

Exercise #10

Rutile, TiO_2, is a structure in which edges and corners of octahedra are shared. The sketch of a partial rutile structure is shown:

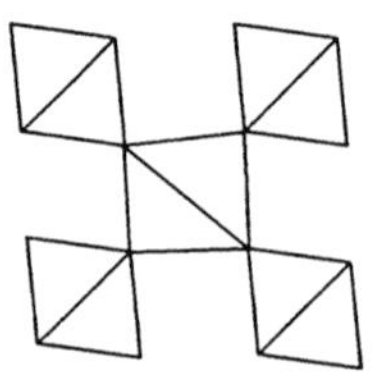

Fig. 8.

Using the hot glue gun, glue five pairs of paper octahedra so that an edge is being shared. Two representations for the pairs glued together are as follows:

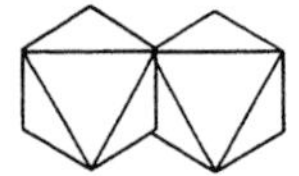

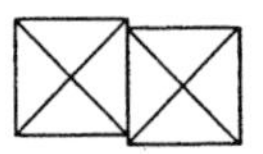

Fig. 9.

Standing two pairs of edge-shared octahedra so that it looks like the part of the drawing when viewed edge up, glue the points of the center set of octahedra to the corners of the edge which is shared for the "outer" pair of octahedra. Note that the center column of octahedra is displaced by half an octahedra from the corner or outer pairs. When the structure shown in the drawing is finished, extra octahedra can be glued between the outer columns replicating the center octahedra. This will make the structure more rigid and make clear the repeating pattern.

The following is a picture of the unit cell as it appears in many inorganic textbooks.

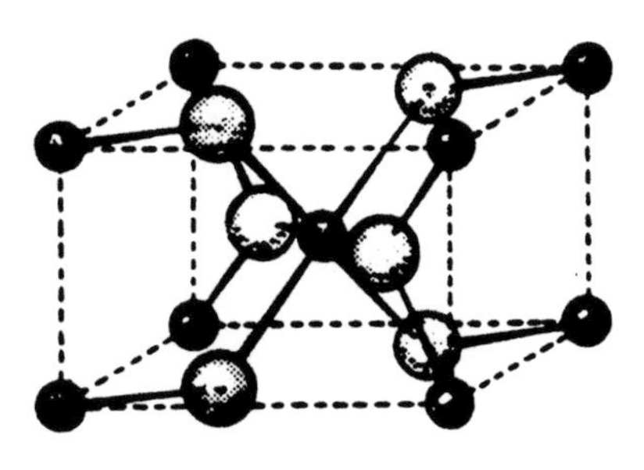

Fig. 10.

Find the unit cell in your model.

How many atoms of titanium and how many atoms of oxygen are in the unit cell?

Auxiliary Experiment

Make a ball and stick model of NaCl. Compare this model to the polyhedral representation of NaCl and the close-packed representation of NaCl.

Questions to be Considered

1. The three structural representations, ball and stick, polyhedral, and close-packing, emphasize different structural aspects. List the helpful features of each structural representation.

2. Locate a repeating pattern in the ReO_3 structure and confirm the stoichiometry.

3. Locate a repeating pattern in the CdI_2 structure and confirm the stoichiometry.

Helpful Background Literature

1. Wells, A. F. "Structural Inorganic Chemistry", 5th ed.; Clarendon Press: Oxford, 1984.

2. Galasso, F. S. "Structure and Properties of Inorganic Solids"; Pergamon Press: Oxford, 1970.

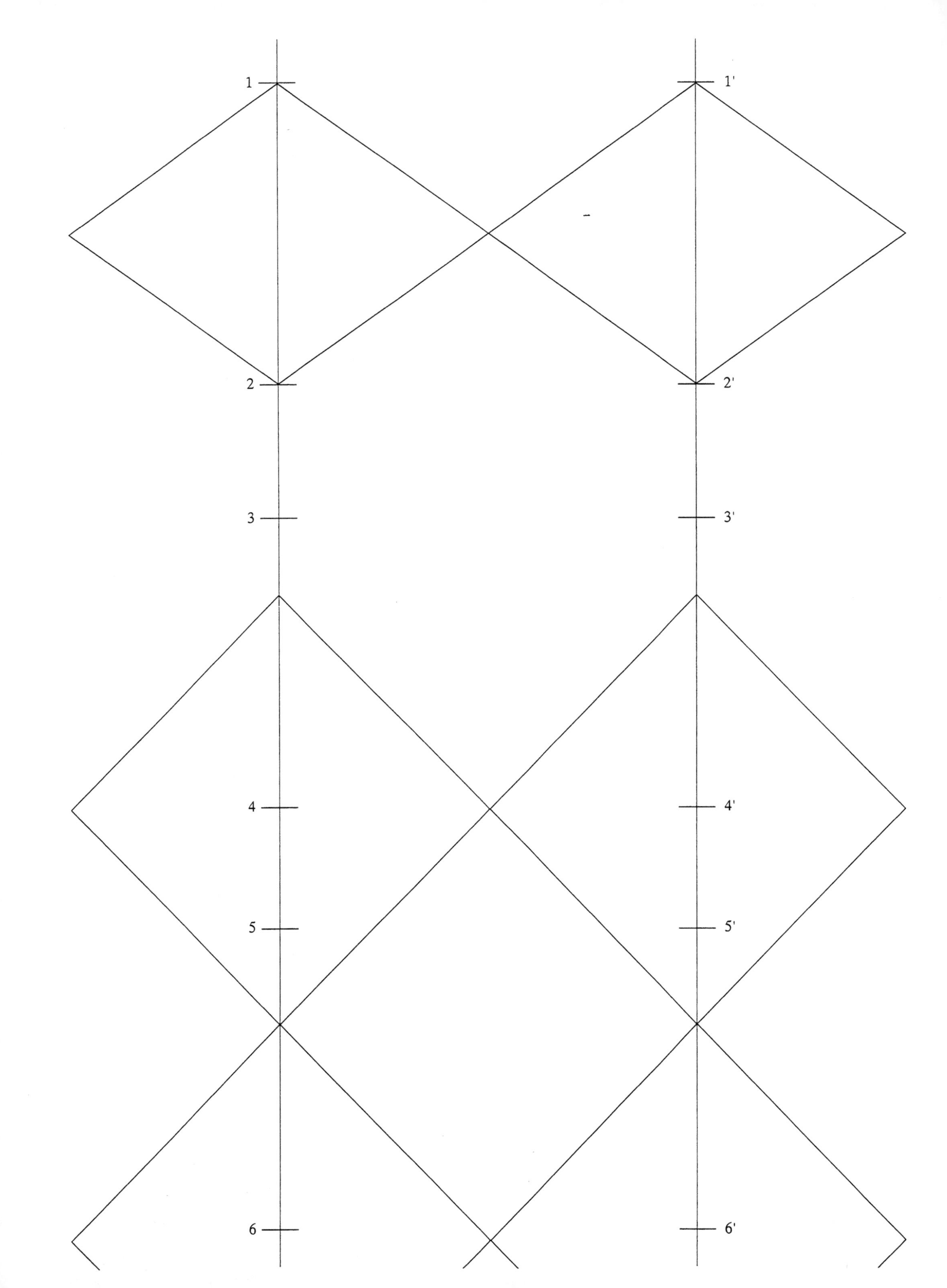
1
2
3
4
5
6
1'
2'
3'
4'
5'
6'

SODIUM PEROXOBORATE

Equipment Needed:
150 mL beaker
125 mm test tubes
Buchner funnel
filter flask
ice bath

Chemicals Needed:
4.75 g borax (sodium borate)
1 g NaOH
5.7 mL 30% H_2O_2
12 mL 0.1 N HCl
10 mL 0.1 M $KMnO_4$
0.2 g KI
Litmus paper
5 mL alcohol
5 mL ether

Time Needed: 2 hours

Safety Notes:

Hydrogen peroxide in the vapor form is an irritant and caustic to the mucous membranes, eyes and skin. If exposed to vapors of hydrogen peroxide, or if the liquid comes into contact with the skin, wash with copious amounts of water.

Hydrogen peroxide and inorganic peroxides are strong oxidizing agents. If organic solvents come into contact with peroxides, do not distill the solvent without adequate peroxide removal or destruction procedures. One of the most frequent causes of laboratory explosions is the distillation of solvents which have been oxidized to form peroxides. It is important to understand the chemistry of peroxide formation.

The allowable concentration of 90% hydrogen peroxide in the air for daily work exposure is 1 ppm. Higher concentrations for short exposures are equivalent to low concentrations for long exposures. In any case, it is recommended that the hydrogen peroxide be used in the hood. Use care, but do not panic. Keep in mind that dilute hydrogen peroxide was a standard disinfectant used for minor wounds for generations.

Purpose

This preparation illustrates the formation of the salt of a peroxy acid from the precursor salt and hydrogen peroxide. The equilibrium nature of the reaction is illustrated. The experiment also provides an opportunity to consider the structure of some inorganic peroxides.

Introduction

Though the usual hydrogen peroxide manufactured and sold for household applications is 3% in water, pure hydrogen peroxide can be prepared. In the pure state, it is a faint blue syrupy liquid freezing at -2°C. Vapor pressure vs. temperature studies indicate a boiling point of about 150°C. Because pure hydrogen peroxide decomposes violently when heated much above 80°C, there are no real experimental points to confirm this extrapolated boiling point. Other than the freezing point, physical properties vary considerably from that of water. The refractive index (1.414 at 22°C vs. 1.333 at 20°C), the density (1.442 at 25°C vs. 1.000 at 4°C), and the viscosity in cp (1.25 vs. 1.00 at 20°C) all indicate higher values for hydrogen peroxide as compared to water.

One commercial method for making hydrogen peroxide is to electrolytically oxidize ammonium hydrogen sulfate to ammonium peroxosulfate. This latter is equilibrated in an acidic aqueous solution to yield ammonium hydrogen sulfate and approximately 6% hydrogen peroxide. Today the preferred industrial route involves dibenzohydroquinone which is oxidized to yield the dibenzoquinone and an approximately 30% hydrogen peroxide solution. This latter method is advantageous in that less energy has to be expended to concentrate the hydrogen peroxide.

Hydrogen peroxide is consumed in the 200 million lbs per year level. The leading use of about 75×10^6 lbs is in the chemical industry for making inorganic peroxygen compounds, organic peroxygen compounds, glycerin, plasticizers, amine oxides, etc. The textile industry consumes some 44×10^6 lbs primarily for bleaching purposes as does the pulp and paper industry (22×10^6 lbs). Water treatment requires 28×10^6 lbs. For example, hydrogen sulfide can can be removed from water systems by oxidizing the H_2S to S using hydrogen peroxide.

For some purposes, the liquid and somewhat unstable hydrogen peroxide is not as useful as a solid peroxide containing active oxygen. Sodium peroxoborate is such a compound. It is used as the monohydrate in dental cleansers and hair lighteners. It is used as a bleach in laundry detergents as well as some bleach preparation. Because peroxoborates are not as effective as other bleaches in cold water, there is a drop in peroxoborate sales with an increasing percentage of low temperature detergent manufacture.

The peroxoborate structure contains the oxygen-oxygen bond characteristic of peroxides. Prior to the X-ray structural determination, the structure of the peroxoborate was uncertain. Sodium peroxoborate is reported to form a heptahydrate, a tetrahydrate, a trihydrate and a monohydrate. The tetrahydrate is not stable above 15°C and at 60°C is converted entirely into the monohydrate. The kinetics for the crystallization of the trihydrate is very slow. The monohydrate is not really a hydrate in the usual sense. The structure of the peroxyborate monohydrate anion is:

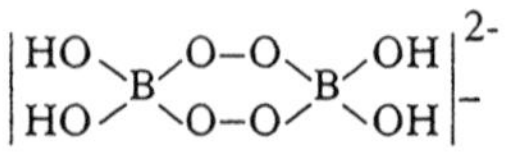

$$\left[(HO)_2B(O{-}O)_2B(OH)_2 \right]^{2-}$$

Fig. 1.

Crystal structure studies have shown that this anion is coordinated with six waters of hydration.

Part of the richness of the borate structures stems from the oxygen sharing capabilities of boron oxides. First, as in the silicates, the oxygen to boron ratio varies as the number of oxygens shared.

orthoborates O : B = 3 : 1
no oxygens shared

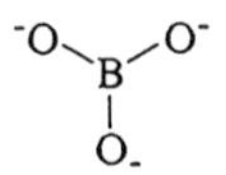

$$BO_3^{3-}$$

pyroborates O : B = 2.5 : 1
1 oxygen shared

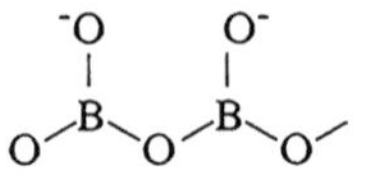

metaborates (rings or chains) O : B = 2 : 1
2 oxygens shared

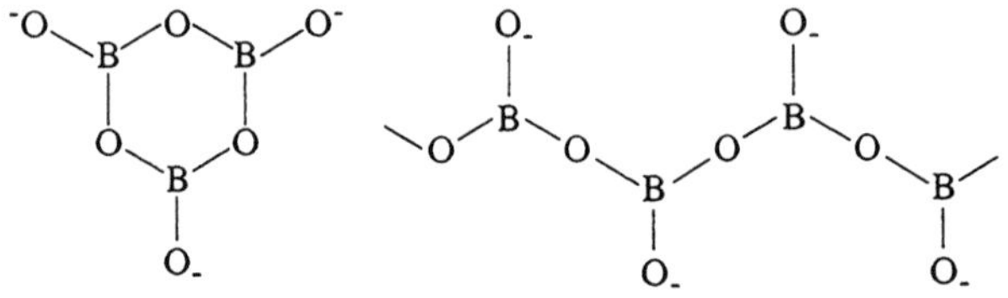

2D nets O : B = 1.5 : 1
3 oxygens shared

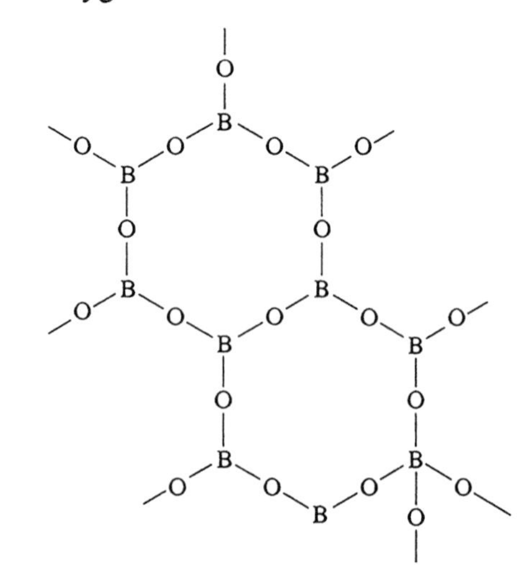

Fig. 2.

In addition to the variety of oxygen to boron ratios varying from 3 : 1 to 1.5 : 1 shown above, some borons are found to be four coordinate. Such is the case in borax and in the peroxoborate monohydrate structure discussed earlier. Borax consists of $B_4O_5(OH)_4^{2-}$ units connected by a system of hydrogen bonding.

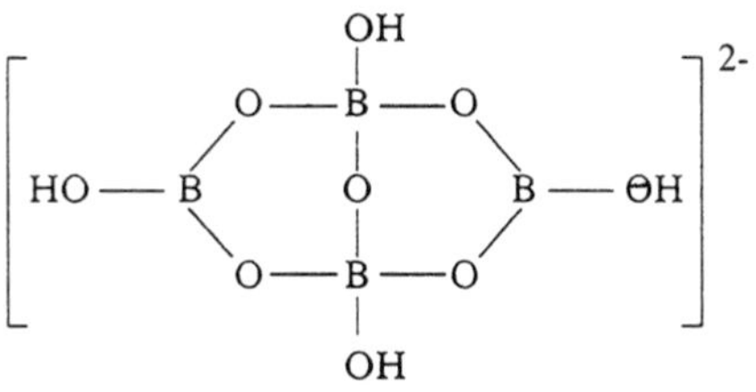

```
                  OH
                  |                      2-
┌           O —— B —— O                 ┐
|          /      |     \               |
|  HO —— B        O       B —— ΘH      |
|          \      |     /               |
└           O —— B —— O                 ┘
                  |
                  OH
```

Fig. 3.

Recognition of this structure within borax indicates that the classical formula for borax, $Na_2B_4O_7.10H_2O$, is better written $Na_2B_4O_5(OH)_4.8H_2O$ Note that two of the borons have the expected 3-coordinated geometry, but two of the borons have been converted to a 4-coordinated geometry.

The dissociation constant for sodium peroxoborate monohydrate as shown in the equation

$$NaBO_3.H_2O \rightleftharpoons NaBO_2 + H_2O_2$$

has been measured to be K = 0.03 mole/l by polarographic studies. This indicates a considerable dissociation of the peroxyborate at equilibrium and suggests that the synthesis might employ LeChatelier's principle to shift the equilibrium to the left.

Experimental Procedure

Prepare a solution of sodium metaborate, $NaBO_2$, by dissolving 4.75 g (0.0125 mole) of borax, $Na_2B_4O_7.10H_2O$ and 1 g (0.025 mole) of sodium hydroxide in 20 mL of warm water. Cool this mixture in ice and then slowly add 5.75 mL of 30 per cent hydrogen peroxide diluted with 15 mL of water. Keep the flask in ice and stir for about 15 minutes, or until the crystallization of sodium peroxoborate appears to be complete. If the crystals do not appear after 15 minutes, try seeding the solution with a seed crystal obtained from one of the other students in the class. Alternatively, add 30% hydrogen peroxide dropwise until precipitation initiates. Do not add more than 20 drops. This will help initate the precipitation by a "salting out" effect. Filter the crystals under suction, wash with alcohol and ether, and dry. On the basis of chemical analysis, one might write the formula as $NaBO_2.H_2O_2.3H_2O$ or $NaBO_3.4H_2O$, but the crystal structure indicates that the formula is more accurately written as $Na_2B_2(O_2)_2(OH)_4.6H_2O$.

Qualitatively test approximately 0.1 g samples of the sodium peroxoborate you have prepared. If a balance is not conveniently available, 0.1 g can be estimated to be about the volume of solid sample approximating the volume of a small pea.

To the 0.1 g sample of the sodium peroxoborate, add 6 mL of 0.1 N HCl. To this acidified solution, add 0.1 molar $KMnO_4$ solution dropwise. Observe the results. Write equations for the reactions. Was Mn^{2+} or Mn^{4+} formed?

To another 0.1 g sample of the sodium peroxoborate, add 6 mL of 0.1 N HCl. To this acidified solution, add dropwise a fresh solution of 1.65 g KI in 100 ml water (0.165 g

KI in 10 mL water if the KI is made up individually by each student). Observe the results. Write equations for the reactions.

Questions to be Considered

1. If the solution of sodium metaborate prepared in this experiment is evaporated, borax will crystallize out. Explain.

2. A solution of borax in water is quite strongly alkaline. Explain. Why can borax be regarded as an acid salt?

3. How would you attempt to prepare solid $NaBO_2$?

4. Compare the structural formula for sodium peroxoborate with the formula of perchloric acid.

5. Assume that you have a 0.1 molar solution of sodium peroxoborate. What percent would be decomposed to hydrogen peroxide?

6. Write a balanced equation describing what happens in the permanganate test and the iodide test.

7. Describe an analytical procedure for determining the degree of hydration of the peroxoborate.

Helpful Background Literature

1. Wells, A. F. "Structural Inorganic Chemistry", 4th ed.; Clarendon Press: Oxford, 1975, pp. 857 – 858.

2. Hansson, A. "On the Crystal Structure of Hydrated Sodium Peroxyborate", *Acta Chem. Scand.*, 1961, *15*, 934 – 935.

3. Kern, D. M. "A Polarographic Study of the Perborate Complex", *J. Am. Chem. Soc.*, 1955, *77*, 5458 – 5462.

MEASUREMENT OF HIGH TEMPERATURES

Equipment needed: Two furnaces capable of 1000°C
Thermocouple and readout device
Platinum resistance thermometer
Readout device for Pt thermometer. A multimeter capable of reading 0.01Ω can be used in place of a Wheatstone bridge for instructional purposes.
Optical pyrometer

Chemicals needed: None

Time needed: 1.5 hours

Safety notes: Take care not to burn yourself on hot surfaces.

Purpose

This experiment is to familiarize the student with methods of high temperature measurement.

Introduction

The familiar mercury thermometer is limited by the boiling point of mercury (356.58°C). In order to measure high temperatures, other alternative methods must be used.

One of the most generally used is the thermocouple. The discussion in the section on *Heating* which deals with thermocouples should be read. In most commercial thermocouple readout devices in use today, the ice junction is eliminated by a reference built into the electronic instrumentation. Depending on the accuracy desired and the cost one wishes to incur, the temperature can be detected by instruments ranging from an analog meter which reads the temperature directly to sensitive voltmeters which accurately read the emf output of the thermocouple. The temperature can then be determined from a table of emf vs. temperature for that particular type of thermocouple.

As shown in Table II of *Heating*, the platinum-platinum/10% rhodium thermocouple has an upper temperature range of 1760°C as opposed to 1370°C for the Chromel/Alumel thermocouple. It also has the advantage in that it does not corrode. The disadvantage is the cost of platinum ($412 per troy ounce in 1995 with gold at $376 per troy ounce) and the relatively small voltage change per degree in comparison to Chromel/Alumel. When the voltage change is 0.005 millivolts per degree rather than 0.02 millivolts per degree, a more sensitive readout is needed in order to achieve an accurate temperature reading.

In general, metals increase in resistance with increasing temperature, while semiconductors decrease in resistance with increasing temperature. The resistance of platinum and nickel wires change rather uniformly with rise in temperature. This, in addition to the non-corrosive nature of platinum and the protective corrosion layer established on nickel wire lends these wires to be used as resistance thermometers.

Platinum wire which is 99.99% pure shows a change of resistance with respect to temperature of 0.00385 ohms/ohm/°C. A slightly different platinum wire gives a response of 0.00392 ohms/ohm/°C, and is called the American curve. Commercial platinum resistance probes are limited in temperature range by the insulation system used. The ceramic constructions are good to 750°C – 850°C, whereas the glass and silicone coated platinum wire is only useful to 350°C. Commercial platinum resistance probes can be obtained for around $100.

Since the change in resistance per degree of platinum resistance thermometers are in the milliohm range, the accuracy of determining temperature is limited by the method used to measure the resistance. With the modern electronic ohmmeters, it is in principle possible to measure 0.01 ohms reasonably well. The readout may be 0.001, but the changes of resistance with the precise nature of contacts (e.g. the way a banana plug is inserted into its socket) render accurate resistance measurements in the milliohm range impractical by direct reading ohmmeters. Accurate resistance measurements in these low ranges are best done by a bridge circuit of some type. As shown in *Heating*, the elementary bridge circuit compares the resistance of the wire to that of a standard resistance. Commercial units are available which convert the resistance readings from a platinum resistance thermometer to a digital readout of temperature with an accuracy of 0.01°C. These instruments cost in the range of $2000 to $3000.

Optical pyrometers are available for a variety of temperature ranges as well as a variety of prices. The older instruments used a hot wire which was heated by a current from a battery. The temperature of the wire was controlled by a rheostat which in turn was calibrated to read a temperature. The hot wire was seen in a scope which enabled the comparison of the color of the hot furnace with that of the wire. The newer optical pyrometers are infrared sensors with digital readouts. Thermocouples in the instrument measure the infrared radiation emanating from the hot source. There are a wide variety of models available with different sensitivities and ranges.

Accurate temperature determinations using an optical pyrometer require the knowledge of various emissivities. For example, the accurate temperature determination of a molten metal requires adjustment of the instrument for that particular metal. Stray light or other infrared sources must also be considered. For general temperature determinations with less expensive instruments, provisions for these fine adjustments are not provided and the temperature measurement is much simpler.

Experimental Procedure

The Chromel-Alumel Thermocouple

In teams of two or three, measure the temperature of ice water and boiling water using a Chromel-Alumel thermocouple. Insert the thermocouple into the muffle furnace and measure the temperature. Turn off the furnace and allow it to cool. Measure the temperature with time. Compare the temperature measured to that of the readout on the oven if there is one and to that of the platinum resistance thermometer. Compare the initial temperature and that of the other readings to that obtained with an optical pyrometer.

The Platinum Resistance Thermometer

Calibrate the platinum resistance thermometer by measuring the temperature of ice water and boiling water. Insert the thermometer into the muffle furnace along with that of the thermocouple and measure the initial temperature. Compare the temperature obtained with that from the thermocouple and the readout on the furnace if there is one. Turn off the power to the furnace and observe the change of temperature as a function of time.

Optical Pyrometer

Read the temperature of the muffle furnace using the optical pyrometer. Compare the reading obtained with the readout shown on the thermocouple detector on the muffle furnace.

Questions to be Considered

1. Critically assess the advantages and disadvantages of each of the high temperature measuring schemes.

2. Under the conditions of the experiment, which method would you guess is the most accurate?

3. In what way can the platinum resistance thermometer reading be made more accurate?

4. Why can't the platinum resistance thermometer be put directly into water to make a temperature measurement?

5. Look up the thermocouple output tables in a handbook of chemistry. On the basis of the output changes per degree, range of temperatures measurable by each thermocouple, and other factors, indicate how you might select a thermocouple for several practical applications.

Helpful Background Literature

1. Omega Engineering Company, "Temperature Measurement Handbook".

PREPARATION OF PHOSPHORS

Equipment Needed:	agate or mullite mortar and pestle quartz boat glass rod tube furnace (1000° – 1050°C) quartz tube fitted for nitrogen flow copper furnace for nitrogen stream (approx. 400°C) alkali filled blow off flask thermocouple and detector U.V. lamp Buchner funnel
Chemicals Needed:	1 g electronic grade ZnS (99.999% with respect to metals contamination) 0.5 g reagent grade NaCl 10 mL concd HCl 2 mL ethyl alcohol $CuCl_2$ solution (0.21 g/L to 2.1 g/L) $AgNO_3$ solution (0.16 g/L to 1.5 g/L)
Time Needed:	2.5 hours
Safety Notes:	Take care to make sure that the gas train has no obstruction to gas flow. Glass apparatus cannot be pressurized to more than several atmospheres without some danger of bursting.

Purpose

This preparation illustrates the importance of trace amounts of impurities in solid state chemistry. Ordinary zinc sulfide cannot be used in this experiment because of the trace impurities which act as "killers" to prevent phosphor formation. Likewise, only low levels of the activating impurities are effective in phosphor formation.

Introduction

Luminescent solids are used in fluorescent lights, oscilloscope tubes and TV tubes. These solids prove useful because on excitation with some non-visible excitation source they emit visible light. The excitation may be by electron beams, ultraviolet light, infrared light, radioactive decay, or electric fields. The most important group of solids which exhibit these properties are alkaline earth (especially zinc and cadmium) sulfides, selenides, silicates and halides.

Luminescence is not exhibited by pure solid ZnS. On proper doping with small amounts of activators, the luminescent properties are observed. The condition of "proper" cannot be overemphasized. Luminescence is observed for concentrations of doping ions around 1 ppm. Maximum luminescence is generally observed at around 100 ppm. At higher concentrations of doping ions the luminescence drops off. A part per million is a small amount. Keep in mind that a ppm means a part per million by weight.

As with activators, very small amounts of many impurities, especially nickel and iron, act as poisons. For this reason the host solid such as zinc sulfide must be carefully purified. The purification steps used to prepare phosphor grade sulfide or selenide are illustrative of the techniques used to purify inorganic compounds and would seem to be pedagogically useful as a laboratory exercise. However, the level of purity required is so high that it is not feasible to try to purify small batches. Apparently, impurities are introduced into the sulfide by the handling necessary to remove others. Because it is impractical for each person in the laboratory to purify several pounds of sulfide, commercial phosphor grade zinc sulfide will be supplied.

The phosphor grade zinc sulfide can be activated in small quantities by mixing with 2 to 20 percent sodium chloride as flux and adding a few drops of dilute copper chloride solution to give the required parts per million of copper impurity. The copper doping gives a green phosphor. Silver doping with silver nitrate gives a blue phosphor. The introduction of silver ion into the divalent lattice requires valence compensation. One purpose of the sodium chloride flux is to provide this valence compensation.

$$ZnS + x\,AgCl \rightarrow ZnAg_xSCl_x$$

This same effect can be brought about by adding some trivalent ions such as Al^{3+} or Ga^{3+}. If copper goes in as Cu^{+} it must also be compensated; as Cu^{2+} compensation is not needed.

It is important to carefully dry the sample before firing. In the presence of moisture, zinc sulfide can react to form zinc oxide and hydrogen sulfide. A dark color after firing may well indicate that the sample was wet.

Zinc sulfide is ordinarily prepared as the cubic zinc blende. The hexagonal wurtzite is formed at around 1020°C, but is unstable with respect to the blende. It is reported that wurtzite will convert to the blende by being ground in a mortar. If the sample is raised to above 1000°C during firing, the transition from blende to wurtzite will occur. On cooling, the reverse transition is expected to occur. Wurtzite can be prepared by quenching. For example, sublimation of zinc sulfide from above 1000°C to a cold finger in an inert atmosphere (H_2, H_2S, N_2, SO_2, etc.) results in the wurtzite form.

Experimental Procedure

Prepare a solution of copper(II) chloride (or silver nitrate), unless it is already done for you, so that two drops of the solution will give a doping ion concentration of 10 to 100 ppm for 1 g of zinc sulfide. Add two drops of the doping solution to 0.25 – 0.5 g of reagent grade sodium chloride. Evaporate off the water in a drying oven, but take care that rust particles are not introduced. In order to avoid introducing unwanted impurity ions, use an agate or mullite mortar and pestle which is carefully cleaned with hydrochloric acid to remove all traces of nickel and iron. Because nickel is a poison, a nickel spatula should not be used. Use a porcelain or glass spatula instead of the metal spatula. Thoroughly mix the doped sodium chloride with 1 g luminescent grade zinc sulfide. Add a few drops of alcohol and grind into a paste. Place the paste in a quartz boat and dry in a drying oven.

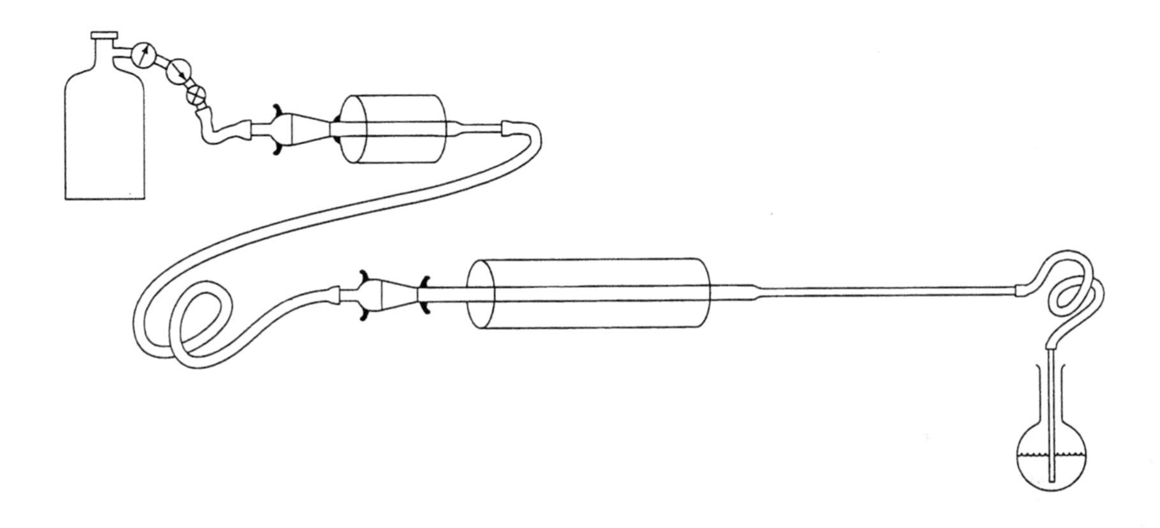

Fig. 1. Apparatus for Preparation of the Phosphor

Heat a glass rod and pull a small hook at one end. This rod can be used for pushing the quartz boat into the tube furnace and for pulling it out. Slide the quartz insert in the tube furnace as far as feasible toward the entry side. Remove the glass joint with the nitrogen tube and slide the quartz boat containing the ZnS-NaCl-$CuCl_2$ up to the edge of the furnace. Replace the nitrogen inlet tube and flush the tube with nitrogen for at least five minutes. It is desirable to pass the nitrogen through a hot copper furnace to remove traces of oxygen. Check to see that the copper furnace is heated to 400°C. This temperature is not critical. A desirable range is between 300°C and 450°C. After the initial flush, slow up the nitrogen flow so that it is just barely bubbling through the alkaline sodium hydroxide filled blow-off at the exit side of the tube furnace. Slide the quartz insert into the tube furnace so that the boat is in the hot zone. Monitor the temperature by measuring the emf of the thermocouple. The temperature of the tube furnace should be 1000° – 1050°C. Keep in mind that the recommended operating range for quartz is 900°C. The extreme temperature that quartz can withstand for short periods without distortion is 1100°C. After 15 to 30 minutes of heating, pull the quartz insert out so that the boat is out of the hot zone. Allow to cool in the nitrogen stream before withdrawing the boat from the tube.

Check the product for luminescence with a U.V. lamp. If a cathode ray simulator is available, use this to check for luminescence. Wash the sodium chloride out by stirring the fired product in 10 mL of distilled water. Filter on a small Buchner funnel and wash the insoluble residue with several 3 mL portions of water. Dry the solid residue and re-check the luminescence.

Auxiliary Experiment

One to two percent of manganese ion (Mn^{2+}) gives a yellow orange phosphor. Weigh 0.5 g electronic grade zinc sulfide and mix with 0.05 g NaCl doped with 0.018 g anhydrous $MnCl_2$. Fire the mixture for 35 minutes at 875°C. Take care to fire under nitrogen as described for copper doping.

Questions to be Considered

1. If one wanted to prepare a doping solution of silver nitrate so that one drop per gram of phosphor is equivalent to 20 ppm, how should this solution be prepared?

2. What are the purpose or purposes of the flux?

3. A one to two percent Mn^{2+} dope can be used to form an orange phosphor. Write a detailed description for the preparation of this phosphor.

4. What is the lower temperature range at which a copper furnace is effective? Why is a slightly or considerably higher temperature desirable? For a copper furnace constructed of glass, what controls the useful upper temperature?

Helpful Background Literature

1. Wells, A. F. "Structural Inorganic Chemistry", 4th ed.; Clarendon Press: Oxford, 1975; p. 164.

2. Symposium, Cornell Univ. "Preparation and Characteristics of Solid Luminescent Materials"; John Wiley and Sons, Inc.: New York, 1948; pp. 23 – 31.

COPPER(II) OXIDE

Equipment Needed:
- 100 mL beaker
- Buchner funnel
- Whatman #4 filter paper for Buchner funnel
- filter flask
- wash bottle
- medium porosity sintered glass funnel
- crucible
- scissors
- hot plate
- muffle furnace
- hot glue gun

Chemicals Needed:
- 5 g $Cu(NO_3)_2.3H_2O$
- 4 mL concd. NH_4OH
- posterboard

Time Needed: 2.5 hours

Safety Notes: Ammonia, though not acutely toxic, is unpleasant. All work involving concentrated ammonia should be carried out in a hood.

Purpose

This experiment illustrates the preparation of a compound which is a useful catalyst. It also illustrates the use of high temperatures in inorganic synthesis. The structural exercise serves as a reminder that structures are just as important as stoichiometry in learning about inorganic compounds.

Introduction

Copper(II) oxide is found naturally as the rare mineral Tenorite (Cornwall, Vesuvius and Keweenaw Point, MI) and as the more common amorphous or colloidal mineral melaconite which is found at or near copper mines.

Copper(II) oxide can be prepared in the laboratory by one of several methods. Copper(II) nitrate or carbonate can be calcined (heated). For example, when the nitrate is heated, the first decomposition product is the green basic salt $Cu(NO_3)_2.3CuO$ which on further heating to 850^oC is decomposed to CuO. It can also be prepared by oxidation of copper metal or by heating copper(II) hydroxide. This latter method will be used for this experiment.

When copper(II) salts are precipitated with base in cold solutions, a blue gelantinous precipitate of $Cu(OH)_2$ is formed. In crystalline form, it can be dried at 100^oC without decomposition. The gelantinous form, however, decomposes with heating to CuO. The blue gelatinous precipitate can easily be converted to the black oxide by boiling the water suspension. The black oxide is formed directly when base is added to the solution of copper(II) ions above 80^oC.

Ammonium hydroxide is the base of choice for this experiment. The ammonium salts which might be coprecipitated with the copper(II) hydroxide can be driven off by heating whereas the alkali metal salts cannot. For example, ammonium nitrate decomposes at 170° – 180^oC. On gentle heating the products are N_2O and H_2O. On rapid heating the reaction is:

$$5\ NH_4NO_3 \rightarrow 4\ N_2 + 9\ H_2O + 2\ HNO_3$$

Since nitric acid can catalytically accelerate the reaction, the decomposition of bulk ammonium nitrate can be explosive. The ammonium nitrate formed in the scale of this experiment is not dangerous. If the experiment is scaled up a great deal, care should be taken to avoid rapid heating.

When the copper salt is selected as the starting material for this experiment, the nature of the ammonium salts formed must be considered. Ammonium chloride melts at 520^oC but has 10 mm of vapor pressure at 210^oC and 760 mm at 338^oC. The vapor is partially dissociated into NH_3 and HCl. Ammonium sulfate decomposes at 357^oC in the atmosphere forming a melt of $(NH_4)_2SO_4$ and $(NH_4)HSO_4$ with a loss of NH_3. The $(NH_4)HSO_4$ melts at 251^oC and boils at 490^oC. Because of the higher temperature needed to boil off the $(NH_4)HSO_4$, the final heating period when copper sulfate is used requires a higher temperature for longer periods of time than when the nitrate is used.

One of the side reactions, especially when an inadequate amount of base is added, is the formation of basic salts. Understanding the chemistry of the basic salts of copper will help in understanding how to prevent it from appearing in the final preparation of CuO. Copper sulfate has been reported to form a large number of basic salts with ratios of CuO : $CuSO_4$ from 9 : 2 to 1 : 1. A typical formula for one of these salts is $3CuO.CuSO_4.4H_2O$. A similarly large number of basic copper chlorides have been reported. Some 18 compounds of CuO : $CuCl_2$: H_2O ranging from ratios of 1 : 1 : 1 to 8 : 1 : 12 have appeared in the literatures. Probably the best known of these basic salts is $Cu_2(OH)_3Cl$. Not as many basic salts have been described for copper nitrate. One is the basic salt $Cu_2(OH)_3NO_3$ which is found in nature as the mineral gerhardtite. If the original copper hydroxide precipitate is filtered in order to separate it from the bulk of the salts, and the precipitate resuspended in water and heated with a bit of ammonium hydroxide, some of the basic salt which carries over with the precipitate will be converted to the hydroxide which then undergoes dehydration to the oxide. However, the equilibrium is not completely displaced and the precipitate always contains some basic salt.

It is important, therefore, to select a basic salt which can be calcined to the oxide. Copper oxide sulfate is stable at high temperatures whereas the basic salts of chloride and nitrate are not.

If a large excess of ammonium hydroxide is added, the copper ammonia complex ion, $[Cu(NH_3)_4]^{2+}$ is formed. The equilibrium is almost entirely on the side of the complex ion. However, on boiling, the small concentration of copper(II) hydroxide dehydrates to the copper(II) oxide and the equilibrium is slowly displaced. Because this reaction is slow, another method is to precipitate the copper(II) hydroxide with a very slight excess of ammonium hydroxide, filter out the copper(II) hydroxide to separate it from the bulk of the ammonium nitrate, resuspend the copper(II) hydroxide in distilled water and heat to dehydrate to the oxide. If the copper(II) hydroxide has gotten crystalline, it will not dehydrate readily to the oxide at the boiling point of water. In order to aid the dehydration and to insure the conversion of any co-precipitated copper(II) nitrate to copper(II) hydroxide, an additional amount of ammonia can be added to the suspension of copper(II) hydroxide. If the ammonia is not in large excess, the copper(II) hydroxide will not be completely converted to the complex ion.

Copper(II) oxide itself has an equilibrium oxygen pressure of 1.963×10^{-25} mm at 450°C and 118 mm at 1000°C. The decomposition creates a solid solution of copper(I) oxide in copper(II) oxide. As long as the vapor pressure of oxygen is below the partial pressure of oxygen in the air (about 150 mm), the decomposition in air will be incomplete. As the temperature is lowered, the Cu_2O in air will be reoxidized to CuO.

Because copper reacts easily with oxygen, the copper(II) oxide can be prepared by direct oxidation. The reaction, however, slows down once the surface of the metal is coated with oxide. For this reason, the preparation must utilize very thin foils as the oxidation substrate. A practical use for the oxidative reaction of copper is the use of copper as an oxygen getter for purifying inert or reducing gases for laboratory use. For example, small amounts of oxygen impurities in argon, nitrogen or hydrogen can be removed by passing the gas through a heated column packed with copper wool or copper chips. Column temperatures of 200°C to 600°C are used, the higher temperatures being necessary for faster flow rates. If Pyrex glass columns are used, heating should be kept below 500°C. When the surface of the copper is darkened by an oxide coating, the column is regenerated by reducing with hydrogen. The reduction of the oxide takes place between 135°C and 200°C depending on the previous history of the oxide. When finely divided, copper reacts much more readily with oxygen than when in the massive state. BASF Inc. markets a BTS catalyst which consists of 30% copper in a very finely dispersed form stabilized on a carrier which is formed into small pellets. Oxygen in amounts of less than 1.5% can be removed at 70°C to 250°C. Above 250°C the activity of the catalyst is reduced by the sintering of the copper particles. The finely divided copper is so active, the reduced catalyst is pyrophoric and will spontaneously burn in air. Because of this, gas streams containing more than 1.5 to 2% oxygen must be pretreated to reduce the oxygen level before passing over BTS catalyst. Since the reaction is exothermic ($Cu + 1/2\ O_2 \rightarrow CuO$, $\Delta H = -37.1$ kcal), care must be taken not to externally heat the column near the 250°C maximum since a temperature rise will always accompany the getter action. The same type of care must be taken during reduction. The column is heated to 120° – 140°C with hydrogen being gradually bled into the nitrogen stream. As the reduction proceeds, the nitrogen is slowly displaced with hydrogen. This is done to keep the temperature in the column from rising above 200°C from a rapid exothermic reduction reaction.

Copper(II) oxide is hygroscopic. When heated to high temperatures, the crystallinity seems to increase and the material becomes less hygroscopic. It also absorbs gases less readily and loses some of its catalytic qualities. The X-ray powder patterns before and after heating show no change in pattern indicating that there is no change in the structure on the atomic scale. The structure of cupric oxide is a distorted PtS structure. The copper and oxygen atoms are both four-coordinated. The copper and four oxygens, instead of being a regular square planar arrangement as in the idealized PtS structure, form two O-Cu-O angles of 84.5^o and two of 95.5^o.

It might be mentioned that copper(I) oxide which is formed from base and copper(I) ions, (the copper(I) hydroxide being even more unstable than copper(II) hyroxide) has quite a different structure. It consists of two interpenetrating anti-fluorite structures. In the anti-fluorite structure, the oxygen atoms are at the face centered cubic positions with copper in all the tetrahedral holes. For the Cu_2O structure, picture two of these anti-fluorite ball and stick models interpenetrating each other with no bonds connecting one framework with the other. Copper(I) oxide is familiar as the precipitate in the Fehling's test for aldehydes. A yellow precipitate first forms which on heating is converted to red crystals. Since the X-ray powder pattern of both the yellow solid and red crystals are identical, it is thought that the yellow form is just a finely divided form of Cu_2O.

Experimental Procedure

Dissolve 5 g of copper(II) nitrate trihydrate in 20 mL of water. Dissolve 3 mL of concd. ammonium hydroxide in 25 mL of water. Slowly add the copper(II) nitrate solution to the ammonium hydroxide solution with vigorous stirring. Filter the precipitate formed using Whatman #4 filter paper or its equivalent in a standard Buchner funnel. Wash the precipitate with about 10 mL of water. Transfer the pasty blue precipitate to a 250 mL beaker. Add 100 mL water and heat to 90° – 100°C on a hot plate for about 5 minutes. Add 1 mL of concd NH_4OH and continue heating between 90° – 100°C for another 10 minutes. Vacuum filter the black precipitate through a medium porosity sintered glass funnel and wash with 5 mL water. Transfer to a size number 0 porcelain crucible and place in a drying oven for about 15 minutes to evaporate off the excess water. Put the crucible with the dry or semi-dry solid into a furnace heated to 600° – 650°C for one hour. (A half hour is the minimum recommended heating period.) Remove the crucible from the furnace, cool for about five minutes and transfer the solid to a weighed sample bottle.

Construction of PtS Model

While waiting for the CuO to finish reacting in the muffle furnace, make a model of the PtS structure.

Cut at least 17 one-and-a-half inch squares (1.5″ x 1.5″) out of posterboard. Twice the number will make a better model. In order to make the model easier to construct, the squares should be uniform. The exact size is not critical, but all the squares must be as close in size to each other as possible.

Put four dots using a pencil or pen on each square as shown in Fig. 1.

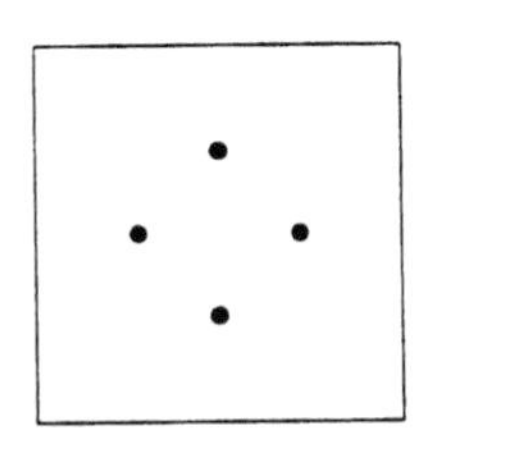

Fig. 1.

Cut from the corners to each dot as shown by the lines. (See Fig. 2.)

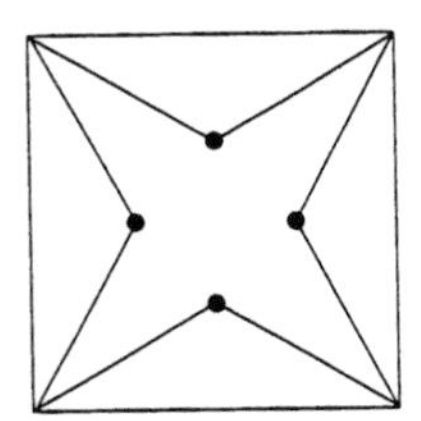

Fig. 2.

Using a piece of polyethylene sheet as a working surface, glue the corners of the four pointed "stars" in the manner shown in the instructor's PtS model. Use the hot glue gun provided. If the glue sticks to the polyethylene sheet, use a knife or a spatula to loosen the "stars" before the glue cools down. The vertical pieces will have to be held in place for a short while until the glue cools and hardens.

The platinum is four-coordinated (square planar) and is represented by the center of the four pointed "star". The sulfur is four-coordinated in a different way (tetrahedral) and is located by the point at which the tips of the "stars" come together.

Questions to be Considered

1. CuO has the distorted PtS structure. Describe the arrangement of the copper and oxygen atoms.

2. Copper(II) hydroxide is amphoteric. It reacts with two molecules of NaOH. Write the reactions which indicate its amphoteric properties.

3. Schweizer's reagent is formed by dissolving copper hydroxide in aqueous ammonia. Four moles of ammonia coordinate with copper. Write the formula for Schweizer's reagent. Schweizer's reagent is capable of dissolving cellulose. This is one way rayon can be prepared.

4. Commercial ammonium hydroxide is labeled to be 28% NH_3 with a density of 0.90. What is the molarity of this ammonium hydroxide? How many milliliters would be required to react exactly with 5 g $Cu(NO_3)_2 \cdot 3H_2O$.

5. If NH_4OH is added to a copper(II) chloride solution, a light blue non-filterable precipitate is formed. If NH_4OH is added to a copper(II) sulfate solution, a light blue filterable precipitate is formed. Offer a possible explanation.

6. If the copper(II) sulfate solution is added slowly with vigorous stirring to a dilute ammonia solution, a dark blue color initially results which turns into a blue precipitate on further addition of the copper(II) sulfate. This precipitate is preceptively darker than the one formed by adding ammonia to the copper(II) sulfate solution. Offer a possible explanation.

7. The precipitate formed by adding ammonia to a copper(II) chloride solution does not decompose to the oxide on heating to 100°C. On separating the precipitate and heating to 600°C in a furnace, pure copper oxide is prepared. Offer a reasonable explanation.

8. Describe the infinite three dimensional structure of PtS in words. How easy would it be for someone else to construct the model representing the PtS structure from a verbal or written description which you might provide?

Helpful Background Literature

1. Greenwood, N. N.; Earnshaw, A. "Chemistry of the Elements"; Pergamon Press Ltd.: Oxford, 1984, pp. 1372 – 1376, 1382.

2. Latimer, W. M.; Hildebrand, J. H. "Reference Book of Inorganic Chemistry" 3rd ed.; The Macmillan Co.: New York, pp. 109 – 112.

3. Wells, A. F. "Structural Inorganic Chemistry", 5th ed.; Clarendon Press: Oxford, 1984, p. 1120.

STRUCTURAL INORGANIC CHEMISTRY VIA MODELS - 3

Session Three of Three

Equipment Needed:	18 trigonal balls (dark blue) 20 tetrahedral balls (black) 50 bonds for above balls 1 paper octahedron 1 paper tetrahedron 20 triangles to make icosohedron 12 pentagons to make pentagonal dodecahedron 6 squares to make cube 4 layers of clear and blue spheres (See type B, Fig. 8) 1 D stix kit 1 tin of marbles 1 hot glue gun (per 3 students)
Chemicals Needed:	None
Time Needed:	3 hours
Safety Notes:	None

Introduction

In *Sessions One and Two* of the study of Inorganic Structures with Models, the concept of representing structures with ball and stick models, with polyhedral models, and with close-packed spheres was introduced. An introduction was made to the property of nets and polyhedra.

The rationale for studying solid state structures of inorganic compounds is that so much of inorganic materials exist in the solid state. Consider the elements themselves. At room temperatures, there are approximately 10 gaseous elements, 3 liquids, and 80 solids. In most cases, when reactive gaseous elements (oxygen, fluorine, chlorine, etc.) react with the solid elements, solid products such as oxides, fluorides and chlorides are formed. Since the field of inorganic chemistry is the study of all the chemical elements, it becomes predominantly the preparation and study of crystalline compounds. These crystalline compounds can be described as regular repetitions of groups (molecules or discrete ions) or as infinite structures in one-, two-, or three-dimensions. Our knowledge of the actual structure of solids has only been available since the discovery of X-ray diffraction by crystals in 1912. Furthermore, structural analysis was a slow tedious process until the advent of fast computers in the 1960s.

Consider for a minute the historic development of chemical formulas and structure.

One of the earliest tools in the elucidation of a chemical formula was the use of elemental analysis. Elemental analysis of a compound consisting of discrete, finite molecules gave information on the relative numbers of atoms of different elements, and this combined with a knowledge of the molecular mass revealed the actual numbers of atoms that form a discrete group (molecule).

The structural problem consists of two parts.

(1) The *topology* of the molecule describes how the atoms are joined together, that is, which atoms are joined to which and by how many bonds.

(2) The *geometry* defines the relative positions of the atoms in space; it is concerned with bond lengths, bond angles, and dihedral angles.

A knowledge of the geometry of a molecule obviously implies a knowledge of its topology; but, until the development of physical methods such as spectroscopy and diffraction, the chemist could, by experiment, study only the topology of molecules. In simple cases this follows directly from a knowledge of the combining powers to the atoms. For example, if there are to be four links from each C and one from each H atom, C_2H_6 can only be $H_3C - CH_3$.

The topology of more complex molecules was deduced from the way in which they could be synthesized from or degraded into simpler units. It cannot be deduced from the chemical formula if there are alternative formulations satisfying the valences of the atoms (e.g., $CH_3CH_2CH_2CH_3$ or $HC(CH_3)_3$). These two forms of C_4H_{10} may be called topological isomers since they differ in the basic skeleton of connected carbon atoms in contrast to geometrical isomers such as the *cis* and *trans* forms of, for example, $R - N = N - R$ where R is an atom or group.

Initially, only inspired guesses could be made about the geometry of molecules and then only about the local arrangements of bonds around certain atoms. The arrangements suggested usually corresponded to the most symmetrical ones; for example, four tetrahedral bonds from carbon and six octahedral bonds from Co^{3+} and Cr^{3+}.

Supporting evidence for these stereochemistries came from studies of geometrical and optical isomers. It happened that the above guesses were correct and led, for example, to an elaborate structural chemistry of carbon compounds that has proved substantially correct in broad outline though necessarily lacking all precise geometric details. This development of structural organic chemistry was possible because most organic compounds (other than polymers) exist as the same finite molecules in all states of aggregation. The structure of the molecule was deduced from its behavior in solution, and the solid is simply an aggregate of molecules with essentially the same structure. No such development of structural inorganic chemistry was possible for two main reasons. The first is that inorganic chemistry is concerned with approximately one hundred elements, many with much less symmetrical bond arrangements than those mentioned above. The second is that whereas reactions are often carried out in solution (sometimes in the solid state or vapor state) most inorganic compounds are intractable insoluble solids at ordinary temperatures.

As mentioned, organic chemistry is primarily a study of discrete molecular compounds. The bonding and geometrical shape can be quite adequately described by considering a single molecule. There are some inorganic compounds which are molecules in this same sense. $TiCl_4$ and SiH_4 are examples. SiO_2 and TiO_2, however, are not molecular compounds. Consider the difference in the properties of CO_2 which is molecular while SiO_2 (e.g. quartz) is an insoluble solid. Quartz is built up of tetrahedral SiO_4 groups and TiO_2 contains octahedral TiO_6 groups. These two compounds are examples of inorganic derivatives which do not form discrete inorganic molecules but instead form three-dimensional structures.

The importance of thinking about the solid state is underscored by the confusion between gram molecular weight and gram formula weight in the minds of many a student in general chemistry. Because a mole of a substance can be either a GMW or GFW, and most compounds encountered by the beginning student are molecular in nature, a mole of NaCl is thought by an unsophisticated student to be Avogadro's number of molecules of NaCl. Since there is no such thing as a molecule of NaCl, the distinction between GMW and GFW is important.

Geologists for years named minerals without knowing their structure. As the structures of crystalline minerals were elucidated, other similar crystalline structures came to be classified by the mineral first studied by X-ray crystallography. We, therefore, say that many of the alkali halides have a sodium chloride structure or that a whole family of compounds have the perovskite ($CaTiO_3$) structure.

The structures of inorganic compounds can be subdivided into one-dimensional, two-dimensional, and three-dimensional structures. In order to understand inorganic structures, it is helpful to understand the nature of polyhedra and to appreciate the topology of one-dimensional chains, two-dimensional planar nets, and three-dimensional frameworks. Further considerations on the nature of sphere packing are also important. These subjects which were introduced in the first two Models experiments will be discussed here in a greater depth.

Experimental Procedure

Exercise #1

In Exercise #3 of *Session One*, the five regular solids were discussed. These were found to have faces of regular polygons (n-gons) of the same kind with the same number of polygons meeting at each (and every) vertex, i.e., same connectedness (p). Each can, therefore, be given the symbol (n, p). Plato demonstrated that there can be only five of these regular polyhedra (enclosed, finite solids).

Review the representation of these solids by sketching an octahedron looking down a line (not looking at the line) joining

(a) opposite vertices
(b) the mid-points of opposite edges
(c) the mid-points of opposite faces

For a tetrahedron, sketch the representation looking down a line joining

(a) a vertex with the midpoint of the opposite face
(b) the mid-points of two non-adjacent edges.

Given the length of the edge of the octahedron as ℓ, what is the distance between opposite vertices? (In Exercise #1 of *Session Two*, αUF_5 was described as octahedra sharing opposite vertices. Your answer will give the uranium-uranium distance in an idealized αUF_5 structure.)

Given the length of the edge of the octahedron as ℓ, what is the distance between the center of the octahedron to an edge along a line perpendicular to an edge? (In Exercise #1 of *Session Two*, NbI_4 was described as NbI_6 octāhedra sharing opposite edges. Your answer should enable the calculation of the niobium-niobium distance in the idealized NbI_4 structure)

Exercise #2

The pentagonal dodecahedron is not commonly observed in chemical systems. The shape has been observed in a number of gas hydrates in which the oxygens (from H_2O) form the dodecahedral framework, and the gases (Cl_2, Br_2, $CHCl_3$) occupy positions in the voids.

The icosohedron is important in boron hydride chemistry. $B_{12}H_{12}{}^{2-}$, $CB_{11}H_{12}{}^{1-}$, and $C_2B_{10}H_{12}$ have icosohedral structures. Many of the boron hydrides with less than 10 boron atoms have structures which are segments of the icosohedron.

Complete the following table for all the regular polyhedra.

Table I

Polyhedron	n, p	# Vertices	# Faces	# Edges
Tetrahedron				
Octahedron				
Cube				
Dodecahedron				
Icosahedron				

Euler, in the eighteenth century, derived the relationship

$$\text{Vertices} + \text{Faces} = \text{edges} + 2$$

Show how the Euler relationship holds for the data compiled in Table I above.

If the vertices of a polyhedron are oriented to the faces of another polyhedron, they are said to be reciprocal to one another. Which of the regular polyhedra are in reciprocal relation to each other?

Exercise #3

There are some less-regular polyhedra than the regular solids which are of chemical importance. One group of these polyhedra includes those whose faces are all regular polygons, but the polygons are of more than one kind, and all vertices are equivalent.

The thirteen Archimedean semi-regular solids are formed by symmetrically cutting off (truncating) the vertices of regular solids so that the faces of regular polygons are produced. For example, a truncated tetrahedron has one triangular face and the hexagonal faces meeting at each new vertex. The truncated octahedron has one square face and two hexagonal faces meeting at each vertex.

Another large group of semi-regular solids include the regular prisms and antiprisms. Some of these chemically important solids will be discussed in greater detail during the exercises.

There are some interesting relationships involving polyhedra. A general topological proof is concerned only with the type of face (n-gon) and the number (p) of edges meeting at each vertex. From Euler's relation discussed in Exercise #2, the following can be shown:

For 3-connected polyhedra:

$$3\ f_3 + 2\ f_4 + f_5 + 0\ f_6 - f_7 - 2\ f_8 - \ldots = 12 \qquad [1]$$

For 4-connected polyhedra:

$$2\ f_3 + 0\ f_4 - 2\ f_5 - 4\ f_6 - \ldots = 16 \qquad [2]$$

For 5-connected polyhedra:

$$f_3 - 2\ f_4 - 5\ f_5 - 8\ f_6 - \ldots = 20 \qquad [3]$$

"Connected" means the number of faces (or edges) meeting at the vertex. (Not all solids are uniformly connected. For example the trigonal bipyramid is both 3- and 4-connected.) As mentioned above, the f_n term is the number of faces with n edges, for instance, f_3 is the number of triangular faces. From these equations it follows that there are only *three* 3-connected polyhedra with all faces of the same type, namely $f_3 = 4$ (tetrahedron), $f_4 = 6$ (cube), and $f_5 = 12$ (pentagonal dodecahedron). The octahedron is the only 4-connected polyhedron. There is only one 5-connected polyhedron with all faces of the same type, the icosahedron with $f_3 = 20$. These equations do not require the faces to be regular. The most symmetrical forms of the polyhedra with all faces of the same type are called the Platonic solids or regular solids. It also follows that there are no polyhedra having 6 (or more) edges meeting at every vertex. The corresponding equation for the 6-connected polyhedra would be:

$$0\ f_3 - 4\ f_4 - 8\ f_5 - \ldots = 24$$

An example of a semi-regular polyhedron is a cuboctahedron. A cuboctahedron results from truncating the 8 vertices of a cube to the mid-points of the cube edge as shown:

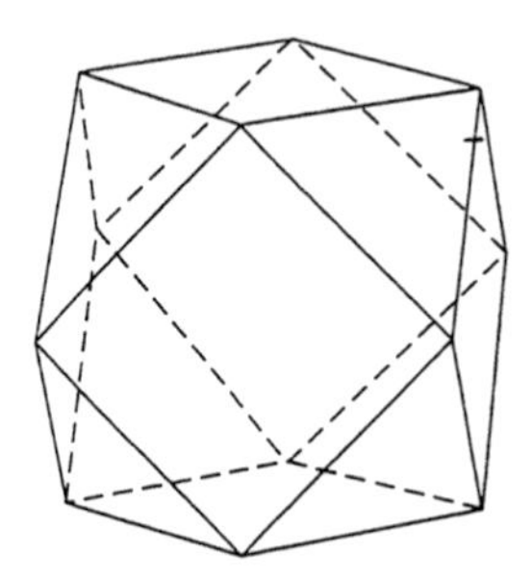

Fig. 1.

This results in one of the thirteen Archimedean semi-regular solids. For the cuboctahedron it can be seen that:

Each vertex is 4-connected

There are six square faces. Therefore $f_4 = 6$

There are eight triangular faces. Therefore $f_3 = 8$

Because the polyhedron is 4-connected the second equation [2] holds:

$$2\,(8) + 0\,(6) - 2\,(0) - 4\,(0) \ldots = 16$$

$$16 = 16$$

Make a model of a cuboctahedron using the D-stix kit. You will need 24 white and 6 blue sticks, and 12 six-sleeved connectors. Leave the spare sleeves of each 6-connector pointing into the square faces of the cuboctahedron. Insert a blue stick as one diagonal of each square face, ensuring that only one blue stick comes to (from) each vertex.

Inspect your model neglecting the blue diagonal sticks.

Now replace each blue diagonal stick with a shorter red stick and identify the polyhedron formed.

The relationship between a cuboctahedron and an icosahedron suggests a possible mechanism for the conversion of *ortho-* to *meta*-carborane (two isomers of $B_{10}C_2H_{12}$). This interesting reaction of $B_{10}C_2H_{12}$ takes place in the solid state at a temperature of 450°C.

In *ortho*-carborane, two adjacent vertices of an icosahedron are the positions of carbon atoms and the remainder of the vertices are the positions of boron atoms. In *meta*-carborane, the carbons are separated by one boron atom.

Justify how the conversion of *ortho*-carborane to *meta*-carborane can occur by the icosahedron to cuboctahedron to icosahedron mechanism. Could *para*-carborane, the carbon atoms at opposite vertices of the icosahedron, be formed by using the same mechanism?

The closest packing of equal spheres of one kind represent the crystal structures of the rare gases and the majority of metals. The vertices of the cuboctahedron represents the positions of the 12 nearest neighbors of a central sphere in cubic close-packing (ccp) of equal spheres. This relationship may be seen by standing the cuboctahedron on one of its triangular faces.

Exercise #4

It is possible to think of many patterns which give rise to two-dimensional planar structures. Consider the variety of lace patterns, tile floors, and wallpaper designs.

One way of classifying two-dimensional nets is by their *connectedness*. For example, the following two-dimensional nets illustrate different connectedness. Which is an example of the graphite layer?

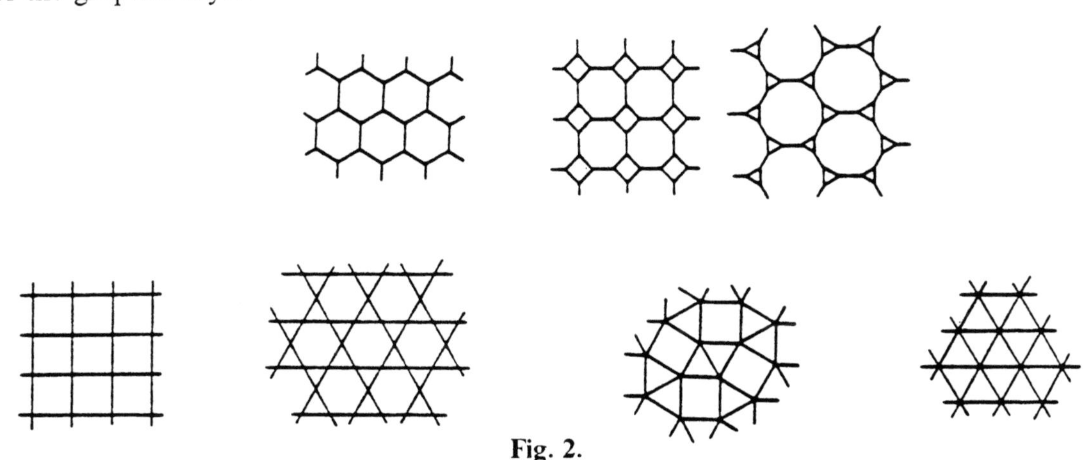

Fig. 2.

Recall in *Session One*, Exercise #5, that the repeating pattern of the BN layer was determined. Reinforce that learning experience by finding the smallest repeat unit, and the number of points (intersections) it contains. What is the bond angle of this planar net?

The equation describing 3-connected nets is:

$$3\,\Phi_3 + 4\,\Phi_4 + 5\,\Phi_5 + 6\,\Phi_6 + 7\,\Phi_7 + 8\,\Phi_8 + 9\,\Phi_9 \ldots = 6$$

$$\text{or } \Sigma\, n\, \Phi_n = 6$$

where Φ_n represents the fraction of the polygons that have n-edges. For the graphite net, all the polygons are hexagons. Therefore $\Phi_6 = 1$, and the equation is seen to hold true.

The analogous equation for 4-connected nets is $\Sigma\, n\, \Phi_n = 4$, for 5-connected nets is $\Sigma\, n\, \Phi_n = 10/3$, and for 6-connected nets is $\Sigma\, n\, \Phi_n = 3$. Check these equations using the nets shown above.

However, the net need not be planar but can be puckered. The bond angles can also vary from 90° to 109.5^o. Several chemical examples are known for the puckered, 3-connected net (or layer) consisting of six-rings with chair form. Once a six-ring chair structure is formed, it can be seen that the third bond on each of the ring members can be e (equatorial) or p (polar). Of the 13 ways of arranging these extra-annular rings, only three lead to periodic layers. These are p^0, p^1, and p^2. The p^0, or the e^6, is the symbol which will be used to indicate that all of the extra-annular connections will be equatorial. The p^2 is the symbol which indicates that two of the extra-annular connections are polar. In Fig. 3, indicate which of the rings is p^0 and which is p^2.

Fig. 3.

Construct a model with tetrahedral bond angles (black balls), in which only three sticks are used at each black ball. *Keep all rings chair-shaped.* As mentioned above, this does not define the configuration of the layer. It is necessary to specify how the third link from each point is related to the mean plane of the ring. Prepare a two dimensional net with repeating pattern in which there is a p^2 arrangement. The two figures below indicate layers with p^0 and p^2 layers. Which is which?

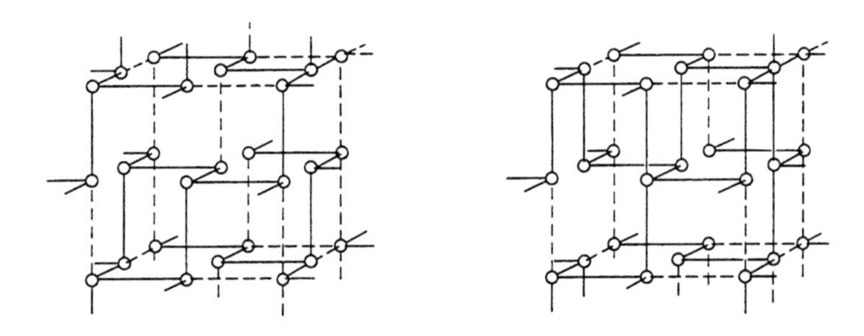

Fig. 4.

The p^0 layer is that found in crystalline As, Sb, and Bi. The p^2 layer is found in black phosphorous and in GeS, SnS, GeSe, and SnSe. In black phosphorous, the layer is deformed from the idealized structure made with the black tetrahedral balls. Two of the bond angles are 102° and one is 96.5^o. Make a model of the p^o layer.

Exercise #5.

The simplest three-dimensional framework is the primitive cubic lattice (Fig. 5.)

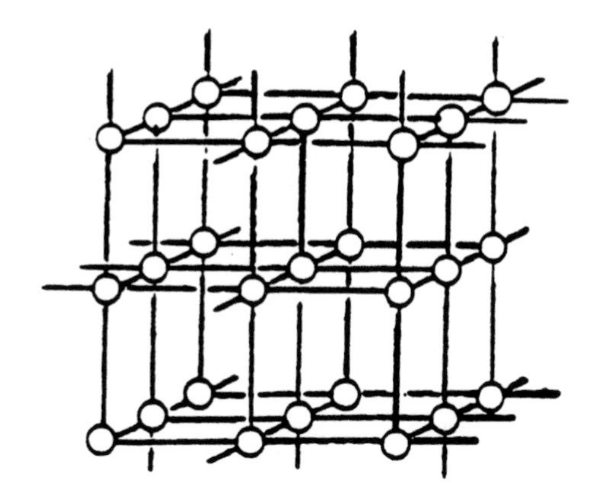

Fig. 5.

This is a six-connected framework consisting of circuits of four points. Starting from any intersection the shortest route back to the starting point is a circuit of four links. Unlike the two-dimensional nets, there is no known equation establishing a relationship between p and n for three-dimensional structures.

The following table only illustrates the empirical relationship between p, the number of connections in the three-dimensional structure, and n, the size of the circuits. These relations hold true only for the simplest and most symmetrical three-dimensional frameworks.

Table II

p	n (3D)	n (2D)
3	10	6
4	6	4
6	4	3

Make a 3-connected, three-dimensional representation of the silicon framework in $ThSi_2$. Using 18 blue trigonal balls, reproduce the unit cell shown in the illustration. Note that even though each three-connected point is trigonal planar, a three-dimensional net can be generated by rotating the trigonal points by 90°.

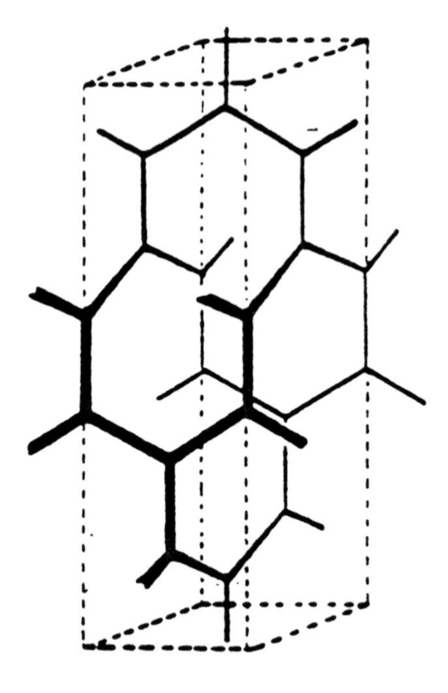

Fig. 6.

In the unit cell of Si atoms in $ThSi_2$, note the size of the smallest rings and the relation to the planar 6-gon net.

There are two forms of P as examples of two different 3-connected systems: the finite polyhedron – P_4 (white P) and the 2D (6,3) layer (black P) illustrated in the previous exercise. A 3D framework structure is not known for any crystalline form of elemental phosphorus. However, silicides of metals provide examples of all three types of structure, Si^- behaving in this respect like P, with which it is isoelectronic.

> $BaSi_2$ or $Ba_2[Si_4]$ contains tetrahedral $Si_4{}^{4-}$ groups.
>
> $CaSi_2$ contains $[Si_2]_n{}^{n-}$ layers similar to those of As (p^0 layer).
>
> $ThSi_2$ and $SrSi_2$ (under pressure) have a 3D Si framework of the type built above.

It should also be observed that P_2O_5 forms related structures of all these three types.

Exercise #6

If there was insufficient time in *Session One* to do Exercise #6 which demonstrated a three-dimensional, 4-connected diamond structure, do so at this time.

Indicate, using this structure, how it might be used to represent the zinc blende structure which is the cubic form of ZnS.

Relate the 3D net representation of ZnS to that of sphere packing. In the latter representation, S can be thought of as close-packed with Zn occupying half of the tetrahedral holes.

Make a model of the hexagonal high pressure form of diamond. This can represent the wurtzite structure of ZnS if alternate spheres are considered to be sulfur atoms and zinc atoms.

Exercise #7

Another four-connected framework is that of PtS, PdS, PtO, AgO, and CuO. A model of this structure is or will be made in the CuO experiment. Once again, square planar Pt, Pd, Ag or Cu are used to form a three dimensional net. Review this structure if the model has already been made. Otherwise, come back to read this section on nets after you have made the model.

Exercise #8

However complex the units that are to be joined together, the problem may be reduced to the derivation of systems of points each of which is connected to some number (p) of others. In the simplest systems, this number is the same for all points:

> $p = 1$: there is one solution only, a pair of connected points.
>
> $p = 2$: the only possibilities are closed rings or an infinite chain.
>
> $p = 3$: the possible systems now include finite groups (for example, polyhedra) and arrangements extending indefinitely in one, two, or three dimensions (p-connected nets).

It is not necessary to include singly connected points ($p = 1$) in nets since they can play no part in extending the net. Also, 2-connected points may be added along the links of any more highly connected net ($p > 2$) without altering the basic system of connected points. We, therefore, do not include either 1- or 2-connected points when deriving the basic nets though it may be necessary to add them to obtain the structures of actual compounds from the basic nets. For example, 2-connected points (representing -O- atoms) are added along the edges of the 3-connected tetrahedral group of four P atoms to form the P_4O_6 molecule, and additional singly connected points (representing O = atoms) at the vertices to form P_4O_{10}.

Still retaining the condition that all points have the same connectedness, we find that there are certain limitations on the types of system that can be formed. For example, there are no polyhedra with all vertices 6-connected, and there is only one 6-connected plane net. These limitations are summarized in Table I, which may be considered in two ways.

> (i) Together with simple linear systems ($p = 2$), the *vertical* subdivisions correspond to the four main classes of crystal structure, namely, molecular crystals, chain, layer, and 3-dimensional (macromolecular) structures The division into polyhedra, plane, and 3D nets is also the logical one for the systematic derivation of these systems.

(ii) However, from the chemical standpoint we are more interested in the *horizontal* sections of Table II. If we wish to discuss the structures that are possible for a particular type of unit, for example, an atom or group that is to be bonded to three others, we will have to select from the vertical groups the systems having p = 3.

Table III
Systems of p-connected_points

p	Polyhedra	Infinite periodic nets: Plane nets	Infinite periodic nets: 3D nets
3	↑	↑	↑
4	│	│	│
5	↓	│	│
6		(one ↓ only)	│
7			│

Give the names of the polyhedra referred to in Table III above and sketch the plane nets for p = 3 and 4. Give chemical examples of the nets where possible.

Which chemical systems can be used as examples of the 3-connected, 4-connected, and 6-connected 3D structures?

There are three simple solutions to the equation

$$\Sigma n\Phi_n = 6$$

corresponding to 3-connected planar nets constructed of equal number of two types of polygons, namely

$$\Phi_3 = \Phi_9 = 1/2$$
$$\Phi_4 = \Phi_8 = 1/2$$
$$\Phi_5 = \Phi_9 = 1/2$$

Draw the simplest solution to the $\Phi_4 = \Phi_8 = 1/2$ net. If this is too simple, try the $\Phi_3 = \Phi_9 = 1/2$ net.

Exercise #9

Determine the basic repeat unit and the formula for the three patterns illustrated below.

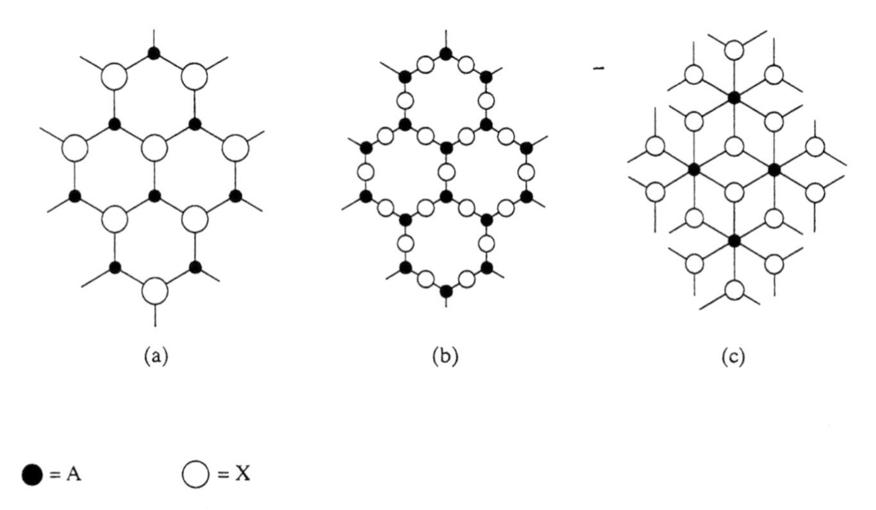

Fig. 7.

Exercise #10

In *Session Two*, Exercise #7, partial filling of octahedral holes in close-packed structures was discussed. Table II in that section indicated the types of compounds representative of different partial hole filling. In this section, structures in which various fractions of the total number of tetrahedral interstices are occupied will be considered. Although it is possible to describe a number of simple structures in this way, many crystals which can be represented by considerations of partial hole filling have appreciable covalent character, that is, the metal atom is forming four-directed (tetrahedral) bonds. For this reason, the two forms of ZnS were discussed under 3D 4-connected nets, and included other structures (PbO, red HgI_2) under "structures built from tetrahedra." The simpler structures which are related in a purely geometrical way as partial tetrahedral hole filling are shown in Table IV.

Table IV

Structures with Tetrahedral Coordination of A and Close-Packing of X Atoms

Fraction of tetrahedral holes occupied	Sequence of c.p. layers AB	ABC	Formula	coordination numbers M and X
All	-	Li_2O (antifluorite)	M_2X	4 : 8
3/4	-	O_3Bi_2 O_6Sb_4 O_3Mn_2	M_3N_2	4 : 6
1/2	Wurtzite β-BeO	Zinc blende PtS, PbO	MX	4 : 4 4 : 4
3/8	Al_2ZnS_4	Al_2CdS_4 Cu_2HgI_4	M_3X_4	4 : 6 4 : 4 *
1/3	β-Ga_2X_3	γ-Ga_2X_3	M_2X_3	
1/4	γ-$ZnCl_2$	HgI_2 ZnI_2 SiS_2 OCu_2	MX_2	4 : 2
1/6	Al_2Br_6	In_2I_6	MX_3	
1/8	$SnBr_4$	SnI_4 OsO_4	MX_4	4 : 1

* 3/4 of tetrahedral metal holes in zinc-blende structure are randomly occupied.

The antifluorite structure of Li_2O may be described as a close packed assembly of O^{2-} ions in which Li^+ ions occupy all the tetrahedral interstices.

Construct the Li_2O structure with the small blue (tetrahedral) balls and the close-packed layers. A small bit of wax will have to be used to position the second layer of tetrahedral balls between each layer.

Observe the relative positions of tetrahedral holes in a hcp assembly and show why no compounds adopt the hcp structure in which *all* tetrahedral holes are occupied.

Exercise #11

There are systems in which atoms of two (or more) kinds form the close-packed assembly. Such atoms must be of similar size, for example, Ba^{2+} and O^{2-} (in complex oxides), Cs^+ and Cl^- (in complex halides), or Cl^- and OH^- (in hydroxyhalides). Small metal atoms (ions) occupy some or all of the octahedral interstices between the anions. Several examples of structures of this type will be illustrated.

Let us consider structures built from close-packed layers consisting of atoms of two kinds, X and O, which are present in the ratio 1 : 3. The two simplest layers of this kind in which the X atoms (represented by the dark circles) are not adjacent are shown in the figure below. (Layers with larger repeat units result from combining strips of these layers.) To form such layers the atoms X and O must have similar sizes. An important difference between the two layers (a) and (b) is that it is not possible to stack (a) layers so as to produce octahedral holes surrounded entirely by O atoms (represented by the white circles) without bringing X atoms into contact, whereas this is possible with the (b) layers.

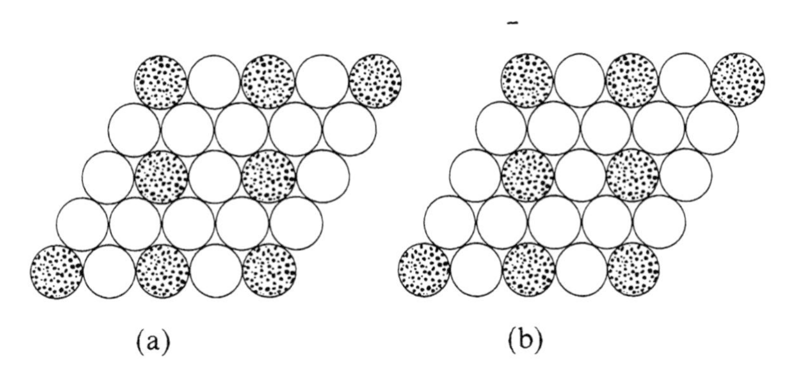

Fig. 8.

Layers of type (a) are found in intermetallic compounds and in hydroxyhalides (the layer being, for example, $Cl(OH)_3$) where contacts between X atoms are permissible. In complex oxides and halides, on the other hand, the layers are of the type BaO_3 or $CsCl_3$, for example, and contact between Cs^+ or Ba^{2+} ions of adjacent layers would reduce the stability of the structure. In such compounds only (b) layers are found, and we confine our attention here to these two large groups of compounds.

Using two portions of XO_3 layer check:

(i) that there is only one arrangement of the layers relative to one another that provides octahedral interlayer positions surrounded entirely by O atoms (O_6 holes) and that there are no X – X contacts

(ii) confirm that there is only one such Y position for every XO_3 in the structure.

The formula of a complex oxide (or halide) depends on the fraction of the octahedral holes occupied by Y atoms:

Table V

Fraction occupied	Formula
all	XYO_3
2/3	$X_3Y_2O_9$
1/2	X_2YO_6

Stack a number of layers in hcp and insert the Y atoms. The YO_6 octahedra form vertical columns in which each octahedron shares two opposite faces. This is the $CsNiCl_3$ or $BaNiO_3$ structure. Note that the number of octahedral holes surrounded by six oxygen atoms equals 1/4 the total number of octahedral holes.

Stack three layers in ccp, the relation between layers being that shown in the figure below:

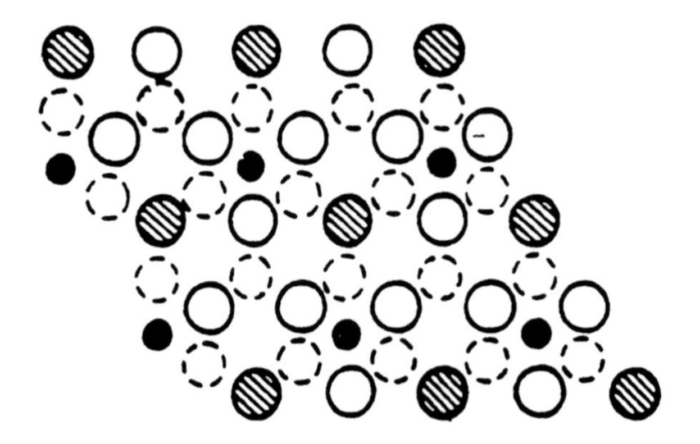

Fig. 9.

In Fig. 9, the shaded circles represent the A atoms, and the open circles represent the X atoms with the dotted circles being the X atoms on the first layer and the full circles being the X atoms in the second layer. The smaller black circles represent the Y atoms in the octahedral holes formed by the X atoms. Using this diagram as a guide, insert four Y atoms above the second layer. The structure is AcBaCb . . . Ascertain whether the YO_6 octahedra share vertices, edges, or faces. In this model the cubic unit cell can be seen with X atoms (blue) at the corners and the Y atoms at the body center. This is the perovskite structure adopted by many complex oxides and fluorides.

There are examples of other layer sequences.

Table VI

Symbol	Number of layers in repeat unit	Examples
hhccc	5	$Ba_5Nb_4O_{15}$
hcc	6	$BaTiO_3$ (hexagonal)
ccch	8	$Sr_4Re_2SrO_{12}$
chh	9	$Ba_3Re_2O_9$
hhcc	12	$Ba_4Re_2MgO_{12}$

(not all of the octahedral holes are occupied in some of these structures.)

The symbols h and c in the above table refer to the layer sequences. The symbol h indicates that there are layers of the same type on both sides of a given layer. The symbol c indicates that the given layer is sandwiched between layers of two different types. For example:

h c h c h c
A B C B A B

h c c h c c h
A B C A C B A B

Fig. 10.

Questions to be Considered

1. Describe a compound as both a polyhedral model and a close-packed model.

2. Describe a compound as both a close-packed model and as a net.

3. Would you consider a two dimensional net consisting of equal number of five membered rings and six member rings to be repeating in nature?

4. Identify several 3-connected semi-regular polyhedra with two different polygons as faces?

Helpful Background Literature

1. Wells, A. F. "Structural Inorganic Chemistry," 5th ed; Clarendon Press: Oxford, 1984.

BRUCITE AND MAGNESIA

Equipment Needed:	sintered glass funnel 220 degree oven muffle furnace (800°C) unglazed (Rose) crucible or nonporous alumina crucible nickel crucible, 50 mL centrifuge
Chemicals Needed:	1 g $MgCl_2.6H_2O$ 28 g KOH
Time Needed:	3 hours
Safety Notes:	Be careful to avoid spills when handling the hot alkali melt.

Purpose

This experiment illustrates the preparation and properties of magnesium hydroxide and magnesium oxide. It also illustrates solid state structures as well as a solid state transformation.

Introduction

Magnesium hydroxide occurs as the mineral brucite. The OH groups in brucite occupy positions of hexagonal close packing with the magnesium ions occupying all the octahedral holes in alternate layers. Another way to visualize the $Mg(OH)_2$ structure is to consider a two dimensional sheet of octahedra with magnesiums at the center of the octahedra and the OH groups located at the corners of the octahedra. These sheets are then stacked so that the OH groups are in a hexagonal close packed arrangement. This structure is generally identified as a CdI_2 structure. The brucite layer is important in that it is involved with silicate groups in minerals such as crysotile, talc, and phlogopite.

Brucite is a mineral found in serpentine along with magnesite and dolomite. It can be prepared by hydrating magnesium oxide, the reverse of the reaction illustrated in this experiment; or it can be precipitated by using an excess of base with magnesium halides.

Magnesium oxide or magnesia is also known as periclase when it is found as a mineral. It exhibits the cubic sodium chloride structure. The large oxygen atoms are cubic close-packed with the small magnesium ions occupying all of the octahedral holes.

In this experiment, the hexagonal magnesium hydroxide will be rearranged in a high temperature solid state reaction to the cubic magnesium oxide. The understanding of the mechanisms of solid state reactions is still in its infancy. Several mechanisms have been postulated for the conversion of the hydroxide to the oxide. In one proposed mechanism (1), water is presumed to be lost from all regions of the crystal. Hydroxyl groups migrate from one layer to another and combine to form water which then diffuses out of the solid array. In another proposed mechanism (2), it was assumed that donor acceptor reactions take place simultaneously throughout the bulk of the solid phase. Mg^{2+} ions migrate to an acceptor site to form MgO with H^+ combining with OH^- ions to form water. The evidence supporting this second mechanism is that topotaxy is observed. Topotaxy is the term used when the macroscopic structure of the parent substance is retained in the final product. Thus the thermal decomposition of nesquehonite ($MgCO_3.3H_2O$) forms MgO with the needle-like structure of the carbonate whereas the thermal decomposition of brucite forms MgO with the hexagonal-platelet shape of the hydroxide. In addition to topotaxy, the MgO structure formed is porous. The argument is that the ion migration is less apt to disturb the macroscopic structure than water migration and that the loss of water will lead to the porosity. A third proposed mechanism (3) interprets the broadening of the O-H band in the infrared at 300°C as an indication of OH-OH interactions on the surface. The water is postulated to form in the bulk by proton tunnelling. If the protons jump from layer to layer, no positive ion migration has to be postulated to maintain electrical neutrality.

One of the interesting industrial applications for magnesium oxide is its use in the forming of Cal-Rods. A Cal-Rod is either round or flattened and is most familiar as the heating element of an electric stove or oven. Because a bare resistance wire would create a shock hazard, the Nicrome heating wire is placed in a concentric position in a stainless steel tube with magnesium oxide filling the space between the heating wire and protective tube. The magnesium oxide must be an electrical insulator but a reasonable thermal conductor. For metals, there is a rule known as the Wiedemann-Franz law which states, on the basis of the free-electron theory, that the ratio of electrical and thermal conductivities at the same temperature is the same for all metals. Thus, if a metal is a poor electrical conductor, it will be a poor thermal conductor. Although the Wiedemann-Franz law does not hold for non-metals, good thermal conductors are generally not good electrical insulators. For the heating rod application, magnesium oxide gives a reasonable thermal conductivity while acting as an excellent electrical insulator.

Properties of Brucite. Because of its layer structure brucite is soft, measuring 2 on the Mho scale. It resembles gypsum and talc in this respect.

At 115°C brucite gives up essentially none of its chemically bound water. In vacuum at 200°C it is dehydrated to MgO only slowly. At 300°C in a vacuum it is 95% converted to MgO in 10 days. In air the dissociation of brucite starts at 400° to 420°C. The reaction is slow. Even at 1000°C the complete conversion to MgO is measured in hours rather than in minutes.

Properties of Magnesia. Because of the three dimensional structure of MgO it is much harder than brucite, measuring 6 on the Mho scale. Artificial crystals of magnesia demonstrate a hardness between that of felspar and that of quartz.

Pure magnesia cannot be melted in an oxygen-hydrogen flame at temperatures sufficient for fusing quartz. It is melted in an electric furnace at temperatures of approximately 3000°C. The grains of magnesia are reported to sinter together at 2250°C. Melting points are reported from 2250° to 2500°C. Part of the difficulty in determining a melting point is the depression in the melting point caused by impurities.

The specific heat between 24°and 100°C is 0.24394. The thermal conductivity for 600 mesh powder over the same temperature range is reported to be 0.00047 cal per square cm degree difference of temperature per second. This compares to values of 0.00029 for lime and 0.00028 for firebrick.

Conductivity studies indicate that fused magnesia is a better insulator than porcelain below 1100°C. Both materials show a decrease in resistance with increase in temperature. At 1100°C there is a crossover but magnesia exhibits a resistance minimum at around 1100°C with the resistance increasing with a further rise in temperature so that the resistance at 1300°C is about the same as that at 700°C. The resistance in ohm-cm for strongly pressed powder is shown in Table I. The significance of this data is that the Cal-Rod contains MgO powder compressed into the stainless steel jacket.

Table I

Temperature (°C)	Resistance (ohm-cm)
650	6.6×10^6
700	3.3×10^6
750	1.8×10^6
800	1.0×10^6
850	6.2×10^5
900	3.7×10^5
950	2.5×10^5
1000	1.6×10^5
1050	1.0×10^5

Experimental Procedure

Magnesium Hydroxide

Place 28.0 g of KOH and 15 mL of water in a nickel crucible, and heat in a 200°C oven until all of the solid is dissolved. Carefully add 1.0 g of $MgCl_2 . 6 H_2O$ to the concentrated KOH solution in the crucible and heat the mixture in the 200°C oven for 45 minutes. After this period of time, remove the crucible from the oven. The $Mg(OH)_2$ is found as a precipitate at the bottom of the crucible. Use a medicine dropper to remove most of the upper layer which consists of the excess alkaline solution. As the next step in removing the excess KOH, add 10 mL of water, dissolving the KOH if it has precipitated on cooling. Place the precipitate and the aqueous KOH solution mixture in an appropriate test tube or centrifuge tube. Centrifuge for three minutes to separate the precipitate. Decant the supernatant solution, and add 10 mL of distilled water to the precipitate in the centrifuge tube. Centrifuge for three minutes. Repeat the wash process a total of three times. Check the final wash supernatant with pH paper to determine if it is neutral. Repeat the wash if it is not. (Alternatively, the solid can be transferred to a sintered glass funnel and washed with three portions of water.)

Divide the product into two parts. Place one portion in an unglazed porcelain crucible for the MgO preparation. Place the other portion on a watch glass and dry in an oven at 100° to 115°C for half an hour. This should be $Mg(OH)_2$. (If the only drying oven in the laboratory is set at 200°C for the first part of this experiment, the $Mg(OH)_2$ may be dried at 200°C for half an hour. Only a small portion of the product will be converted to MgO.) If an X-ray powder pattern equipment is available, do an X-ray powder diffraction pattern of this material to confirm that it is $Mg(OH)_2$.

Magnesium Oxide

Dry the magnesium hydroxide prepared above which was placed in the unglazed porcelain crucible in a 200°C oven for 15 minutes. This prevents splattering which may occur when a wet product is suddenly placed in a hot muffle furnace. Remove the crucible from the drying oven and place in an 800°C muffle furnace for one hour. The reason for using an unglazed porcelain crucible is that the glaze is a lower melting compound and will not maintain its integrity at 800°C. The somewhat more expensive alundum crucible can be used. The nickel crucible used in the first part of the experiment can also be used although the surface of the crucible will become oxidized to NiO at these temperatures. Weigh the product. Determine the X-ray powder diffraction pattern if the equipment is available. Confirm that the product is cubic. Calculate the cell size.

Auxiliary Experiment

Test a small crystal of magnesium hydroxide and magnesium oxide you have prepared with 6 N HCl.

Determine the conditions necessary to drive the reaction in reverse, i.e. to react MgO with water to form the hydroxide.

Questions to be Considered

1. Write balanced equations for the preparations of the two compounds.

2. What is the difference between the CdI_2 structure and the $CdCl_2$ structure?

3. The structure of talc can be represented as follows:

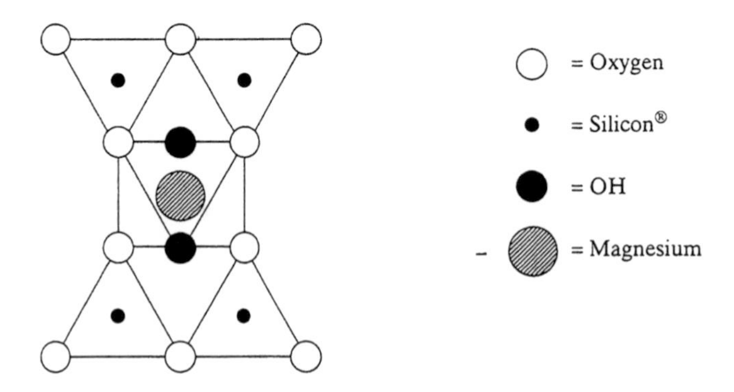

Fig. 1.

Verbally interpret the structure in terms of polyhedra, corner or edge sharing, etc.

Helpful Background Literature

1. Goodman, J. F. *Proc. Roy. Soc.*, 1958, *A 247*, 346.

2. Ball M. C.; Taylor, H. F. W. *Minerolog. Mag.*, 1961, *32*, 754.

3. Freund, F. *Angew. Chem. Internat. Ed.*, 1965, *4*, 445.

4. Gmelin, *Handbuch der Anorganischen Chemie, System Nummer 27, Teil B,* Verlag Chemie, Weinheim/Bergstr., 1937, pp. 11 – 52 (oxide) and 54 – 68 (hydroxide).

5. Mellor, J. W., *A Comprehensive Treatise on Inorganic and Theoretical Chemistry, Vol. IV*, Longmans: London, 1963, pp. 280 – 296.

Appendix I

The following tables give the X-ray powder pattern data MgO and $Mg(OH)_2$. If the X-ray data were collected as a Debye-Scherrer film, the data obtained from the film should be entered into the table below and into your notebook. If the X-ray data were obtained from a defractometer instrument, the instrument will have done some of the calculations for you. Compare your experimental values with those given in the Powder Diffraction Files (PDF).

Table I

Magnesium Oxide (MgO) PDF Data (Index No. 04-0829)

2θ	I/I_1	θ	$\sin \theta$	$\sin^2 \theta$	d	$h^2 + k^2 + l^2$	hkl	a
	10				2.431		111	4.213
	100				2.106		200	
	52				1.489		220	
	4				1.270		311	
	12				1.216		222	
	5				1.053		400	

Table II

Magnesium hydroxide [$Mg(OH)_2$] PDF Data (Index No. 07-0739)

2θ	I/I_1	θ	$\sin \theta$	$\sin^2 \theta$	d	$h^2 + k^2 + l^2$	hkl	a
	90				4.770		001	
	6				2.725		100	
	100				2.365		101	
	55				1.794		102	
	35				1.573		110	
	18				1.494		111	
	16				1.373		103	
	2				1.363		200	
	12				1.310		201	

STRONTIUM ZINCATE(II)

Equipment Needed:	agate mortar and pestle pellet press muffle furnace (1000°C) tube furnace (1000°C) alumina boat large unglazed crucible (e.g., 50 mL alumina or Rose crucible) top loading balance small sample vial reduction valve for nitrogen cylinder
Chemicals Needed:	1.22 g zinc oxide 2.21 + 1.47 g strontium carbonate 1.5×10^{-2} moles of strontium acetate hemihydrate* 1.5×10^{-2} moles of zinc acetate dihydrate 1.5 mL glacial acetic acid acetone nitrogen cylinder
Time Needed:	3 hours
Safety Notes:	Pyrolysis of the acetate salts of strontium and zinc result in organic decomposition products. The furnace in which this pyrolysis is carried out should be in the hood.

Purpose

This preparation illustrates how different starting materials may lead to the same solid state reaction. The discussion also illustrates how differential thermal analysis-thermal gravimetric analysis (DTA-TGA) can shed light on the decomposition processes. The product formed is an example of a four-connected two-dimensional net discussed in

---

*If not available, follow directions for preparation in the Experimental Procedure section.

Session One. Recall that the net was puckered so that the strontium ions can be accomodated between the network layers.

Introduction

Strontium zincate was first prepared by the thermal decomposition of a mixture of strontium carbonate and zinc oxide under nitrogen at 1000°C (1). It has also been prepared by the thermal decomposition of strontium acetate hemihydrate and zinc acetate dihydrate at 1150°C under oxygen (2).

Both methods for the synthesis of strontium zincate will be described. The apparatus for both methods will also be made available. Some of the class should use one method and others should use the other method to compare the relative merits of each recipe.

The path of pyrosynthesis can often be followed by DTA-TGA. TGA, which measures the weight loss as a function of temperature, indicates that the strontium acetate hemihydrate commences to lose water at 180°C. After an initial slow loss of about two-thirds of the water of hydration, the remainder is rapidly lost at 200°C. The fusion of the anhydrous strontium acetate takes place at 321° $\pm$ 1°C. After fusion, the weight loss again commences at about 350°C with rapid decomposition commencing at 430°C with the loss of acetone and the formation of strontium carbonate as the products at temperatures above 500°C.

Zinc acetate dihydrate loses the two molecules of water in one step commencing at around 60°C and finishing slightly above 100°C. The TGA then indicates that the zinc acetate gradually continues to lose weight. This is ascribed to a slow sublimation of the product. At 180°C there is evidence that some of the zinc acetate decomposes to tetrazinc(II) monooxoacetate ($Zn_4O(CH_3COO)_6$) and acetic anhydride. On further heating, the DTA indicates a double endotherm at 252°C and 258°C. These two temperatures have been intrepreted as being the fusion of the tetrazinc(II) monooxoacetate and the zinc acetate. Following fusion, there is rapid weight loss during which the zinc acetate decomposes to zinc oxide, acetone, and carbon dioxide. The tetrazinc(II) monooxoacetate also decomposes to the same products, although in a slightly different ratio.

Since the strontium acetate and zinc acetate decompose to strontium carbonate and zinc oxide, there should not be a great deal of difference whether one starts with strontium carbonate and zinc oxide or with strontium acetate and zinc acetate. There are some variables, however, which should be considered. There is the question of the relative purity of the starting materials. There is the question of how readily one can make an intimate mixture of the two substances. It should also be noted that the zinc oxide formed from the zinc acetate decomposition is porous which may facilitate the solid state reaction. The evaluation of the two synthetic methods requires the X-ray analysis of the powder formed. Strontium carbonate by itself undergoes decomposition at 1260°C. For an incomplete reaction, one would look for the X-ray powder pattern lines for strontium carbonate, strontium oxide and zinc oxide.

Experimental Procedure

Method A: Weigh out 1.22 g zinc oxide and 2.21 g strontium carbonate and place in an agate mortar. Make a mull with acetone and grind thoroughly with the pestle. The grinding should be continued until most of the acetone has evaporated. Press the mixture into one or more pellets using the 1/2 inch pellet press applying enough pressure to make pellets which hold together and place in an alumina boat. Slide the boat into a tube furnace and fire at 1000°C under a nitrogen flow for 12 hours.

Method B: Weigh out 1×10^{-2} moles of strontium acetate hemihydrate and 1×10^{-2} moles of zinc acetate dihydrate.

If the strontium acetate hemihydrate is not available in the storeroom, prepare some from strontium carbonate. Weigh out 1.47 g of strontium carbonate (1×10^{-2} moles). Stoichiometry indicates two moles of acetic acid for each mole of strontium carbonate. For a stoichiometric amount, 1.14 mL of glacial acetic acid is needed. Since acetic acid is volatile, a slight excess will insure that enough is available to react with all of the strontium carbonate. However, do not use more than 1.5 mL. Add the glacial acetic acid to the strontium carbonate and add about 1 mL of water. Bubbling will indicate that a reaction is taking place. Allow the reaction to proceed for a while because it is not a fast one. Place the product on a steam bath until the excess acetic acid and water have been removed. Weigh the product and calculate its molar amount.

Carefully weigh an equal molar amount of zinc acetate dihydrate. Place both reactants in an agate mortar and mull with acetone. Grind thoroughly until the acetone is gone. Place the reactants in an unglazed crucible and place in a muffle furnace which is at room temperature. Because there will be some bubbling of the reactants as acetic acid is lost, use a large unglazed crucible (e.g., 50 mL alumina). When everyone in the class has placed their reaction mixture into the muffle furnace, start the muffle furnace and heat to 1000°C. Hold the temperature at 1000°C overnight.

Bottle the product. Label properly. Compare the X-ray powder patterns for the products prepared by the two methods. Look for contamination of the strontium zincate by strontium carbonate and zinc oxide.

Questions to be Considered

1. Sketch the $SrZnO_2$ structure.

2. Explain why the $SrZnO_2$ structure is puckered whereas the similar four-connected net of red mercuric iodide is not.

Helpful Background Literature

1. Scholder, R.; Stocker, *Angew. Chem.*, 1954, *66*, 461.

2. Schnering, H. G.; Hoppe, R., *Z. Anorg. Allg. Chem.*, 1961, *312*, 87.

3. McAdie, H. G., *J. Inorg. Nucl. Chem.*, 1966, *28*, 2801 – 2809.

CADMIUM(II) IODIDE

Equipment Needed: 50 mL and 125 mL Ehrlenmeyer flasks
100 mL beaker
hot plate
steam bath
sublimer
vacuum system
small sample vial

Chemicals Needed: 3 g $CdSO_4.8/3H_2O$
4 g KI
50 mL ethyl alcohol
dry ice
aluminum foil for heat shield
dry ice
6 drops 0.1 M $AgNO_3$
5 drops CCl_4
0.5 mL 0.1 M $Fe(NO_3)_2$

0.5 mL 6M *HNO* sub 3
one liter of liquid nitrogen

Time Needed: 3 hours

Safety Notes: Do the evaporations in the hood. Take care not to allow the dry ice to remain in contact with the skin for even a short length of time. It can create bad burns by freezing at -78^oC.

Purpose

This preparation illustrates a basic type of inorganic reaction, that of metathesis. The structure of cadmium(II) iodide is also of interest in that it is of a type exhibited by a number of compounds. It can be visualized by either a close-packed model or a polyhedral model.

Introduction

The cadmium(II) iodide synthesis to be carried out is an example of a metathesis reaction. Characteristic of many inorganic reactions, the rate of the reaction for this exchange reaction is fast. Most of the actual experimental time is spent in isolating and purifying the product.

Cadmium(II) iodide crystallizes in a structure, a single layer of which can be represented by octahedra sharing six edges. A close-packed picture is that of iodide ions hexagonally close-packed with cadmium ions occupying all the octahedral holes in alternate layers. Other compounds with the CdI_2 structure are $CdBr_2$, $FeBr_2$, $CoBr_2$, $NiBr_2$, MgI_2, CaI_2, ZnI_2, PbI_2, MnI_2, FeI_2, CoI_2, $Mg(OH)_2$, $Ca(OH)_2$, ZrS_2, SnS_2, TiS_2, etc. The cadmium iodide structure contrasts with the cadmium chloride structure in that the latter can be pictured as cubic close-packed chloride with cadmium ions occupying all the octahedral sites in alternate layers. Examples of other compounds with the cadmium chloride structure are $MnCl_2$, $FeCl_2$, $CoCl_2$, $NiCl_2$, NiI_2, $MgCl_2$, and $ZnBr_2$.

That cadmium iodide is relatively covalent is indicated by its ease of extraction with alcohol. This covalency is not surprising in view of the position of cadmium on the periodic table.

Because cadmium iodide can crystallize in a number of polytypes, the product obtained after the alcohol extraction does not give the X-ray powder diffraction pattern for crystalline cadmium iodide. It is only after subliming that powder patterns similar to that reported in the literature can be obtained. The sublimation of cadium iodide is not easy. It is slow and a very good vacuum is needed.

Experimental Procedure

Weigh out 3 g $CdSO_4 \cdot 8/3H_2O$ and 4 g KI into a 50 mL Ehrlenmeyer flask. Add 11 mL of water. Boil gently until the solution clears. Decant the solution into an evaporating dish. Place on a hot plate to concentrate. Before the solution reaches the point where there is danger of splattering, remove the evaporating dish from the hot plate and finish evaporating on a steam bath.

Add 25 mL of alcohol to the solid residue and heat on a hot plate. Filter into a 125 mL Ehrlenmeyer flask. Repeat with a second 25 mL portion of alcohol. Save the residue, if any, for further tests. Pour the alcohol solution into a 100 mL beaker and evaporate off the alcohol on a hot plate. Once again remove the last portions of alcohol on a steam bath to prevent splattering.

Place the solid left after the alcohol evaporation into the sublimer constructed with a standard taper joint. Note that the joint in the sublimer is "reversed" with the male joint on the body of the sublimer. The reason for this is so the cold finger can be extracted from the sublimer without danger of getting grease on the sublimed product.

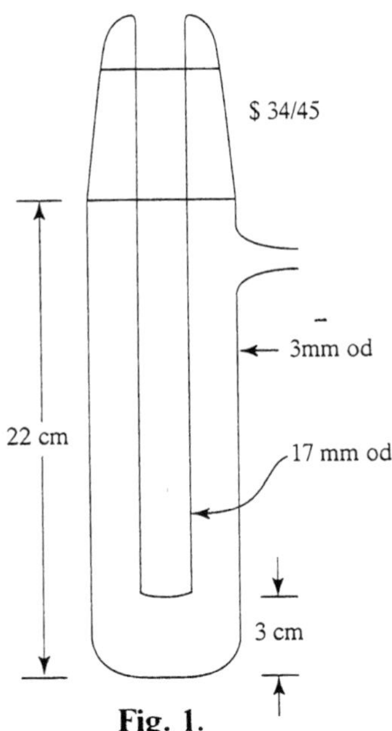

Fig. 1.

Pull a vacuum on the sublimer, place powdered dry ice in the sublimer cold finger and gently heat the bottom of the test tube with a Bunsen burner. If sublimation does not occur, it is probably due to a poor vacuum. After sublimation (not all the product need be sublimed) carefully pour out the dry ice before breaking the vacuum. Remove the cold finger and scrape the solid onto a piece of glassine paper. Place in a 1 dram sample vial which has been weighed with the label. This empty weight which includes the label and lid is the tare weight. Be sure the label has the compound name, notebook page, tare weight, net weight and your name.

There should be little or no potassium iodide left after extraction of the cadmium iodide by alcohol. Any residue left after the alcohol extraction should be tested for iodide. Either method listed below can be used.

1. Acidify with nitric acid so that the solution is approximately 1 N. Test the resulting solution with 5 or 6 drops of 0.1 M $AgNO_3$.

2. Place fifteen drops of the aqueous solution of the residue in a small test tube. Add 2 drops of 6 M HNO_3 in excess of that required to neutralize. Add 3 to 5 drops of CCl_4. Add some 0.1 M $Fe(NO_3)_3$ solution equal in volume to the aqueous portion in the test tube and shake. A violet color in the CCl_4 layer indicates I^-.

Questions to be Considered

1. What would you expect if sublimation is attempted without pulling a vacuum?

2. In the first method described for an iodide test, a pinch of solid is dissolved in 3 mL of water. How many drops of concd. HNO_3 (15 M) is required to make the solution approximately 1 M?

3. Write the balanced equations indicating the reactions which occur in the second method for the iodide test.

4. Which reactant is in excess, $CdSO_4$ or KI?

Helpful Background Literature

1. Wells, A. F. "Structural Inorganic Chemistry", 4th ed.; Clarendon Press: Oxford, 1975; pp. 142, 209, 353, 375.

MANGANESE(IV) OXIDE

Equipment Needed:	evaporating dish glass crucible or boat for tube furnace 200°C hot plate 300°C tube furnace "T" tube for oil bubbler surface thermometer small sample vial
Chemicals Needed:	3.0 mL manganese(II) nitrate solution (50%) tank of oxygen
Time Needed:	3 hours
Safety Notes:	Decomposing manganese(II) nitrate at high temperatures gives off oxides of nitrogen. The hot plate for this decomposition should be in the hood. The exit tube for the tube furnace should be in the hood. Avoid splattering while driving off the last bits of water.

Purpose

This decomposition illustrates the formation of one of a class of oxides with the rutile structure. MnO_2 is also a commercially useful substance.

Introduction

Manganese forms a large number of oxidation states with oxides and oxy anions being represented by manganese oxidation states of 2+ to 7+. Manganese(II) oxide, like magnesium oxide crystallizes with the sodium chloride structure. Manganese(III) is observed in Mn_3O_4 which is an oxide consisting of Mn(II) and Mn(III). Manganese(III) oxide hydroxide, MnO(OH), occurs as a mineral manganite. Mn(IV) oxide, MnO_2, is the most stable oxide. Mn(V) is seen in the unstable $MnO_4{}^{3-}$ ion. For example crystalline $Na_3MnO_4.10H_2O$ is a known compound. Mn(VI) is represented by manganates such as K_2MnO_4. The best known of the manganese oxide anions is the Mn(VII) oxyanion represented by potassium permanganate, $KMnO_4$. Manganese(VII) oxide, Mn_2O_7, is an oily dark green-brown liquid which has a tendency to explode.

MnO_2 is one of the components of a dry cell battery. The Leclanche cell, which was patented in 1866, is still in widespread commercial use. The cell consists of a zinc anode and a carbon electrode surrounded by a paste of MnO_2, ammonium chloride, and graphite. In operation, the zinc is oxidized to Zn^{2+} and the manganese reduced to Mn(III). The alkaline dry cell utilizes the same reactions except that the ammonium chloride is replaced by potassium hydroxide. This results in a cell which provides about 50% more energy than a classical dry cell of the same size.

In addition to its practical use, MnO_2 belongs to a family of metal dioxides which exhibit the rutile structure or a distorted rutile structure. This family consists of Ti*, V*, Nb*, Ta, Cr, Mo, W, Mn*, Tc, Re*, Ru, Os, Rh, Ir, Pt, Ge*, Sn and Pb. Those indicated with an asterisk (*) are polymorphic, however, one of the polymorphs is a structure related the the rutile structure. Manganese dioxide probably crystallizes in this experiment as a mixture of several related structures generally called γ-MnO_2. The X-ray powder pattern is that of the mineral pyrolusite which is sometimes called β-MnO_2. The material prepared probably contains some ramsdellite besides the rutile structure because the pyrolusite slowly converts to ramsdellite on continued heating.

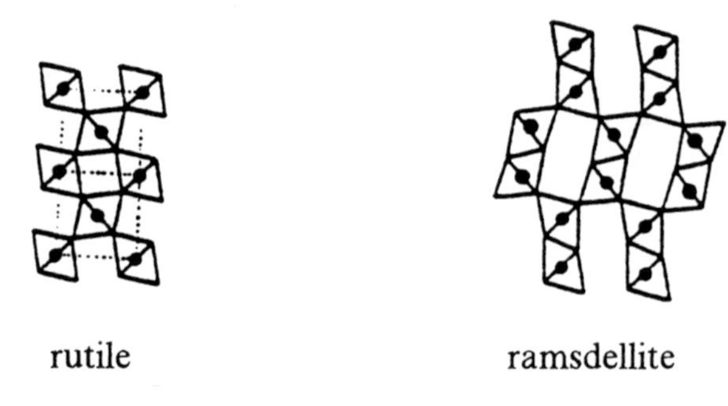

Fig. 1.

The product prepared can be crystallized by hydrothermal means to form the rutile type crystals.

This family also contains other useful oxides. TiO_2 is the white pigment which is basic to paint formulation. PbO_2 is the active substance in the lead acid storage battery. CrO_2 is the ferromagnetic material which is used to make magnetic tapes for tape recorders. As demonstrated in another experiment in this set, SnO_2 is used to make surface conductive glass.

The MnO_2 can be made in several ways. Although it can be made by the direct oxidation of elemental manganese, the easiest way to make the compound is to pyrolyze manganese(II) nitrate. Although manganese(II) nitrate can be obtained as the crystal, it is quite expensive compared to the 50% solution. Under pyrolytic conditions, the nitrate ion oxidizes the Mn(II) to Mn(IV). Even under mild heating, the nitrate ion is a strong enough oxidizing agent to oxidize the Mn(II). The black solid formed on evaporating off the water is Mn_2O_3 which contains both Mn(II) and Mn(IV). On heating this material under oxygen, NO_2 is given off and the pyrolusite structure of MnO_2 is formed. It is interesting that decomposing the $Mn(NO_3)_2$ in air in a muffle furnace often gives rise to the Mn_2O_3 rather than the MnO_2.

Experimental Procedure

Pour out 3.0 mL of the 50% manganese(II) nitrate solution into an evaporating dish. Place the evaporating dish on a hot plate prewarmed to 200°C. When the residue becomes pasty or within about a drop of dryness, transfer the pasty residue to a Pyrex boat and slide into a tube furnace equipped so that oxygen can be flowed over the boat containing the black pasty mixture. Start warming the tube furnace while flowing oxygen through the tube. The NO_2 will start to come off at around 180°C. Allow the tube furnace to continue heating to 300°C. When a period of time has passed with no more NO_2 being evolved, allow the tube furnace to cool, or, alternately, slide the insert out so that the boats are out of the hot zone. When the boats have cooled so they can be handled, remove them from the tube furnace. Scrape the greyish black solid into a vial, label the vial properly, and turn in to your instructor.

The instructor will have X-ray diffraction patterns for pyrolusite and Mn_2O_3.

Questions to be Considered

1. On heating manganese(II) nitrate, brown fumes are observed to be emitted. Write the Lewis structures for the oxides of nitrogen.

2. Write a balanced redox reaction for the pyrolysis of manganese(II) nitrate.

3. Look up the description of the dry cell in a general chemistry text and write the reactions which can take place under different conditions of discharge.

4. Draw the unit cell for the rutile structure. How many oxygens and how many manganese atoms are there per unit cell?

5. Arrange the metals whose dioxides form the rutile structure in a periodic arrangement.

Helpful Background Literature

1. Wells, A. F. "Structural Inorganic Chemistry," 5th ed.; Clarendon Press: Oxford, 1984. pp. 540 – 542, pp. 553 – 556.

2. Rogers, D. B.; Shannon, R. D.; Sleight, A. W.; Gillson, J. L. "Crystal Chemistry of Metal Dioxides with Rutile-Related Structures", *Inorg. Chem.*, 1969, *8*, pp. 841 – 849.

PILLARED CLAYS

Equipment Needed:	magnetic stirrer hot plate magnetic stirrer bar 1 L Ehrlenmeyer flask small dropping funnel two #1 Buchner funnels (50 to 70 mm)
Chemicals Needed:	5.0 g of Bentolite L 8.2 mL of Chlorhydrol solution distilled deionized water 0.1 M $AgNO_3$ in a dropping bottle
Time Needed:	3 hours
Safety Notes:	Heating chlorhydrol drives off gaseous HCl. The boiling, therefore, should not be done on the open desk. Make sure that you carry out this operation in a hood.

Purpose

This preparation illustrates changes in structure of a two dimensional infinite network inorganic compound using rather mild conditions. The separation of the layers in clays is of commercial importance in the preparation of catalysts. The experiment also serves as an introduction to several commercially important classes of minerals.

The use of X-ray powder diffraction patterns is illustrated. From the X-ray powder diffraction of the starting material and the X-ray diffraction of the pillared material, the expansion in lattice spacings is calculated.

Introduction

Clays are a group of minerals which are characterized by being soft, exhibiting easy hydration, and having cation exchange properties. Many, but not all, of the clays are related to kaolinite and pyrophyllite.

Kaolinite consists of a double layer, one of octahedra and one of tetrahedra. The octahedral layer consists of octahedra sharing alternate edges. Refresh your memory by looking up the BiI_3 structure in *Session Two* of the models experiment. The tetrahedral layer consists of tetrahedra sharing three vertices to form rings of six. The fourth vertex of the tetrahedra are shared with certain of the vertices of the octahedra. Fig. 1 illustrates the kaolinite structure. Aluminum is located in the center of the octahedra, and the vertices of the octahedra shared with other octahedra are locations of OH groups. The vertices of the tetrahedra are locations of oxygens and the center of the tetrahedra are locations of silicon. The formula of kaolinite is $Al_2(OH)_4Si_2O_5$.

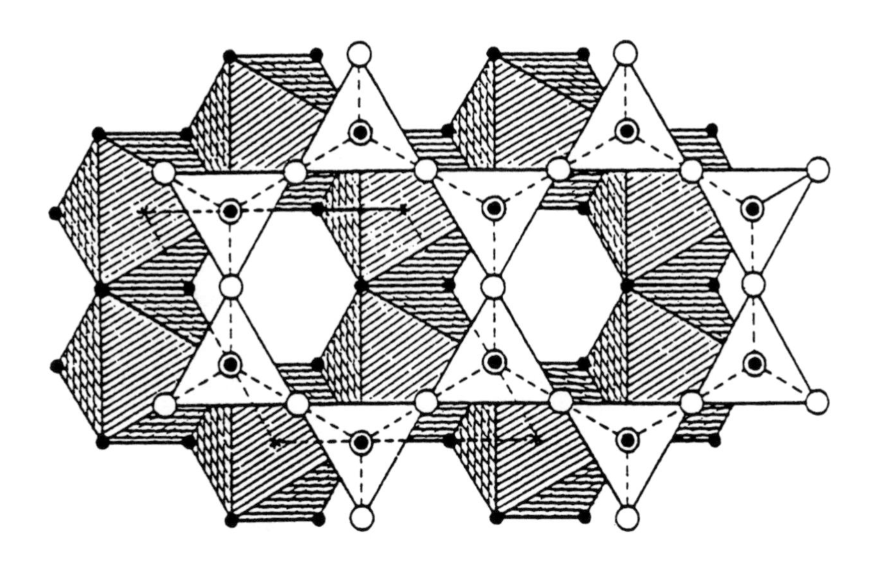

Fig. 1. The kaolinite double layer

The pyrophyllite structure is an extension of the kaolinite structure. If the kaolinite structure is pictured with the tetrahedral layer on both sides of the octahedral layer, then a pyrophyllite structure results. Because the octahedral layer is the "baloney in the sandwich" formed by the two tetrahedral layers, there will be twice the number of silicons in pyrophyllite as there is in kaolinite. The formula for pyrophyllite is, therefore, $Al_2(OH)_2Si_4O_{10}$. Note the decrease in the number of OH groups because of the increased number of octahedral vertices shared by the tetrahedral vertices, these latter being oxygens. A method of writing the vertical cross-section of these layer structures is as shown in Fig. 2.

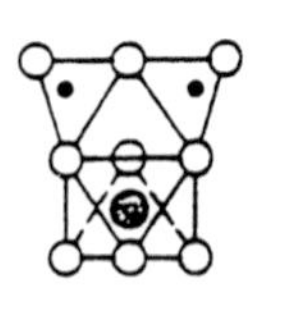

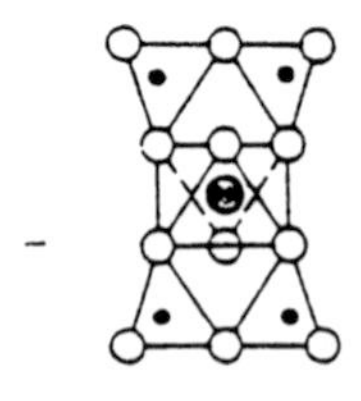

Fig. 2. Kaolinite (left) and Pyrophyllite (right)

In the clay minerals, part of the aluminum atoms in the octahedra can be replaced by other ions. In some cases, some of the silicons in the tetrahedra can be replaced by aluminum. For example, montmorillonite (bentonite) is formed by replacing some of the aluminum atoms in the octahedra with magnesium atoms. The formula is $[Mg_{0.33}Al_{1.67}Si_4O_{10}(OH)_2]^{-0.33}Na_{0.33}$ The number of magnesiums and aluminums total two, which is the number of aluminums in pyrophyllite. Because magnesium is 2+ and aluminum is 3+, charge compensation is needed by an external sodium + ion. The weak charge on the layers and the presence of the sodium ion account for the properties exhibited by clay – softness, cation exchange, and hydration.

Montmorillonite forms thick gelatinous suspensions in low concentrations. It is used as a binder for foundry sand and as a thickener for oil drilling mud. It is one of the important components of Fuller's earth.

In clays, the spaces between layers are occupied by the charge compensating cations and by water. It is possible to introduce other larger species between the layers. This will then push the layers further apart, or pillar the clay. The process of introducing the large pillaring agents is called intercalation. For example, hydroxy-aluminum oligomeric cations can be introduced between the layers of montmorillonite clays and then converted to alumina pillars by subsequent heating. These pillared clays can then be appropriately treated to act as acid cracking and hydrocracking catalysts for petroleum refining.

Clays can be pillared by alumina by using an aqueous aluminochlorohydrate solution which is sold as Chlorhydrol, a proprietary compound. The solution is 50% W/W and is basic aluminum chloride. Analysis gives 23.4% aluminum oxide and 8.25% chloride. This indicates an aluminum to chloride ratio of 1.97:1. The solution is thought to contain the cation $[Al_{13}(O)_4(OH)_{24}(H_2O)_{12}]^{7+}$. This large cation enters the sites between the layers of montmorillonite. On heating, the cation dehydrates and forms the alumina pillars.

In addition to alumina, clays can be pillared by chromia, zirconia, and iron oxide. The interlayer space opened up by pillaring can be as large as 21 Å and can, therefore, accommodate larger molecules than zeolites.

Read the section on X-ray diffraction before starting this experiment. To determine whether the experiment has affected the clay structure or not, the X-ray powder patterns of the clay starting material and the final product need to be determined. If the facilities are not available to determine the powder patterns, determine the expansion of the lattice spacing from the two patterns provided.

Experimental Procedure

Disperse 5 g. of bentonite in the Ca^{2+} form (commercial name Bentolite L) in 500 mL of distilled-deionized water in a 1L Ehrlenmeyer flask. Stir with a magnetic stirrer hot plate to keep the clay in suspension. Add dropwise 8.2 mL of Chlorhydrol solution to the clay suspension. Continue to stir for 2 hours while warming the suspension to 70°C. Filter the clay, and wash until chloride free (test with a few drops of silver nitrate solution). Filtration is relatively rapid when using a coarse sintered glass funnel or glass filtering crucible. Filtration with a fast filter paper such as Whatman 4 is satisfactory, as long as approximately one hour is available for the total filtration process and washing operation. Dry the filtered and washed product overnight in an oven at 120°C.

If it is possible for the class to determine X-ray powder diffraction patterns of the bentonite and the pillared clay, proceed to the separate experiment giving directions for obtaining the data. However, if it is not feasible to do individual powder pattern determinations, look at the sample patterns attached, and do the calculations indicated.

Compare the X-ray data for the bentonite and the pillared bentonite sample. Pillaring occurs along the c direction of these materials. The first reflection (at low 2θ) of clay materials like a bentonite sample is generally an (00ℓ) reflection. This means that this plane is infinite along the a and b directions. Determine the d-spacing for this reflection.

If pillaring of bentonite is successful, then the pillaring agent such as aluminum chlorhydrol will intercalate between the layers of the bentonite and the material will expand. There are several applications of pillared type materials because the expansion causes an increase in surface area and open space in the material. This can lead to enhanced adsorption of the pillared material and an accordion-like appearance of the particles. Two specific applications of such systems are the use of vermiculite which is an expanded clay for packaging materials, and as an adsorbent such as in cat litter products.

Once the bentonite expands, the surface area will increase substantially. In addition, the X-ray d-spacings that have a component along c (all of the reflections where ℓ does not equal zero) will change due to this expansion. The planes that will change the most are the (00ℓ) planes since they have no a or b component, neither of which should influence the expansion. It is very simple then to calculate the amount of expansion of the bentonite by just subtracting the d-spacing value of the same (00ℓ) reflection of the bentonite sample from the d-spacing value of the same (00ℓ) reflection of the pillared bentonite material.

It is important to look at all of the other diffraction peaks and see how these peaks change from the bentonite sample to the pillared bentonite sample. Make a list of each of the X-ray peaks in terms of d-spacings and note the change in d-spacing after intercalation of the pillar.

Auxiliary Experiments

If time and equipment permit, determine the X-ray powder diffraction pattern of the pillared clay sample you have prepared.

Questions to be Considered

1. Postulate how metal cations which might act as catalysts can be introduced into a pillared clay.

2. How might different interlayer spacings be obtained?

3. What value of an increase in d-spacing did you obtain or calculate from the powder patterns in the Appendices. Is this a reasonable value?

4. Did you observe the same expansion for all of the diffraction lines? Why or why not?

5. Are there any reflections that did not shift after intercalation of the pillar? If so, why is this the case?

6. Compare the widths and intensities of the peaks for the pillared and precursor materials. Are there any noticeable differences? Why is this so?

Helpful Background Literature

1. Wells, A. F. "Structural Inorganic Chemistry," 5th ed.; Clarendon Press: Oxford, 1984.

2. Occelli, M. L.; Tindwa, R. M., *Clays Clay Min.*, 1983, *31*, 107.

3. Suib, S. L.; Carrado, K. A.; Skoularikis, N. D.; Coughlin, R. W., *Inorg. Chem.*, 1986, *25*, 4217.

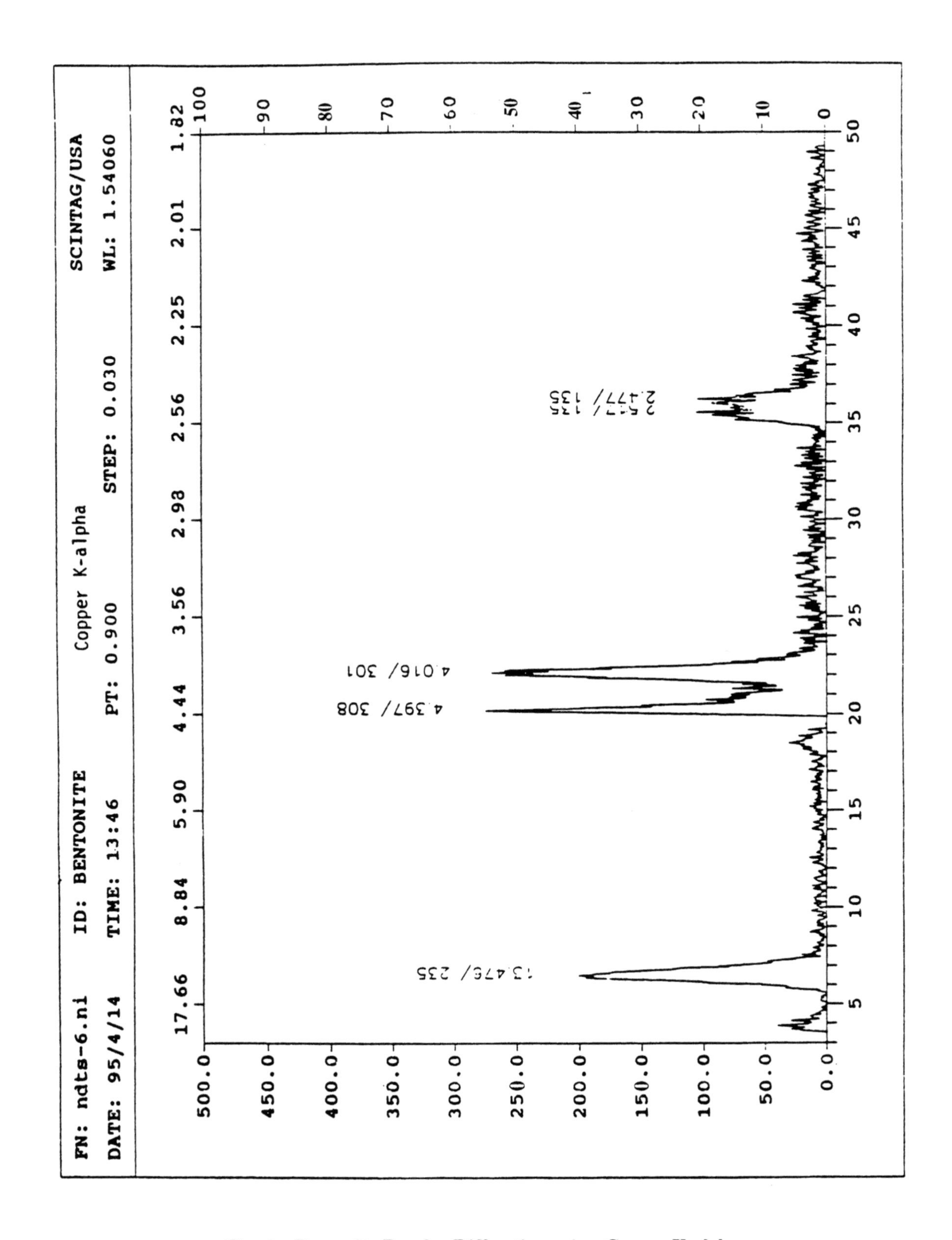

Fig. 3. Bentonite Powder Diffraction using Copper K-alpha

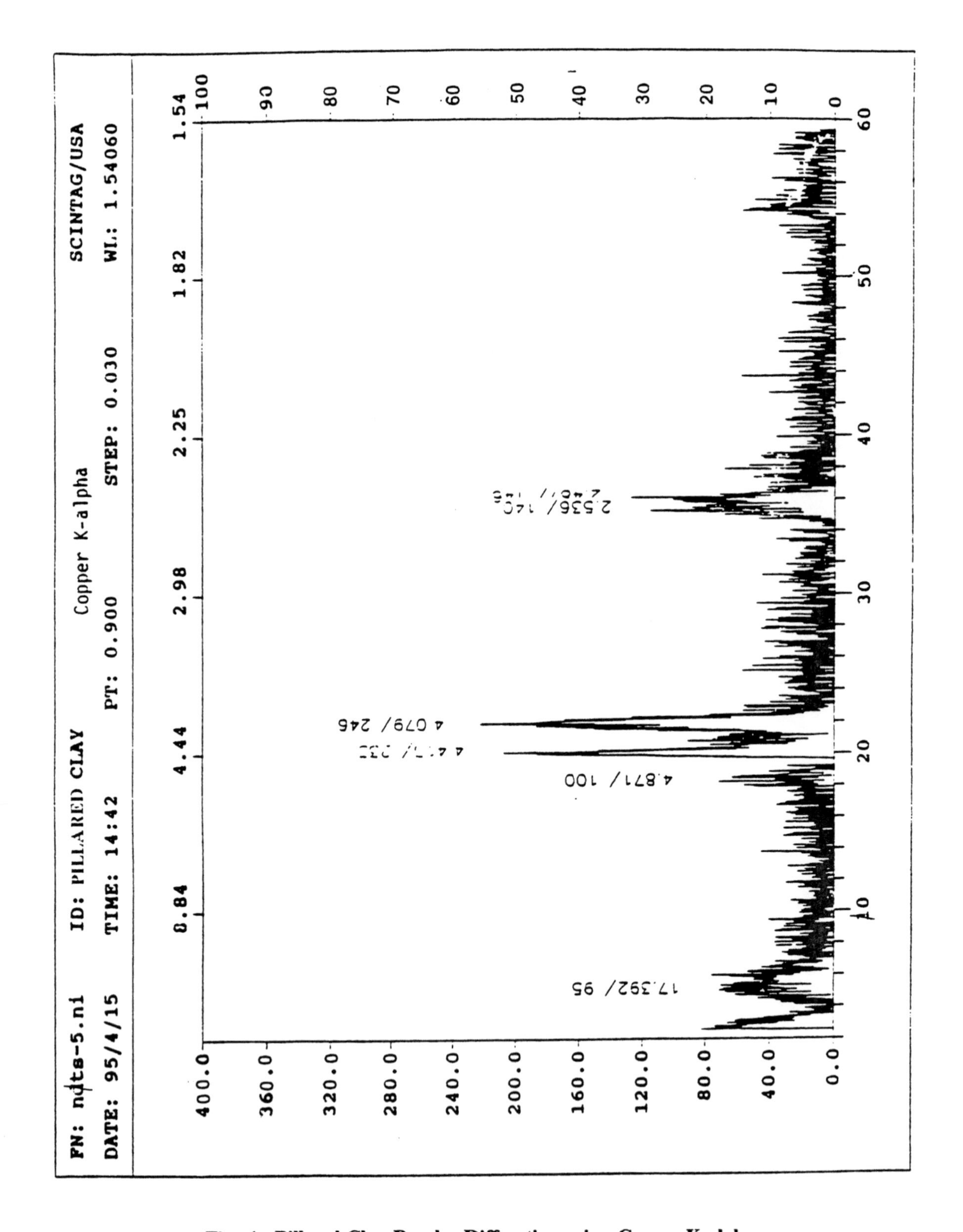

Fig. 4. Pillared Clay Powder Diffraction using Copper K-alpha

POTASSIUM NITRILOSULFONATE

Equipment Needed:	125 mL Ehrlenmeyer flask 50 mL beaker 250 to 400 mL beaker 25mm x 200mm test tube Buchner funnel Filter flask sample vial flowmeter Flowmeter with 1 mm capillary Dishpan 500 mL graduated cylinder
Chemicals Needed:	3 g KNO_2 10 g KOH Tank of SO_2 (0.178 M = 11.4 g) 10 mL alcohol 10 mL ether ice
Time Needed:	3 hours
Safety Notes:	Do not let SO_2 escape into the laboratory atmosphere. Make sure there are no open flames when using ether. Do not dispose of ether in the sink.

Purpose

This preparation illustrates the formation of a novel sulfur nitrogen compound.

The experiment also illustrates how a gaseous reactant can be added in approximate stoichiometric amounts.

Introduction

In General Chemistry, students are introduced to alkali, alkaline earth, and ammonium sulfates. Sulfites and the oxides of sulfur (SO_2 and SO_3) are also discussed. However, the variety of known compounds containing sulfur-sulfur and sulfur-nitrogen bonds are not introduced. This experiment illustrates the formation of a sulfur nitrogen compound and, in general, provides a framework for the consideration of several other types of compounds containing sulfur-nitrogen bonds.

There are amides and imides of sulfuric acid; but, unlike the amides and imides of carboxylic acid, the sulfuric acid analogs are not as well known. Amides are formed when one acid moity is substituted for a hydrogen on ammonia, and imides are formed when two acid moities are substituted on ammonia. Potassium nitrilosulfonate or potassium nitridotrisulfate is formed when three sulfuric acid moities are substituted on ammonia. The free acid cannot be isolated but the salts are reasonably stable. Analogous to carboxylic acid chemistry, if two sulfonic acid groups are substituted on ammonia, the imidosulfonic acid is formed; and, if one sulfonic acid is substituted, the amidosulfonic acid is formed. Amidosulfonic acid is also called sulfamic acid. Sulfamic acid and imidosulfonic acid are stable as the free acid.

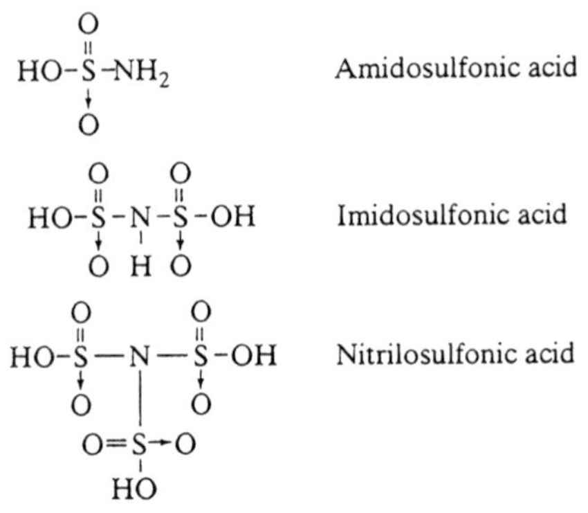

```
   O
   ||
HO-S-NH2            Amidosulfonic acid
   ↓
   O

   O     O
   ||    ||
HO-S-N-S-OH         Imidosulfonic acid
   ↓ |   ↓
   O H   O

   O       O
   ||      ||
HO-S — N — S-OH     Nitrilosulfonic acid
   ↓   |   ↓
   O   |   O
     O=S→O
       |
       HO
```

Fig. 1.

Ammonium nitrilotrisulfonate has been reported[1] to be formed by bubbling SO_2 into aqueous NH_4NO_2 and NH_4HSO_3. On cooling, crystals of ammonium nitrilotrisulfonate are formed. Potassium nitrilosulfonate is formed when potassium nitrite is stirred with a hot solution of potassium bisulfite in an excess of potassium hydroxide. The ammonium salt is slightly soluble in water. The potassium salt is less soluble. At 23°C, 100 parts of water dissolve 2 parts of salt.

It might be expected that a ratio of three moles of bisulfite to one mole of nitrite would be involved in the reaction. However, it has been found that better yields are obtained when bisulfite in excess of the 3 : 1 ratio is used. With mole ratios of 6 : 1 to 4.5 : 1, yields of 64% were obtained. With lower ratios of bisulfite, the yields were less. This indicates that the reaction is best written as:

$$4KHSO_3 + KNO_2 \rightarrow N(SO_3K)_3 + K_2SO_3 + 2H_2O$$

Potassium nitrilosulfonate hydrolyzes in boiling water to yield potassium amidosulfonate. The reaction proceeds by way of potassium imidosulfonate since this compound can be isolated if the hydrolysis time is short. The hydrolysis is slowed if free alkali is present. Appreciable hydrolysis does not take place at 40°C, and recrystallization has been accomplished at 40° – 50°C.

Drying in a vacuum desiccator over concentrated sulfuric acid causes a loss of some of the waters of crystallization. When heated to 100 – 110°C, all of the waters of crystallization are lost. Potassium nitrilosulfonate slowly decomposes to potassium hydrosulfate and potassium imidosulfonate on storage.

In this experiment, 0.178 moles of KOH will be reacted with 0.178 moles of SO_2 to form 0.178 moles od $KHSO_3$. A given amount of a solid can be determined by weighing. A given amount of a liquid can be determined by weighing or by volumetric measurements. Gases are a bit more difficult to measure out. Volumetric measurements are done with vacuum lines. With benchtop equipment, a flow device will enable an estimate of the amount of gas used.

A simple flowmeter can be calibrated and the amount of gas at a given flow rate estimated from the time the gas is allowed to flow. One simple flowmeter is shown:

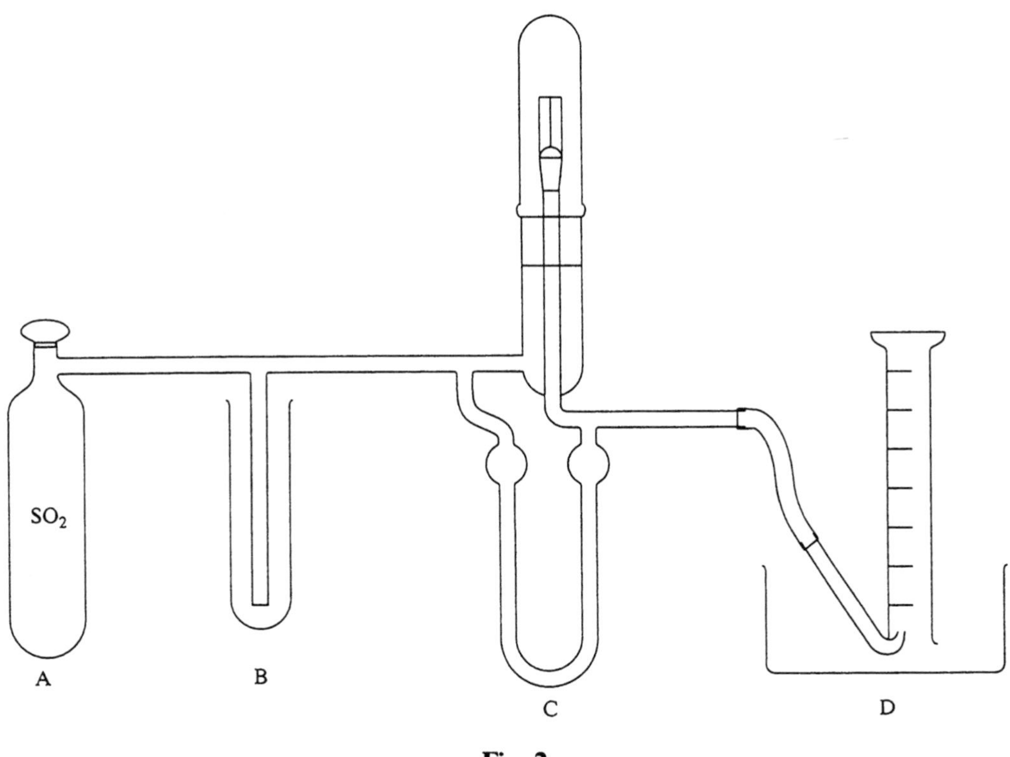

Fig. 2.

A = SO_2 Tank. A lecture bottle should be sufficient for one laboratory section.

B = Pressure Blowoff. This is a safety device in case the bubbling tube becomes accidentally plugged with the solid reactant. The large "test tube" is filled to 17 – 20 cm with mineral oil.

C = Flow Meter. The gas flows through a 1 mm capillary attached to a standard taper 10/30 joint. The pressure drop is measured with an oil manometer. The manometer should be filled about half way with the oil column being about 10 cm high. The expansion volumes on both arms of the manometer are designed to retain the oil in the U-tube in cases when the pressure differential exceeds twice the height of the oil column at rest. The range of the flow can be governed by varying the length of the capillary choke. For this experiment, a 2 to 3 cm piece of 1 mm capillary will give a flow rate of about the right magnitude.

D = Dish Pan. The dish pan is approximately half-filled with water.

E = Graduated Cylinder. A 500 mL graduated cylinder is filled with water and inverted in the dish pan.

Experimental Procedure

The flow meter can be calibrated with the laboratory compressed air. Establish a reasonable flow rate and note the pressure differential with the manometer. This flow is then directed through a tube with a hooked end into the water filled graduated cylinder. The time to displace a given volume of water is determined. The pressure of the gas in the graduated cylinder is then brought to atmospheric by adjusting the water level in the graduated cylinder to be the same as that in the dish pan. Knowing the volume of gas delivered in a measured time will allow for the metering of a given quantity of gas which is flowing at a given manometric pressure difference.

Change the hooked glass tube with a straight glass tube to bubble the SO_2 into the KOH solution. If a lecture bottle of SO_2 is used, make sure it is clamped in a vertical position so that the SO_2 in the gas phase can be removed.

In a 50 mL beaker, dissolve 3 g of KNO_2 in 13 mL of water. Place 10 g of KOH in a 25 mm x 200 mm test tube. Add 19 mL of water and effect a solution. Holding the test tube with a test tube clamp to minimize the possibility that the basic solution will splash out on your hand during the bubbling process, bubble the volume of SO_2 calculated to be 0.178 moles. Keep in mind that rapid bubbling may not allow time for the SO_2 to react. Also very rapid bubbling may splash some KOH solution out of the test tube. After the addition of SO_2, pour the resulting bisulfite solution into a 125 mL Ehrlenmeyer flask. If the bisulfite solution is not warm, heat it on a hot plate. Do not heat above 50°C. Slowly add the KNO_2 solution to the hot bisulfite solution. The solution will become cloudy and, on cooling, a mass of needle-like crystals will form. After standing in the mother liquor (recrystallization mixture) for one hour, transfer the mixture to a 250 mL – 400 mL beaker using 200 mL of water. Make sure that the solution is still basic. If necessary, add a few pellets of potassium hydroxide. Heat to 40° – 50°C. Continue to monitor the solution to insure that it remains alkaline. If the solution becomes acid, the imidodisulfonate rather than the nitrilosulfonate will be obtained. On cooling, the needle-like crystals form once again. Filter the precipitate on a Buchner funnel and wash with some ice water in order to remove the bisulfite. If the bisulfite is not removed, it will catalyze the decomposition of the product. Follow with about 10 mL of alcohol and 10 mL of ether in order to remove water. Dry briefly in a vacuum desiccator.

Place in a sample vial and weigh the product. Report the yield on the basis of nitrite used. Properly label the vial and turn in the product to the instructor.

Auxiliary Experiment

Boil the potassium nitrilosulfonate in water and isolate the potassium sulfamate.

Questions to be Considered

1. Write the Lewis structures of SO, SO_2, SO_3, SO_3^{2-}, and SO_4^{2-}.

2. Suggest how you might distinguish potassium sulfamate from potassium nitrilosulfonate.

3. Write the structure for thiosulfuric acid.

4. Write the reaction showing the formation of methyl sulfate from the action of methyl alcohol on chlorosulfonic acid.

5. Alkyl- and arylsulfonates are industrially important as detergents (linear alkylbenzene sulfonate) or as ion exchange resins (sulfonated polystyrene). Show the difference in the structure of a sulfonate and a sulfate ester.

Helpful Background Literature

1. Zeegers, R. N. G., U. S. 2,686,106, Aug. 10, 1954.

2. Fremy, E. *Ann. Chim. Phys. (3)* 1845, *15*, 408.

3. Raschig, F. *Liebig's Ann.*, 1869, *152*, 336.

4. Sisler, H.; Audrieth, L. F. *J. Am. Chem. Soc.*, 1938, *60*, 1947.

TIN(IV) CHLORIDE

Equipment needed:	1 L filter flask 250 mL dropping funnel 2-hole rubber stopper to fit filter flask assorted rubber stoppers 4 traps for drying train chlorine combustion tube standard taper 19/38 test tube $SnCl_4$ distillation "U" tube ampoule on 19/38 joint 2 small Dewar flasks tweezers vacuum pump with trap
Chemicals needed:	17 g $KMnO_4$ 20 g $CaCl_2$ 40 mL H_2SO_4 7 g granulated tin 160 mL concd. HCl tin foil liquid nitrogen 200 mL chlorobenzene (for whole class) masking tape
Time needed:	3 hours
Safety notes:	Chlorine is noxious. Set up the chlorine generator in the hood. Do not add sulfuric acid to permanganate! Wash out everything in the hood and flush with lots of water.

Purpose

This experiment illustrates that combustion can take place without oxygen. It illustrates with trap to trap distillation the manipulation of inorganic compounds in a vacuum system. The technique of sealing an ampoule under vacuum is demonstrated.

Introduction

Tin, a group IV metal, is silver-white, lustrous, soft and capable of being rolled into thin sheets (tin foil). Even though the element differs greatly from the non-metal group IV elements, many of the compounds which it forms are analogous. For example, tin forms the hydride SnH_4, the halide $SnCl_4$, the dioxide SnO_2, and the sulfide SnS_2. Tin hydride, or stannane, is a gas which has a melting point of —149.9°C and a boiling point of – 52.3°C. It is a more stable compound than the difficulty of its formation would lead one to predict. The dioxide exists in nature as the mineral cassiterite and the disulfide is best known as the yellow precipitate formed in the sulfide separations employed in qualitative schemes. The tin disulfide is used commercially as a pigment.

Tin(IV) chloride is a covalent liquid compound with a specific gravity of 2.226, mp of – 30.2°C, and bp of 113°C. Like carbon tetrachloride, it is tetrahedral with a tin-chlorine bond distance of 2.30 ± .02 Å. This compares to the 1.77 Å carbon-chlorine bond distance in carbon tetrachloride. Because it is a molecular compound, it is not surprising that it is a liquid at room temperatures. When frozen, the solid $SnCl_4$ has the SnI_4 structure. In the SnI_4 structure, the halogen atoms are in ccp array with the tin atoms occupying one-eighth of the tetrahedral holes. Remember from the models experiments that there are two tetrahedral holes per close-packed sphere. The tetrahedral holes occupied are alternatively three-sixteenth and one-sixteenth of the tetrahedral holes between alternate layers of the close-packed halogen atoms. This is an interesting exercise in partial hole filling. See Clark (2), p. 86 for the hole filling patterns.

Tin(IV) chloride is soluble in cold water, alcohol, carbon disulfide, and is decomposed by hot water. Like HCl which is considered covalent in the anhydrous state and completely ionized in water, stannic chloride apparently ionizes in water and slowly hydrolyzes. The hot water merely accelerates the hydrolysis reaction. An interesting property of tin(IV) chloride is that, whereas the anhydrous material is a colorless, fuming, caustic liquid, the hydrate is a white crystalline substance. The solid hydrate $SnCl_4.5H_2O$ called crystalline butter of tin, is a white, opaque, deliquescent crystal which can be obtained by adding no more than an equal volume of water to the anhydrous tin(IV) chloride.

Tin(IV) chloride is generally prepared by direct combination of tin with chlorine. Although not as extensively used as tin(II) chloride which is an important reducing agent, it finds outlets as a mordant in the textile industry, a stabilizer for certain resins and a component in the manufacture of fuchsin, blue print, drugs and ceramics.

The purification procedure in this experiment illustrates, in a crude way, trap to trap distillation in a vacuum line. The more sophisticated procedures involve use of a high vacuum line and control of temperatures by the use of various "slush" baths. The rate of transfer of volatile materials in a vacuum line is among other things, a function of the mean free path of the molecules. A low vacuum with more air molecules decreases the mean free path of the substance being transferred and thereby makes the distillation very slow. On a practical basis, transfers take place readily in a vacuum of 10^{-4} mm mercury pressure or better and very poorly at 10^{-2} mm mercury pressure. Slush baths are organic solvents in which a liquid-solid equilibrium has been established. As the water-ice equilibrium holds constant at 0°C, so liquid chlorobenzene and solid chlorobenzene holds constant at -45.2°C, liquid $CHCl_3$-solid $CHCl_3$ holds constant at -63.5°C, liquid CS_2-solid CS_2 holds constant at -111.6°C etc. By adjusting the temperature to hold back impurities while passing the desired compound through the trap, purification can be achieved.

In order to achieve good vacuum, precision ground stopcocks and joints must be used. These in turn must be sealed and lubricated with a proper grease. Vacuum stopcocks are always hand lapped and are, therefore, non-interchangable. Note that the vacuum stopcocks on the apparatus used have numbers identifying the matching plug and barrel. Note also that all vacuum stopcocks are designed in such a way that the atmospheric pressure is used to hold the plug tightly in the barrel. The grease used for vacuum work must have a very low vapor pressure and must not striate when the stopcocks are turned. Two common high-vacuum greases used are the silicone greases and the hydrocarbon greases. Both are relatively expensive. The Apiezon high-vacuum hydrocarbon greases cost around $60 per 25 g for the better greases. Apiezon N has a vapor pressure of 10^{-9} torr (mm mercury) at 20°C and has a recommended temperature range of 10°C to 30°C. Apiezon T has a vapor pressure of 10^{-8} at 20°C but has a useful range of 0°C to 120°C. Apiezon M is a less expensive grease with vapor pressure and temperature range almost identical to N, but in practice is not as satisfactory as N for stopcocks. It is ideal for joints which are repeatedly connected and disconnected. Silicone high-vacuum greases have vapor pressures less than 10^{-6} torr. The major advantage is that they are more resistant to organic vapors than the hydrocarbon greases.

Experimental Procedure

Direct Combination of Elements

Set up a chlorine generator and a drying train for chlorine. The chlorine is generated by reacting potassium permanganate with concentrated hydrochloric acid. Fig. 1 is a diagram of the apparatus for generating and drying chlorine. Put 17 g of potassium permanganate in a one liter filter flask. If a pressure equalizing dropping funnel is available, adapt it to the top of the filter flask using a rubber stopper. If not, mount a 250 mL dropping funnel using a two-hole rubber stopper adapted for pressure equalization as shown. Set up the entire apparatus in a hood. Chlorine is a toxic gas with a threshhold limit value of 1 ppm or 3 mg/cubic meter of air.

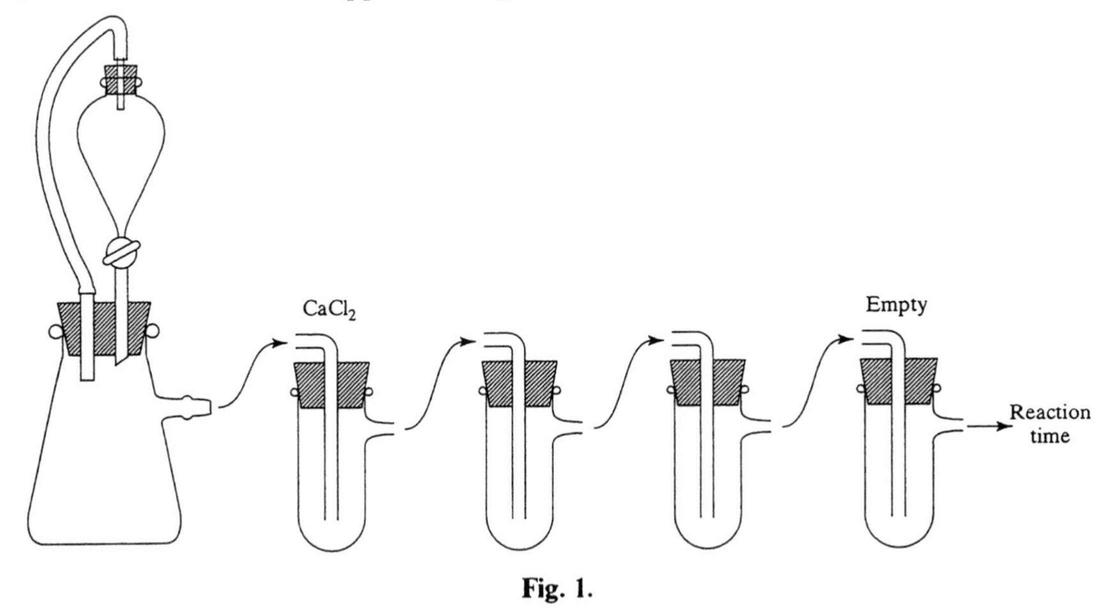

Fig. 1.

If available, use 8 inch side-arm test tubes and assemble a drying train as shown. Otherwise use standard 8 inch test tubes with two-hole rubber stoppers. The first test tube of the drying train is filled with anhydrous calcium chloride. The next two are about half-filled with concd. sulfuric acid. The empty trap at the end acts as a trap for sulfuric acid which may be blown over. The calcium chloride trap not only helps to dry the chlorine but also acts to insure that there is no suck-back of sulfuric acid into the reaction flask. It must be noted that H_2SO_4 and $KMnO_4$ react explosively!

Weigh out 7 g of granulated tin and place between the indentations in the combustion tube. Position the clamp close to the mouth of the combustion tube so that the heat necessary to carry out the reaction will not burn the rubber on the clamp jaws. (See Fig. 2). Direct heating on some clamp jaws will also cause the metal to melt. Attach this combustion tube to the chlorine generator via a one-hole rubber stopper. Place a standard taper 19/38 test tube at the end of the air condenser to receive the product. These combustion tubes are only good for one or two runs. If indentations are not made in the tube, make sure that it is mounted so that the hot molten tin will puddle where it can be heated.

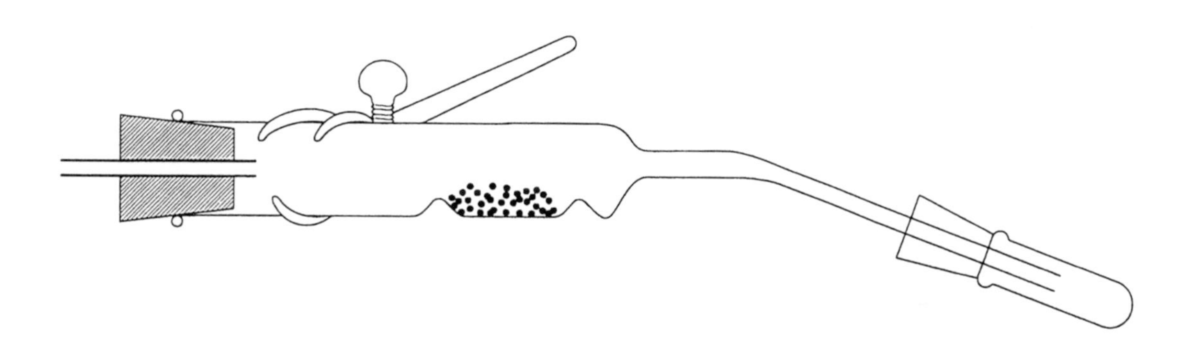

Fig. 2.

With the stopcock of the separatory funnel closed, remove the hose at the top of the separatory funnel and blow into the tube. If both sulfuric acid scrubbing tubes do not bubble, there is a leak in the system. Check all stoppers and tube connections to remedy this leak. *Do not* start to generate chlorine until this is done.

Place 160 mL of concd. HCl into the dropping funnel. Start to add the HCl slowly. After the flow of chlorine has started, heat the combustion tube with a Bunsen burner. The reaction between tin and chlorine will be readily visible as a flame. Continue to generate chlorine by slowly adding HCl to the permanganate. Heat the combustion tube as necessary to keep the reaction going.

When the reaction is complete, stop the addition of HCl. Remove the receiver and add a few pieces of tin foil to the product to react with the excess chlorine. Keep corked. It may take as long as a half hour for the yellow color to fade.

When disposing of the material in the chlorine generator, flush down a hood drain with plenty of water. Cleaning these generators in a sink in the open room allows more chlorine to escape into the room than is comfortable.

Purification

For purification of the anhydrous tin(IV) chloride by trap-to-trap distillation followed by sealing of the pure product in a glass ampoule, proceed as follows. Attach the ampoule to one side of the vacuum manifold and the receiver with stannic chloride to the other side. Arrange to pump the system through the ampoule and through a trap which will protect the pump from any product passing through the ampoule. Freeze the stannic chloride sample with a small Dewar flask of liquid nitrogen. Pull a good vacuum on the system for about 5 minutes. At this point, the ampoule should be flamed briefly with a bushy Bunsen flame in an attempt to desorb all the adsorbed water. Even when the flaming is done, solid stannic chloride hydrate often shows up in the product. This is an indication that water and gases adsorbed on glass surfaces are not easy to remove.

Make a chlorobenzene slush with enough chlorobenzene in a small wide-mouth Dewar flask (enough to cover the ampoule to a point approximately around the vacuum takeoff side-arm). Add small amounts of liquid nitrogen to the chlorobenzene while stirring with a six-inch wooden handled spatula. (The spatula should be long enough to reach the bottom of the Dewar flask.) If too much liquid nitrogen is added at once, the chlorobenzene will freeze solid and it will be difficult to stir to make a slush.

Place the chlorobenzene slush on the ampoule so that the surface of the slush is below the side-arm leading to the vacuum pump. Replace the liquid nitrogen Dewar flask on the impure product with an ice-water bath. Use any appropriate size beaker for the ice-water bath. Be sure to have the trap protecting the pump cooled with liquid nitrogen. As the trap-to-trap distillation takes place, the non-volatile impurities will be held back in the ice cooled container. The stannic chloride will be held back in the ampoule by the -45°C chlorobenzene slush. The more volatile impurities will pass through the -45°C trap and will become trapped in the liquid nitrogen trap.

When the distillation of tin(IV) chloride seems to have been finished, prepare to seal off the ampoule. Take a piece of masking tape and wrap around the top part of the ampoule (just below the constriction) so that the two ends of the tape can be pressed together to form a tab or handle which can be grasped with tweezers. Replace both the ice-water bath and the chlorobenzene slush with liquid nitrogen traps. Grasp the masking tape tab with tweezers which are at least 6″ long. (Shorter tweezers will become very cold.) Seal off the ampoule on the arm away from the pump first. When this has been accomplished, seal off the ampoule side-arm leading to the pump.

Before the sealing operation is done on the ampoule, read the following instructions carefully. It is probably desirable to practice sealing off a piece of glass tubing similar in size to the seal-off points on the ampoule before attempting to seal the ampoule.

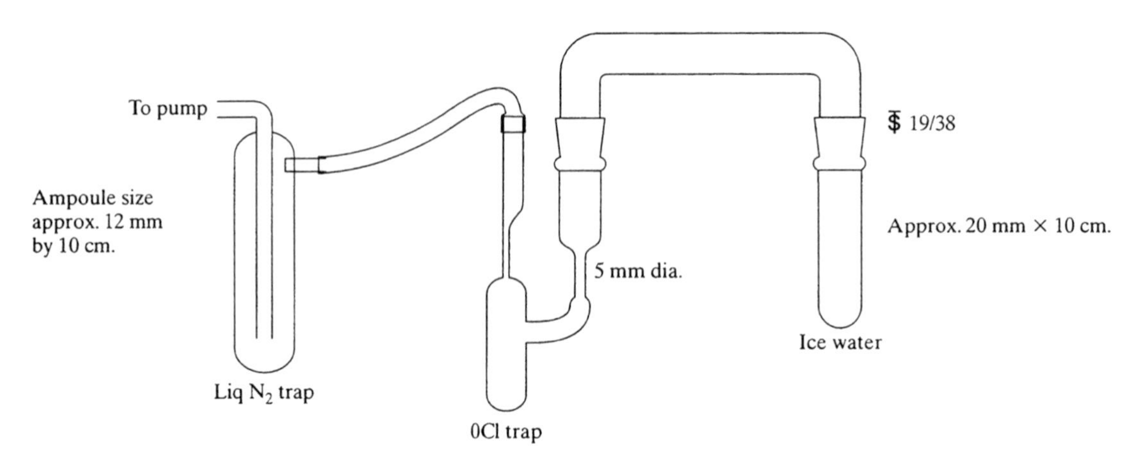

Fig. 3

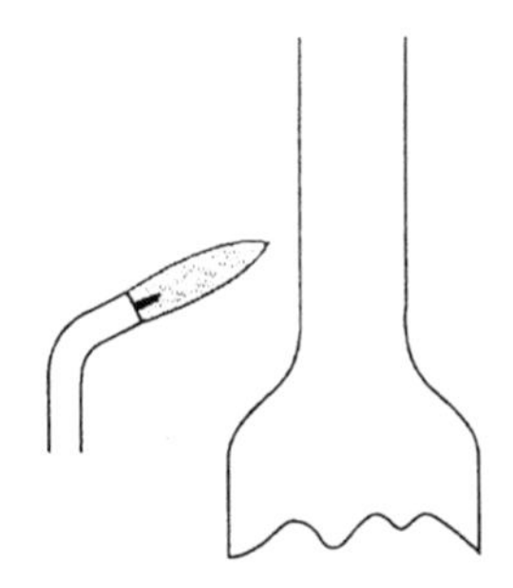

Heat gently and evenly around the circumference

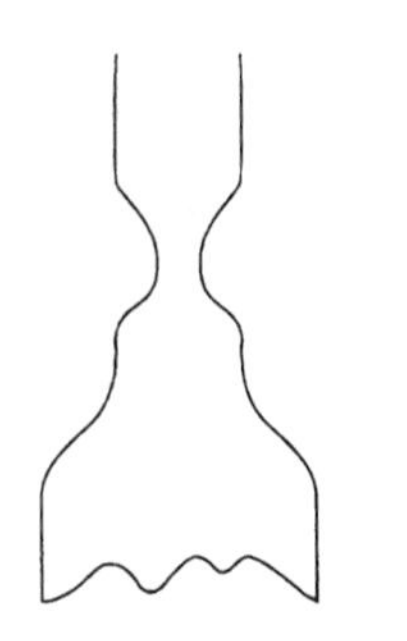

Tube should start collapsing evenly like this

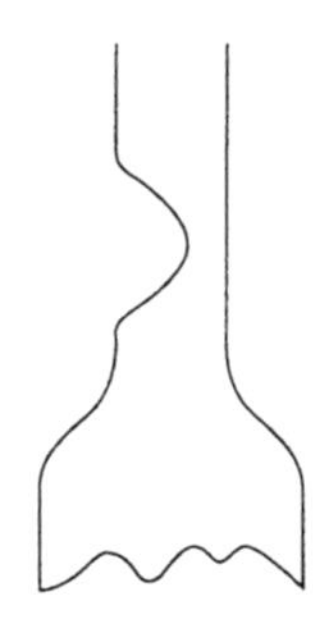

Not like this.

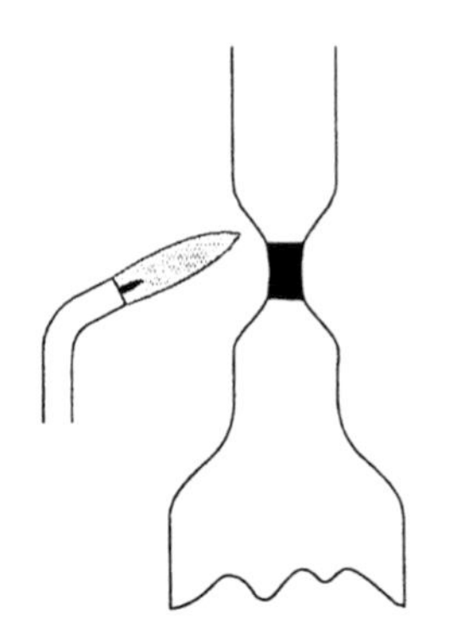

Once the glass has made a seal, move flame above seal.

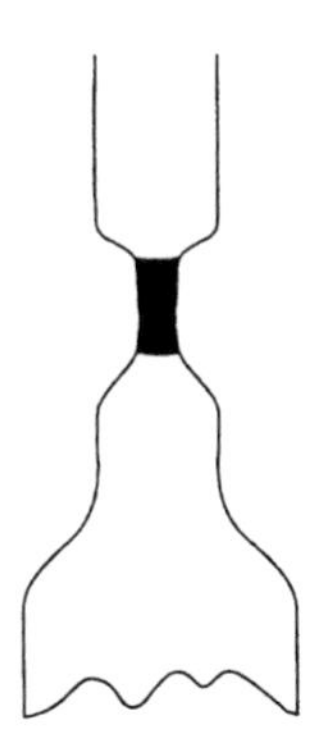

Work to get a length of solid glass. Heat this well so that a capillary pathway is not left. In general keep heat away from ampoule end of solid length.

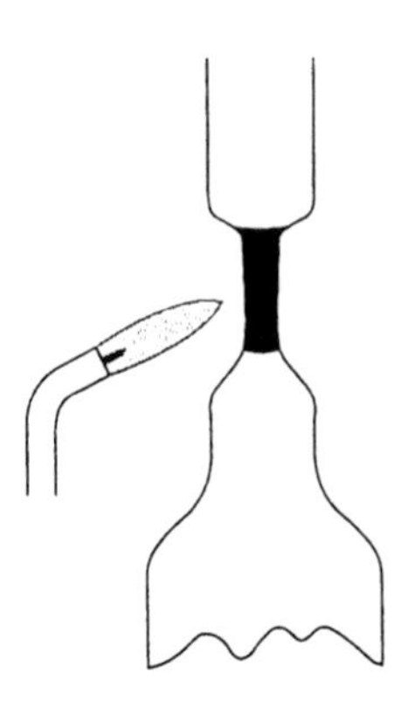

Heat in middle of solid area dropping ampoule slightly. Melt off as much as pulling off.

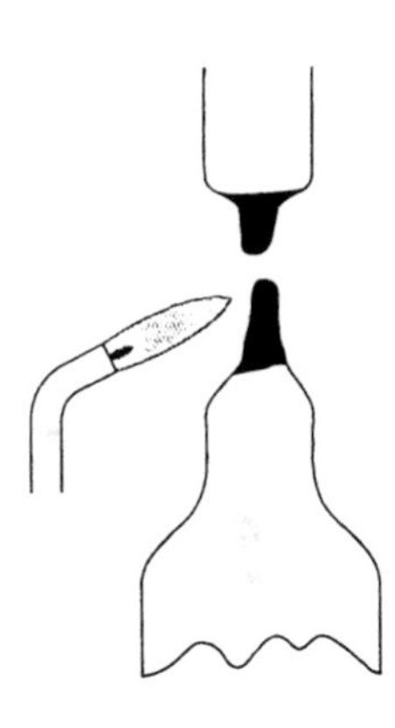

Heat the tip carefully to make sure that the glass is melted closed and no capillary leaks are left.

Fig. 4

Set the newly sealed off ampoule aside for the contents to melt. Properly label and hand in. Return both the standard taper test tube and joint from the ampoule to the instructor. Ground glass joints are expensive and are not considered disposable.

Preparation of the Hydrate

To the residue add 0.5 to 0.75 mL of water for every mL of anhydrous tin(IV) chloride. The amount of water can be estimated; however, if too much water is added, a solution will result rather than the crystalline hydrate. Separate the solid. Leave it in an open beaker in the hood until it stops fuming. Place the solid hydrate sample in a bottle, properly label, and hand in.

Questions to be Considered

1. Write the balanced equation for the reaction.

2. Calculate how many grams of $KMnO_4$ and how many mL of concd HCl will be needed to react exactly with 7 g of tin.

3. Calculate how many mL of water are required for each mL of tin(IV) chloride to exactly form $SnCl_4.5H_2O$.

Helpful Background Literature

1. Wells, A. F. "Structural Inorganic Chemistry," 5th ed.; Clarendon Press: Oxford, 1984; p. 424, p. 915.

2. Clark, G. M. "The Structure of Non-molecular Solids: A Co-ordinated Polyhedron Approach," Wiley: New York, 1972; pp. 85 – 89.

SURFACE CONDUCTIVE GLASS

Equipment needed:
- muffle furnace (600°C)
- 200 mL beaker
- steam bath
- sintered glass funnel
- filter flask
- glass atomizer*
- microscope slide
- firebrick
- tongs
- ohmmeter
- surface resistivity probe (optional)

Chemicals needed:
- 25 g tin(IV) chloride crystals
- 25 mL methanol
- 1 g Sb_2O_3
- 5 mL concd. HCl
- Concd. H_2SO_4**
- Concd. HNO_3**
- acetone**
- alcohol**
- toluene**

Time needed: 3 hours

Safety notes: Make sure that the methanol spray is always directed into a hood. Use a metal box as a "spray catcher." Protect the furnace surfaces which might be exposed to the spray with aluminum foil.

---

*A small sprayer made of glass can be used with the laboratory compressed air with satisfactory results.

**50 mL beaker of this reagent to be used by the whole class.

Purpose:

This experiment illustrates a high temperature formation of a a defect inorganic structure with a practical application.

Discussion

Transparent electrically conducting coatings can be put on glass surfaces. The most common technique is to apply an appropriate tin oxide thin layer on the glass. Tin can exist in either a 2+ or 4+ oxidation state. If a non-stoichiometric oxide layer consisting of a mixture of SnO and SnO_2 is formed, the structure can be pictured as that for SnO_2 with a certain proportion of $O_2{}^{2-}$ ions missing. This allows additional electrons to be associated with the lattice taking the "place" of the missing oxide ions. Another way of picturing this is to consider the tin(II) ion not as Sn^{2+} but as Sn^{3+} with a loosely associated electron. This electron accounts for the semiconductor properties of non-stoichiometric tin oxide films. The conductivity is a function of the number of defect centers. Antimony which has a 3+ oxidation state acts to create an oxygen deficiency in the same sense as partial reduction of tin. A small amount of antimony ions, therefore, increases the defect centers and increases conduction. The non-stoichiometric oxide is formed by hydrolysis of a tin salt on the heated glass. The organic constituent of the spray furnishes a reducing condition. If anhydrous tin salts are used, small amounts of water are added to enable the hydrolysis reaction to take place.

Conductive coatings are useful in a number of ways. One application is for meter faces where electrostatic charges on the glass face might cause serious reading errors of a sensitive meter. Another application is in fluorescent light fixtures where the conductive coating acts as a shield to prevent radio interference by the gas discharge in the fluorescent light. It is also used in electroluminescent panels which have been marketed thus far only in small units as night lights. Corning Glass markets these conductive glasses as Corning E-C glass, Libbey-Owens-Ford as Electrapane and PPG as NESA for non-electro-static application.

The oxide coatings range from 0.1 to 1 micron in thickness. Although transparent, they exhibit iridescence because the film thickness is such as to cause interference of reflected light. This property can be used to estimate the thickness of the film.

Table I

Thickness (mμ)	Order	Colors
78	1	white, yellow, red
232	2	violet, blue, green, yellow, red
387	3	purple, blue, green, yellow, red
542	4	green, red
697	5	greenish-blue, red

The film is adherent, hard, and durable to most chemicals. It exhibits surface resistivities of 100 ohms to a megohm. Surface resistivity is defined as the resistance between two opposite edges of a square of any dimension and is expressed in ohms. For practical purposes, when two point electrodes of an ohmmeter are held close together on a surface, the resistance value read is approximately the same as surface resistivity.

Resistivities can depend on the "art" of application as well as additives. Small percentages of Sb_2O_3, $TeCl_4$, and WCl_6 increase conduction. The presence of copper, manganese and other metals reduce conduction.

One of the problems encountered in application is the development of haze. A number of causes have been proposed. Some of these are too low a temperature, too rapid an application, droplets that are too large, inhomogeneous solutions, bounce back of spray from oven walls or other objects, and too long a heating period. If the temperature is too low, the stannic salt is not properly oxidized to the oxide mixture. If the heating is carried out too long at the high temperatures, sodium in the glass migrates to the surface causing haze in the film.

Experimental Procedure

A muffle furnace located in a hood will be preheated to 600°C. Prepare a spray solution by weighing 25 g of tin(IV) chloride crystal and dissolving it in 5 mL of methanol in a 200 mL beaker. Heating on a steam bath for a short while helps effect the solution. If a gel is present, filter it out using a sintered glass filter. Use half of this solution for the first spray solution. For the second spray solution, add 2 mL of an Sb_2O_3 solution to the other half of the stannic chloride solution. The Sb_2O_3 solution is made by dissolving 1 g of Sb_2O_3 in 3 mL of concd. HCl and diluting with 17 mL of methanol.

Put a microscope slide on a piece of hot firebrick and preheat for 5 minutes. If the firebrick is at room temperature, increase the preheat time to 15 minutes. Pull it out and immediately spray with a fine mist for several seconds. Hold the spray gun 6 inches or more from the slide. The distance depends on the force of the spray. If the spray is strong, the slide may be blown off the firebrick. It will also cool the slide too much. If the slide is cooled by the spray or is allowed to cool appreciably before spraying, the tin chloride will not be oxidized to tin oxide. The best results can be obtained by pulling the slide out on the furnace cover and immediately spraying. However, care must be taken not to spray in toward the oven. This causes a bounce back of the spray. See the discussion of haze formation above. It is often helpful to situate the furnace at an angle in the hood so that the spray can be directed into the hood but not into the furnace. If the spraying is continued more than several seconds, the glass will have cooled an appreciable amount. Put the sprayed glass back into the oven for 2 1/2 minutes. This reheat time is a function of the oven temperature. Once again refer to the discussion on haze formation. Apply a second spray and reheat for 2 1/2 minutes. Be careful not to get too much spray or its decomposition products into the laboratory atmosphere.

Remove the slide and its firebrick support from the oven and allow them to cool. Replace the firebrick in the furnace so that it will be hot for the next student.

Measure the surface resistivity using an ohmmeter. In order to obtain an approximate surface resistivity, the probes of an ohmmeter can be held close together on the surface. Alternatively two probes, consisting of n cm sheets of brass or copper fixed to an insulating board n cm apart can be used to measure the surface resistivity. The value of n to measure surface resistivity of glass of the sample size prepared should be 0.5 cm to 1 cm. The two probes should be connected to posts sufficiently far apart so that alligator clips can be used to connect the probes to the ohmmeter.

Repeat the above procedures using the tin(IV) chloride solution doped with antimony(III) oxide. Test the surface resistivity of this glass and compare it to the surface resistivity of the glass without the antimony(III) oxide dopant.

Test the durability of the coating by measuring the resistivity of the dry slide before and after washing with each of the following solvents: toluene, acetone, alcohol, and water. Check for durability of the coating in concd. HCl, H_2SO_4, and HNO_3. Dip separately for 30 seconds in each of these concentrated acids, rinse with water, and dry. Measure the resistivity to ascertain whether these acids are harmful to the conductive coating. The acids and solvents will be found in beakers in the hood. Since everyone in class will be using these same beakers, keep contamination to a minimum by dipping only dry slides.

Alternate Procedure

Make a study of the experimental variables so that the optimum conditions for preparing a conductive surface can be determined. Start with the above described procedure. Try one or more of the following variations:

1. Vary the furnace temperature from 500°C to 700°C.

2. Vary the heating rate, both preheat and postheat.

3. Vary the number of applications.

4. Vary the amount of Sb_2O_3 solution added.

5. Vary concentrations and length of time of exposure to the acids used to test the durability of the conductive layer.

Questions to be Considered

1. SnO_2 crystallizes in the rutile structure. What other compound or compounds also crystallize with the rutile geometry? What is the polyhedral representation of the rutile structure? What is the rutile unit cell?

2. What is the principle gaseous reaction product formed when the spray hits the hot glass surface?

3. If part or all of the alternate procedure is not done in the laboratory, use it as an exercise in experiment design. What experiments would you do and how would you present your results so that it would be of greatest value to a production manager?

COPPER(I) CHLORIDE

Equipment Needed:
- 1 L acid bottle
- sintered glass funnel
- vacuum desiccator
- seal-off ampoule
- vacuum pump
- pipe cleaner for 6mm glass tube

Chemicals Needed:
- 5 g $CuCl_2.2H_2O$
- 4.5 g Na_2SO_3
- 2 mL concd. HCl
- 40 mL glacial acetic acid
- 30 mL absolute ethanol
- 40 mL ether
- $CaCl_2$ desiccant for desiccator
- parafilm to seal vial

Time Needed: 3.0 hours

Safety Notes: Ether is very flammable. Make sure there are no flames when washing with ether. When using a water aspirator with ether, the sinks can fill with ether vapor, and this vapor can catch fire if there are open flames nearby.

Purpose

This experiment illustrates the preparation of an unstable oxidation state. An expertly prepared sample will remain white. A poorly prepared sample will be green. If the sample is not carefully dried, the white product will turn green in a short while.

Introduction

Copper exhibits two common oxidation states, 1+ and 2+. The 2+ oxidation state is more common and generally more stable. Copper(I) chloride can be prepared, but it is sensitive to air oxidation. This experiment illustrates the preparation of an unstable inorganic compound and reviews the solid state structure of copper(I) chloride.

When inorganic compounds are very unstable in air or easily affected by moisture, they must be handled in inert atmosphere glove boxes or glove bags or in vacuum lines. For example, cesium metal spontaneously ignites in air. Diborane is a gas which also spontaneously ignites in air. Vacuum lines are the technique of choice in manipulating these materials.

Compounds which are a little less sensitive to air or moisture can be prepared by careful and rapid manipulations of ordinary laboratory procedures. Compounds can be protected from air oxidation during filtering if they are kept covered with argon or carbon dioxide (these gases are more dense than air and will protect the solids in a funnel) or by covering a precipitate during suction filtration with a rubber dam.

It is often the case that compounds are much more sensitive to oxidation when wet. The oxidation of copper(I) chloride to copper(II) salts can be easily observed by a color change in the material. Well-prepared copper(I) chloride when thoroughly dried is colorless. Less-expertly prepared copper(I) chloride will be greenish or will soon turn green.

Copper(I) chloride like copper(I) bromide, copper(I) iodide, and silver iodide crystallize at room termperatures in the zinc-blende structure. In the zinc-blende structure, the metal atom is surrounded tetrahedrally by the non-metal atoms. Another way of describing the structure is to say that the copper atoms are occupying half of the tetrahedral holes created by cubic close-packed chloride ions. At 407°C, copper(I) chloride transforms to the wurtzite structure. In the wurtzite structure the copper atoms are still tetrahedrally surrounded by chlorine atoms. The chlorine atoms are, however, hexagonally close-packed in the wurtzite structure as opposed to the cubic close-packing for the zinc blende structure.

Copper(I) chloride vapor at 450°C is shown to be a trimeric molecule (Cu_3Cl_3) by mass spectrometric and vapor pressure studies. Electron diffraction indicates that it forms a six-membered ring.

Experimental Procedure

Dissolve 5 g $CuCl_2.2H_2O$ in 5 mL of water. Dissolve 3.5 g Na_2SO_3 in 25 mL of water. Make up a dilute sulfurous acid solution by dissolving 1 g of Na_2SO_3 and 2 mL of concd. HCl in 1 liter of water. Use the one liter acid bottles assigned to you. Add the copper(II) chloride solution to the sodium sulfite solution with stirring. The solution is first colored, but as the reduction proceeds, the colorless copper(I) chloride will be observed to start precipitating. Pour the reaction mixture containing the precipitated copper(I) chloride into 500 mL of the dilute sulfurous acid solution. After an initial stirring, allow the copper(I) chloride precipitate to settle and decant off as much of the supernatant solution as possible. Quickly transfer the residue into a sintered glass funnel set up for a suction filtration. Filter and wash several times with some of the remaining sulfurous acid solution *making sure that the liquid layer remains above the product at all times.* If allowed to suck dry, the product could be oxidized by the air. Wash successively with four 10 mL portions of glacial acetic acid, three 10 mL portions of absolute ethanol and four 10 mL portions of ether. (Hint: The acetic acid and ethanol can be disposed by flushing down the sink. The ether must be put in an ether waste bottle in the hood. It is a good idea to empty the filter flask of the acetic acid and ethanol before doing the ether wash.) Finally, allow the product to be sucked dry after the final ether wash.

Alternative One. The product is then further dried for 15 minutes in a vacuum desiccator containing $CaCl_2$ as a dehydrant. Flush a pre-dried pre-weighed sample vial using a nitrogen stream. It is convenient to do this by pulling a one to two mm capillary on one end of a 7 or 8 mm glass tube. Adjust the nitrogen stream so that it is very slow. Transfer the dried product into the vial and again flush the vial with nitrogen just before capping it. (If the nitrogen stream is too strong, there is the risk of blowing the product out of the vial.) Seal the cap with Parafilm or wax, properly label, and turn in the product to the instructor.

Alternative Two. The "dry" powder after the ether wash is put into an ampoule using the top portion of the ampoule as a funnel. If some of the product adheres to the narrow seal-off tube, run a pipe cleaner down the tube to push the product into the ampoule. If the seal-off tube is not clean, the product will burn into the glass during the seal off operation changing the composition and the coefficient of expansion of the glass. This will cause the seal off site to be prone to breakage. A one-hole rubber stopper with a glass tube is then inserted into the mouth of the ampoule and attached to a vacuum pump. The ampoule should be pumped for a while to insure that all of the water and solvent are removed. The ampoule is then sealed off under vacuum. See the stannic chloride experiment for instructions on sealing ampoules under vacuum. The cuprous chloride sealed in an evacuated ampoule should then be properly labeled and turned in to the instructor.

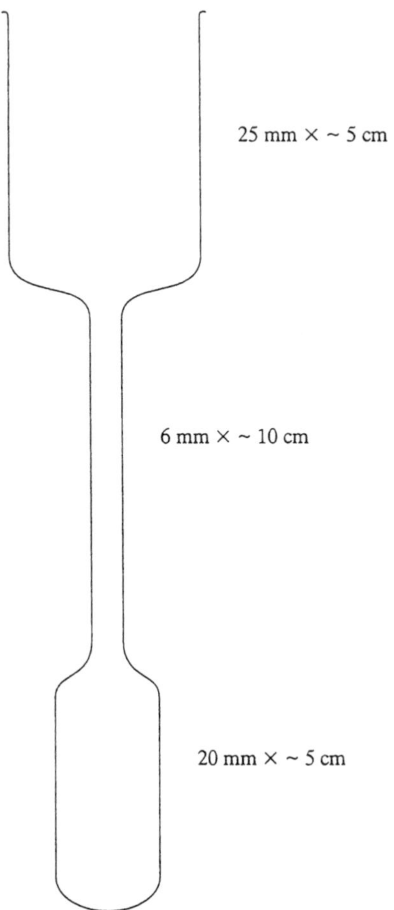

Fig. 1.

Questions to be Considered

1. Write balanced equations for the synthesis.

2. Calculate the yield.

3. Is argon or nitrogen the better inert gas for filling the bottle? Why?

4. Draw the structure of the copper(I) chloride trimer in the vapor phase. Would you expect the bond angles to be the same?

5. How can one tell whether or not the product is contaminated with NaCl or Na_2SO_3?

6. By what other methods may CuCl be prepared?

7. Why is the product finally washed with alcohol and then ether?

8. What explanation would you give for the color of $CuCl_2$ and the colorless nature of CuCl?

9. Describe the difference between the wurtzite and zinc blende structures.

10. If wurtzite and zinc blende are being represented by four-connected three-dimensional nets, what is the best way to characterize the zinc blende structure?

Helpful Background Literature

1. Fernelius, W. C., et al., "Inorganic Syntheses, Vol. II", McGraw-Hill: New York, 1946; pp. 1 – 4.

2. Cotton, F. A. and Wilkinson, G., "Advanced Inorganic Chemistry", 5th ed.; Wiley-Interscience: New York, 1988; pp. 757 – 760.

3. Wells, A. F. "Structural Inorganic Chemistry", 5th ed.; Clarendon Press: Oxford, 1984; p. 410, p. 443.

MERCURY(II) TETRATHIOCYANATOCOBALTATE

Equipment Needed:	magnetic stirrer hot plate 50 mL beaker 250 mL beaker sample vial
Chemicals Needed:	1.4 g cobalt(II) sulfate hydrate ($CoSO_4.7\ H_2O$) 1.5 g ammonium thiocyanate (NH_4SCN) 1.35 g mercury(II) chloride 99% ($HgCl_2$) parafilm to seal vial
Time Needed:	2.0 hours
Safety Notes:	It is undesirable to have mercury ions in the environment. Take care not to dispose of mercury ions in the sink. The product prepared should be disposed of properly. See the teaching assistants for the proper procedure at your institution.

Purpose

This experiment illustrates the preparation of a compound suitable for use in calibrating a magnetic susceptibility apparatus. It also illustrates a solid with an interesting variation of a familiar structure.

Introduction

A number of properties must be met in order for a compound to be utilized as a standard in calibrating a magnetic susceptibility apparatus. It must be able to be prepared pure in a reproducible fashion. It must have a moderate susceptibility ($\chi_g \simeq 10^{-5}$) so that the middle of the scale can be calibrated. It must have stability and a predictable change with changing conditions. These changes which occur must be minor. These conditions are satisfied with mercury(II) tetrathiocyanatocobaltate in that it is a compound stable in moist air, the χ_g changes predictably at temperatures near room temperature, and the solid packs reproducibly in a Gouy tube.

Because of these desirable characteristics, the magnetic susceptibility of mercury(II) tetrathiocyanatocobaltate has been carefully determined (1). Three different samples were prepared and measurements made with different packings on a Gouy balance. The values obtained were 16.45×10^{-6} ($\pm 0.2\%$), 16.40×10^{-6} ($\pm 0.2\%$), 16.48×10^{-6} ($\pm 0.3\%$), 16.38×10^{-6} ($\pm 0.2\%$), 16.44×10^{-6} ($\pm 0.4\%$), and 16.49×10^{-6} ($\pm 0.4\%$). It was concluded that, within the experimental error of 0.4%, there was no difference between the determinations using different preparations or between different tube fillings. This careful work (1) concludes that the $CoHg(SCN)_4$ has the susceptibility of 16.44 (± 0.08) $\times 10^{-6}$ at 20^oC. It obeys the Curie-Weiss law, $\chi_g \propto (T + 10)^{-1}$, where T is in degrees absolute.

The cobalt in $CoHg(SCN)_4$ is surrounded by four nitrogens from the thiocyanato groups in a tetrahedral configuration. The mercury is surrounded by four sulfur atoms from the thiocyanato groups also in a tetrahedral configuration. There are thus thiocyanato bridges between cobalt atoms and mercury atoms. The tetrahedral coordination of all of the metal atoms results in a diamond network with the metal atoms occupying the carbon positions in diamond.

Experimental Procedure

Dissolve 5.6 g of cobalt(II) sulfate hydrate (cobalt(II) sulfate heptahydrate) and 6 g of ammonium thiocyanate (NH_4SCN) in 10 mL of water in a 50 mL beaker. Place on a hot plate which is at 120° to 150^oC. The idea is to get it to the boiling point without boiling so vigorously that the salt solution splatters out of the beaker. Dissolve 5.4 g of mercury(II) chloride ($HgCl_2$) in 60 mL of water. Boil and filter off any insoluble material. Boil the filtered solution once again. While stirring with a magnetic stirrer, add the hot solution of the cobalt(II) sulfate and ammonium thiocyanate. Continue boiling for about two minutes with continued stirring. Decant the supernatant solution and wash several times with distilled water by decantation. Dry the product at 120^oC. Put into a tared vial; determine the yield; and turn in to the instructor.

Questions to be Considered

1. Locate the SCN in the diamond network.

2. Rather than a diamond network, could this have been described as being analogous to the zinc blende or wurtzite structure? Which one would it be, zinc blende or wurtzite? Is there any advantage over describing it as one of these rather than as diamond?

3. Calculate the percent yield.

4. Assume that the percentages of carbon, nitrogen and sulfur are going to be determined. What theoretical amount do you calculate for each of these elements? If carbon

is to be analyzed by conversion to carbon dioxide, what techniques can be used to separate and to determine the exact amount of carbon dioxide?

5. Using the Curie-Weiss law, calculate what you would expect for the magnetic susceptibility at 20°C.

Helpful Background Literature

1. Figgis, B. N.; Nyholm, R. S. *J. Chem. Soc.*, 1958, 4190 – 4191.

2. Wells, A. F. "Structural Inorganic Chemistry", 5th ed.; Clarendon Press: Oxford, 1984; p. 126.

SYNTHESIS OF A SUPERCONDUCTOR

Equipment needed:	Furnace capable of 1000°C Thermocouple and readout device quartz crucible Tube furnace with alundum plates to fit in quartz tube or alundum boat in which the 3/4″ pellets can be placed on edge. Depending on the length of the boat, several pellets can be placed in one boat. Reduction valve for oxygen Exit bubbler for oxygen train
Chemicals needed:	1.13 g yttrium(III) oxide 5.23 g barium nitrate 3.31 g copper(II) carbonate, basic cylinder of oxygen
Time needed:	3 hours
Safety notes:	Take care not to burn yourself on hot surfaces.

Purpose:

This experiment illustrates a defect perovskite structure which shows superconductive properties.

Introduction

The phenomenon of superconductivity, that of a material with zero resistance, was first observed in mercury in 1911. A large number of other elements and alloys were subsequently shown to exhibit this property, but the early examples required temperatures in the liquid helium range (< 4.2 K). Because of the great technical advantages in having electromagnets operate without resistance, experimental generators have been built of superconductive wires by Westinghouse and Mitsubishi. Superconductive magnets have been built and commercialized for devices such as NMR. The major hindrance to the widespread use of superconductivity is the cost of maintaining the liquid helium or supercritical helium temperatures.

For many years, the "holy grail" was to develop a material with a superconductive transition temperature above 77 K or − 196°C, the boiling point of liquid nitrogen. This would make the cryogenic technology much less expensive and open up a number of technically attractive uses for superconductivity.

In 1973, the superconductive transition temperature (T_c) was observed at 23.3 K with the development of niobium-tin alloy (Nb_3Sn).

In 1983, two researchers at the IBM Zurich Research Laboratory, K. Alex Muller and J. Georg Bednorz began to examine metal oxides as possibilities for obtaining higher T_c materials. Some members of a class of perovskites were known to exhibit superconductivity at 13 K. Their initial studies were disappointing, but in 1986 Bednorz observed that a barium-lanthanum-copper oxide showed a T_c at 35 K. When they published their results in *Zeitschrift fur Physik B* there was much skepticism. However, two groups, Shoji Tanaka at the University of Tokyo and Ching-Wu "Paul" Chu at the University of Houston quickly confirmed the findings of the Zurich group and began research on similar compounds. By replacing barium with strontium, three groups (Tokyo, Houston, and Bell Labs) demonstrated that the T_c could be raised to 40 K. In January of 1987, Chu, Wu and coworkers announced that they were able to make a material with a T_c of 90 to 100 K. This was the yttrium-barium-copper oxide with the approximate composition $Y_{1.2}Ba_{0.8}CuO_4$. Subsequent studies of this material by Robert Hazen at the Geophysical Laboratory of Carnegie Institution of Washington D. C. showed that it consisted of two phases. The black phase, which was the superconducting phase had the formula $YBa_2Cu_3O_x$ where x was typically between 6.5 and 7.0.

The structure was shown to be related to perovskites with some oxygen vacancies. The copper atoms are in the octahedral holes and the yttrium and the barium in the cuboctahedral holes. Review the structure of perovskite which was covered in the model laboratory session. In $BaTiO_3$, the titanium atoms are in the octahedral holes and the barium atoms are in the cuboctahedral holes. In the 1 − 2 − 3 superconductor, so called because of the ratio of the metal atoms, 1 Y, 2 Ba, and 3 Cu, all the oxygens in the plane of the yttrium atom are missing. This means that the copper atoms near the yttrium are not surrounded by 6 oxygens in an octahedral arrangement, but are surrounded by 5 oxygens in a square pyramid. The yttrium atoms are in eight-coordination and are located between layers of copper and oxygen atoms which form the base of the square pyramid. The structure is not regular, but shows defects throughout. Some of the oxygen defect sites in the yttrium plane are occasionally occupied by oxygens. Sometimes oxygens are found in other unexpected positions in the structure.

The structure can also be described as a pile of three cubes. The bottom cube and the top cube have barium in its center. The yttrium is in the center cube. The corners of each cube locate the positions of the copper atom. Not all of the copper atoms are surrounded by six oxygens in an octahedral arrangement because of the oxygen vacancies observed in the plane of the yttrium atoms.

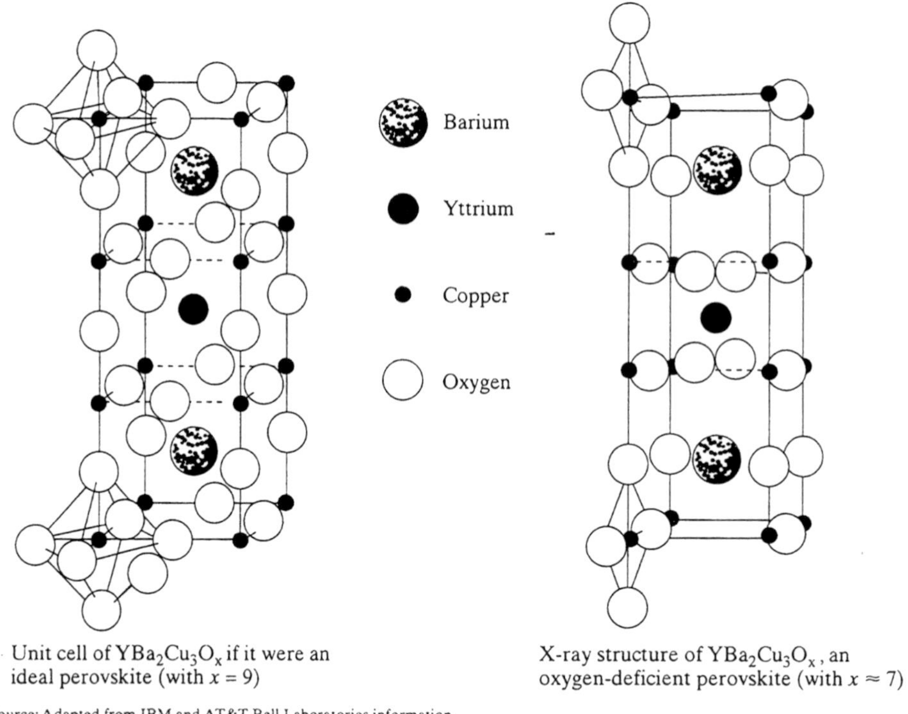

Unit cell of $YBa_2Cu_3O_x$ if it were an ideal perovskite (with $x = 9$)

X-ray structure of $YBa_2Cu_3O_x$, an oxygen-deficient perovskite (with $x \approx 7$)

Source: Adapted from IBM and AT&T Bell Laboratories information

Fig. 1.

Experimental Procedure

$YBa_2Cu_3O_x$ Superconductor

The yttrium, barium and copper are reacted in the ratio of 1 to 2 to 3 moles. Some recipes use yttrium(III) oxide (Y_2O_3), barium carbonate and copper(II) oxide. The reaction works a bit better with yttrium(III) oxide, barium nitrate and copper(II) carbonate. Start with 1.13 g of yttrium(III) oxide (1×10^{-2} M). This will require 2×10^{-2} M of $Ba(NO_3)_2$ and 3×10^{-2} M of copper(II) carbonate, basic.

There is a problem with weighing out a calculated amount of the barium nitrate and the copper(II) carbonate, basic. Barium nitrate tends to be hygroscopic. It should be weighed without standing in air for too long a period of time. If an excess of 0.2 g is weighed beyond the amount calculated for 2×10^{-2} moles, any error due to water in the salt will be insignificant. A slight excess of barium should not be harmful to the experiment. Copper(II) carbonate, basic is generally not stoichiometric. Stoichiometric copper(II) carbonate is not stable at room temperatures at atmospheric pressures. Hydroxycarbonates are more stable. Two well known minerals of hydroxycarbonates are azurite, $Cu_3(OH)_2(CO_3)_2$, and malachite, $Cu_2(OH)_2CO_3$. The commercial copper(II) carbonate, basic is $CuCO_3.Cu(OH)_2$ with a stoichiometry essentially that of malachite. The weight of copper(II) carbonate, basic needed to produce 3×10^{-2} M of CuO can be calculated from the malachite formula, but because of the possibilities of waters of hydration, it is better is determine the gravimetric ratio by heating a sample of copper(II) carbonate, basic to 650°C in order to convert it for practical purposes to copper oxide. A porcelain crucible can be used for this determination. The gravimetric ratio thus determined is then used to weigh out the required amount of the copper(II) carbonate, basic. (This may have been done for you by the teaching assistants in the course in which case the gravimetric factor will be provided.)

The weighed samples of yttrium(III) oxide, barium nitrate, and copper(II) carbonate, basic are ground well in a clean mortar. Place the ground sample in an unglazed porcelain or quartz crucible and fire at 950°C in a muffle furnace in air for two hours. Because the decomposition of the carbonate and nitrate will cause the mixture to bubble, the reaction mixture should be reground after cooling.

Place the ground reaction mixture in a pellet press to make a pellet 3/4 inch in diameter. Use a pressure of about five tons (indicated on the press) to make the pellet. Be sure that the five tons are read on the scale which corresponds to the 3/4 inch pellet press. Place the pellet in a quartz tube heated in a tube furnace under slowly flowing oxygen. This refiring and slow cooling is essential to develop the superconductive property. The temperature should be held at 950°C for about an hour and then decreased from 950°C to 250°C over a period of three to four hours (or longer.)

The mutual repulsion of a permanent magnet and a superconductor is called the Meissner effect. The magnetic flux of the magnet will not penetrate the superconducting material. The magnet will, therefore, be levitated if the material on which it is resting becomes a good strong superconductor. Place the pellet at the bottom of a clear Dewar flask or an insulated container made from styrofoam coffee cups. If the latter is used, a small plastic bottle or a combination of a piece of plastic and glass wool can be used to raise the pellet near the top of the container. Care must be taken to make sure the pellet is not tilted. Carefully place a small magnet on top of the pellet. Slowly pour liquid nitrogen into the Dewar flask or styrofoam cup. The magnet will levitate if the pellet becomes superconductive.

Long term exposure to room temperature moisture may cause the superconductive pellet to crumble. This can be retarded by keeping the pellet in a jar or by coating the pellet with a thin coating of wax. Dip the pellet in a beaker of molten paraffin wax of the type used to seal jams and jellies. Allow it to drip so that only a thin coating is formed. Test the coated disk to see that the superconductive properties are retained. A coating of clear Krylon spray can be used as an alternative to waxing the pellet.

Questions to be Considered

1. What gases are generated in the initial firing process?

2. How might the amount of water in barium nitrate be determined?

3. How else might the amount of copper in the commercial basic copper carbonate be determined?

4. Determine the coordination numbers and geometry of all the metal cations.

Helpful Background Literature

1. Ellis, A. B. "Superconductors, Better Levitation through Chemistry", *J. Chem. Ed.*, 1987, *64 (10)*, 836 – 841.

2. Matsen, F. A. "Three Theories of Superconductivity", *J. Chem. Ed.*, 1987, *64 (10)*, 842 – 846.

3. Harris, D. C.; Hills, M. E.; Hewston, T. A. "Preparation, Iodometric Analysis, and Classroom Demonstration of Superconductivity in $YBa_2Cu_3O_{8-x}$", *J. Chem. Ed.*, 1987, *64 (10)*, 847 – 853.

12-TUNGSTOPHOSPHORIC ACID

Equipment Needed:	125 mL Ehrlenmeyer flask graduated cylinder Bunsen burner top loading balance sintered glass funnel small sample vial scissors for class use
Chemicals Needed:	5.0 g WO_3 2.0 g NaOH pellets 1.0 g Na_2HPO_4 6 mL concd. HCl ice bath tin foil, aluminum foil, iron wire and copper wire
Time Needed:	2 hours
Safety Notes:	Heating concd. HCl drives off gaseous HCl. The boiling, therefore, should not be done on the open desk. Make sure that you carry out this operation in a hood.

Purpose

This preparation illustrates the formation of a molecular structure from an infinite three-dimensional structure.

Introduction

The tungsten(VI) trioxide (tungstic acid), WO_3, has the distorted ReO_3 structure. It consists of tungsten atoms octahedrally surrounded by oxygen atoms. Each corner oxygen is shared with another octahedron but not at 180°. Since all six corners are shared, the stoichiometry is MX_3. The ReO_3 structure was made in the model experiment. If you have forgotten what it looked like, refer to *Session Two*, Exercise #2, part 2.

When tungsten(VI) trioxide is reacted with sodium hydroxide, the salt, sodium tungstate, is formed. When sodium tungstate is reacted with disodium hydrogen phosphate, 12-tungstophosphoric acid (phosphotungstic acid) is formed.

12-tungstophosphoric acid has the anion $PW_{12}O_{40}{}^{3-}$. The acid is a hexahydrate containing $H_5O_2{}^+$ ions. The hexahydrate can, therefore, be written $(H_5O_2)_3(PW_{12}O_{40})$ which is a more accurate depiction than $H_3PW_{12}O_{40}.6H_2O$. 12-tungstophosphoric acid has an interesting structure. Three tungstens octahedrally surrounded by oxygens share adjacent edges. The oxygen at the intersection of the shared edges coordinates tetrahedrally with the phosphorous atom. Ions of this type are called Keggin ions and are represented by ions such as $(HSiW_{12}O_{40})^{3-}$, $(H_2BW_{12}O_{40})^{3-}$, and $(CoW_{12}O_{40})^{5-}$ in addition to the $(PW_{12}O_{40})^{3-}$.

Keggin ions have a variety of applications ranging from catalysis to medicine (3-6).

The color of 12-tungstophosphoric acid varies with the oxidation state of the tungstens. It ranges from white to a deep blue. For this reason, this series of compounds has sometimes been referred to as heteropoly blues. Sodium 12-tungstophosphate crystallizes with a large number of waters of crystallization. The common salts reported contain 19, 13, and 11 waters. One of the reasons given for the large numbers of waters of crystallization is the large size of the ion with negative charges residing on the oxygens at the outer corners of the octahedra.

Experimental Procedure

Weigh out 5.0 g of WO_3 and 2.0 g of NaOH into a 125 mL Ehrlenmeyer flask. Add 15 mL of distilled water to dissolve the salts. Boil the aqueous mixture until a solution results. Note the color. Add 1 g of Na_2HPO_4 to this hot solution. When a solution once again results, carefully add 6 mL of concd. HCl. A precipitate of 12-tungstophosphoric acid forms immediately. Cool in an ice bath. Filter using a sintered glass funnel. Recrystallize by adding 10 drops of water and heating the mixture in a hot water bath (or steam bath). Add water one drop at a time until solution is effected. (Do not add more than 2 to 3 mL total.) Remove from the hot water or steam bath and cool in an ice bath. Filter and dry the white crystals. Bottle about half the yield in a sample bottle, label appropriately, and turn in to the laboratory instructor.

Most of the loss in yield occurs in the recrystallization procedure. Make sure that the minimum amount of water is used. Cooling the filtration apparatus with ice before filtering also helps.

Do not use a metal spatula in working up the product to be turned in. The metal may partially reduce the tungsten to form a blue product rather than the white crystals for the tungsten(VI) product.

With the remaining product that was not bottled, test the above statement that the metal spatula will reduce the tungsten to form a blue compound. Besides the spatula, try several other metals such as tin (tin foil), aluminum (aluminum foil), iron (wire) and copper (wire).

Auxiliary Experiments

Prepare the barium salt.

Prepare the blue tungsten(V) product.

Obtain an X-ray powder pattern on the product you have prepared.

Questions to be Considered

1. Sketch the ReO_3 structure.

2. Some of the older literature give the formula for a compound as 3 Na_2O. P_2O_5.24 WO_3.42 H_2O. The analysis of a compound this large is prone to error. Knowing the formula of the Keggin ion, what errors might have been made in the proposed formulation? Assume that the sodium salt of the Keggin ion can have 19 waters of hydration.

Helpful Background Literature

1. "Gmelin's Handbuch der Anorganischen Chemie", Verlag Chemie: Berlin, 1933. Volume on Wolfram, pp. 365 – 366.

2. Phillips, M. A. "Preparation of Phosphotungstic Acid and of Sodium and Barium Phosphotungstates" *J. Soc. Chem. Ind.* (London), 1950, *69*, 282 – 4.

3. Pope, M. T. "Heteropoly and Isopoly Oxometalates", Springer-Verlag: Berlin, 1983.

4. Day, V. W.; Klemperer, W. G. *Science* (Washington, D.C.), 1985, *228*, 533.

5. Pope, M. T.; Muller, A. *Angew. Chem., Int. Ed. Engl.*, 1991, *30*, 34.

6. Pope, M. T.; Muller, A. "Polyoxometalates: From Platonic Solids to Anti-Retroviral Activity" Kluwer Academic Publishers: Dordrecht, The Netherlands, 1994.

COPPER COMPOUNDS

Equipment needed:	steam bath 85°C drying oven ice bath sintered glass funnel (medium porosity) 250 mL Ehrlenmeyer flask small sample vial
Chemicals needed:	5 g thiourea 1 g Cu turnings or wire 55 mL concd. HCl 15 mL acetone 1.0 g KCl 5.5 g $CuCl_2.2H_2O$ Parafilm for sealing vial
Time needed:	3 hours
Safety notes:	Carry out the reactions with concd. HCl in the hood.

Purpose

The first part of the experiment illustrates the formation of a complex halide. Complex halides and complex oxides are solid state compounds with two or more different metal atoms along with halogen or oxygen. The second part of this experiment illustrates that an unstable oxidation state can be stabilized by complexation.

Introduction

The red $KCuCl_3$ is a complex halide. This is a distinct solid state compound and not a double salt of KCl and $CuCl_2$. $KCuCl_3$ consists of an infinite chain of distorted octahedra of $CuCl_6$ sharing opposite and adjacent edges. It is a double chain of octahedra which can be pictured as a portion of the $CdCl_2$ layer. The chains are held together by the positive K^+ ions. It is a distorted $NH_4[CdCl_3]$ structure which was studied in the models experiment.

A large number of complex halides are known. These are characterized by two different positive ions and a halogen negative ion. The structures for these compounds can vary even for compounds with similar formulas. It would seem that $KCuCl_3$ and $KMgF_3$ might have similar structures. However the former is a linear chain and the latter a three-dimensional structure. $KCuCl_3$ consists of a double chain of octahedra whereas $CsCu_2Cl_3$ consists of a double chain of tetrahedra. In addition, some complex halides contain a discrete $MX_m{}^{n-}$ ion which packs, with a positive ion, into a lattice similar to simpler salts. For example, in $NaSbF_6$ the Na^+ and symmetric octahedral $SbF_6{}^-$ ions form the cubic NaCl structure.

As stated, the $KCuCl_3$ structure is related to the $NH_4[CdCl_3]$. The latter structure is distorted in the $KCuCl_3$ structure in that the octahedra of chlorine about each copper is not symmetrical. Four chlorines are at equivalent distances and two are slightly further away.

The red $KCuCl_3$ is readily hydrolyzed in moist air to a mixture of $CuCl_2.2H_2O$ and $K_2CuCl_4.2H_2O$. The blue (green when moist) $CuCl_2.2H_2O$ structure is that of distorted octahedra sharing opposite edges. The blue-green $K_2CuCl_4.2H_2O$ structure can be described as a body centered analogue of the antifluorite structure in which the positive ions are potassium and the negative ions are the same distorted octahedral $(CuCl_4.2H_2O)^{2-}$ aggregates found in $CuCl_2.2H_2O$.

Double halides exhibit different degrees of stability. Some complexes are stable in water. Others are unstable and can be separated into their component simple halides by recrystallization from water. Carnallite, $KMgCl_3.6H_2O$ is an example of a complex halide of this latter type.

The preparation of the copper 1+ oxidation is illustrated by the CuCl experiment. In that experiment, Cu^{2+} ion is reduced to Cu^{1+} ion using $SO_3{}^{2-}$ ions. In the present preparation copper metal is slowly oxidized by hydrochloric acid. This should not occur according to the reduction potentials of H^+ and Cu^+, where E^o (H^+ to H_2) = 0.000 and E^o (Cu^+ to Cu^o) = +0.518 V, but the presence of thiourea forms a stable complex of the copper 1+ and drives the reaction to completion.

Copper(I) compounds are generally colorless due to their d^{10} electronic configurations while copper(II) compounds (d^9) are highly colored. In the present experiment you will prepare a colorless stable copper(I) complex, $Cu(thiourea)_3Cl$. The highly colored copper(II) compound, $KCuCl_3$ is illustrated in the first part of this experiment. The red color of the $KCuCl_3$ is somewhat unusual for copper(II); most compounds of this ion are blue or green.

Experimental Procedure

Preparation of Potassium Trichlorocuprate(II)

Dissolve 1.0 g (0.014 moles) of potassium chloride in 3.7 mL of distilled water. In a fume hood, dissolve 5.5 g (0.033 moles) of copper(II) chloride dihydrate in 50 mL of concd. HCl in a 250 mL Ehrlenmeyer flask. It may take several minutes of stirring for complete dissolution of the copper salt, but do not heat the solution as this will cause a decrease in the concentration of the hydrochloric acid. Pour the potassium chloride solution into the copper chloride solution *slowly, essentially dropwise, and with rapid stirring.* Mark the Ehrlenmeyer flask for identification and place it in an ice bath for one half hour. Filter the red, needle-like, crystals on a medium porosity sintered glass filter using a faucet aspirator to provide suction. As soon as the solid is free of excess mother liquor, spread the damp precipitate on a watch glass and dry in one of the large drying ovens at a temperature not exceeding 85°C.

Place the product in a small tared sample vial and seal it with Parafilm under dry nitrogen. Label the vial properly and submit it to the lab instructor.

Preparation of Tris(thiourea)copper(I) Chloride

Dissolve 5 g of thiourea in 25 mL of hot water, add 1 g Cu (turnings or fine wire) and then 5 mL concd. HCl in the hood. Heat on the steam bath while the copper dissolves with liberation of H_2 (note – copper in HCl does not readily dissolve). Filter the hot solution and allow to cool slowly. White opaque crystals separate. Filter and wash with acetone and air dry. Bottle, weigh, and record your yield.

Questions to be Considered

1. Write the balanced equations for the reactions you have performed.

2. Calculate the percent yield.

3. What information should be on the label? Why is the tare weight important?

4. $KCuCl_3$ is readily hydrolyzed in moist air, forming a mixture of cupric chloride dihydrate and $K_2CuCl_4.2H_2O$. Write the balanced equation for this reaction.

5. Is helium or nitrogen the more desirable gas for filling the vial? Why?

6. The complex $RbCaF_3$ is found to crystallize in the perovskite structure. This is, therefore, an example of a three-dimensional repeating structure. Describe the relative locations of the rubidium, calcium and fluorine atoms in $RbCaF_3$.

7. Compare the structures of $KCuCl_3$ and $KMgF_3$. Draw the double octahedral chain of $KCuCl_3$ and confirm the stoichiometry from the model.

8. The $CsCu_2Cl_3$ consists of a double chain of tetrahedra. Draw a single chain of tetrahedra sharing opposite edges. Draw two of the single chains joined laterally to form a double chain. Confirm that the tetrahedra are sharing three edges. Confirm the stoichiometry from the model.

Helpful Background Literature

1. Wells, A. F. "Structural Inorganic Chemistry," 5th ed.; Clarendon Press: Oxford, 1984; p. 1138, pp. 214 – 215.

POTASSIUM TRIFLUORONICKELATE(II)

Equipment needed:	balance with $\pm$ 5 mg sensitivity mortar and pestle 20 mL nickel crucible muffle furnace (750°C) 250 mL beaker sample bottle
Chemicals needed:	0.6 g NiF_2 0.7 g KHF_2
Time needed:	3 hours
Safety notes:	Handle the KHF_2 carefully. HF is destructive to tissue. It has an anesthetizing effect so that the attack on the tissue is not felt. If a white spot appears on the skin, wash with copious amounts of water and seek immediate medical attention.

Purpose

This experiment illustrates the formation of a complex halide. $KNiF_3$ and $KMgF_3$ are isostructural, i.e., they both crystallize with the perovskite structure. The K_2NiF_4, which is a by-product of this synthesis, has a structure related to the perovskite structure.

Introduction

A complex halide is any solid phase which contains two or more different metal atoms and usually one kind of halogen atom. The general formula for the simplest type of derivative is $M(1)_aM(2)_bX_c$. The actual formulas of complex halides are of all degrees of complexity, as may be seen from the following selection: $CsAgI_2$, $KCuCl_3$, $TlAlF_4$, K_2PtCl_6, Cs_3CoCl_5, Na_3AlF_6, Na_3TaF_8, $Cs_3Tl_2Cl_9$, and $Na_5Al_3F_{14}$.

A few general observations will be made here before describing the structure of some of these salts. First, similarity in formula type does not mean that the compounds have similar or related structures. Thus the salts $KNiF_3$ $KCuCl_3$, and $CsAuCl_3$ have quite different structures. Second, it is not always correct to use the simplest empirical formula. The "double" chloride, $CsAuCl_3$, should actually be written as $Cs_2Au^+Au^{3+}Cl_6$ for it contains both univalent and trivalent gold. Third, a complex halide of formula A_mBX_n does not necessarily contain a discrete BX_n group. The structural features of the BX_n group depends entirely on the bridging or non-bridging capacity of the halogen atom in a given compound or structure. As illustrated in the model experiments, a halogen atom can act as a terminal ligand or can bridge two or more metal atoms depending on the final structure which the compound adopts.

The stabilities of ternary halides may vary from the very unstable to those which are quite stable. Potassium trichlorocuprate(II), $KCuCl_3$, is an unstable red crystalline compound which readily decomposes to $CuCl_2$ and K_2CuCl_4 in moist air. $KNiF_3$, on the other hand, can survive a treatment in boiling water. There is a second form of potassium nickel fluoride, with stoichiometry K_2NiF_4, which has intermediate stability to water. It survives cold water treatment but decomposes completely to $KNiF_3$ if boiled in water for a few minutes.

The synthesis of both $KNiF_3$ and K_2NiF_4 illustrated in this experiment involves (modified) solid state reactions. The most common type of reaction studied in elementary chemistry involves reactions of solutes in a convenient solvent. Also familiar are examples of heterogeneous reactions in which interactions are observed between solid-gas, solid-liquid, liquid-liquid, and gas-liquid phases. Solid-solid reactions are generally unfamiliar to the student. Yet in these days of "solid-state" devices, these reactions are of more than passing interest.

The general technique in a solid state reaction is to heat the fine, intimately mixed reactants to high temperatures. For some reactants, in order to remove the partially reacted mixture, it is necessary to regrind and refire the product several times. Problems are encounterd in reactions with crucibles, volatilization of one of the reactants, removal of contaminants, and the separation from excess starting materials. In this reaction, potassium hydrogen fluoride (KHF_2 or KF.HF) is reacted with nickel(II) fluoride (NiF_2). A one to one molar ratio would produce $KNiF_3$. A two to one molar ratio would yield K_2NiF_4.

In order to achieve an intimate contact between two unlike reactants, it is desirable that the particle size be small and that they are well mixed. Small samples are efficiently mixed and ground in a "Wig-L-Bug". Containers of plastic, stainless steel or agate are available for this mechanical mixer. Classically, samples are crushed and ground using a mortar and pestle. A mechanized mortar and pestle arrangement can also be used for this operation on larger samples.

A serious problem which is encountered is the selection of a crucible. A porcelain crucible is unsatisfactory. It is attacked by this reaction mixture and significant amounts of potassium aluminum silicate are found in the reaction product. Quartz or glass vessels would be equally unsatisfactory since K_2SiF_6 or its decomposition products would be formed. Nickel crucibles are reasonably priced and exhibit fair resistance to fluoride attack. They are essentially inert to NiF_2 but are slowly attacked by KHF_2. For this reason, K_2NiF_4 can be made in a nickel crucible, but generally shows a darkened color due to the products formed by attack of the crucible and its oxide coating.

Another problem is the purification of the product. If one reagent or the other is used in excess, it must be removed from the desired product. For this purpose selective solubility, volatility, or decomposition might be used. If none of these methods are applicable, it is necessary to weigh out exact stoichiometric amounts of the reactants and hope for a complete reaction. In this experiment, an excess of NiF_2 in the product cannot be removed by treatment with water. However, because it is a very fine powder, small amounts can be suspended in water and can be decanted with some of the finely divided product. An excess of KHF_2 produces some K_2NiF_4 mixed with the desired $KNiF_3$. The separation of these two compounds is simple since a short period of contact with boiling water converts all the K_2NiF_4 into $KNiF_3$. This would seem to indicate that a large excess of KHF_2 should be used. This, however, is not desirable since increasing amounts of KHF_2 would result in increased attack on the crucible. Weight ratios of $KHF_2 : NiF_2$ of 1.5 : 1 is the suggested upper limit. The closer one can achieve the molar ratio of 1.25 : 1, the better. The object is to get close to the molar equivalent ratio and at the same time try to insure that any error be on the side where the impurity can easily be separated or removed.

The decomposition of the solid K_2NiF_4 to the solid $KNiF_3$ in boiling water is an interesting reaction. $KNiF_3$ has the perovskite structure. The potassium and fluoride ions occupy positions of cubic close-packing with the potassium ions spaced out so that they are coordinated by twelve fluoride ions. The nickel atom is coordinated by six fluorides. These structures were discussed in the model experiments and will be available in the laboratory for your inspection. The K_2NiF_4 is illustrated below.

The K_2NiF_4 is representative of a number of compounds exhibiting the same structure. Some of these compounds are K_2ZnCl_4, Rb_2UO_4, Sr_2RuO_4, Ba_2PbO_4, Sr_2FeO_3F and La_2NiO_4. These compounds are said to have the K_2NiF_4 structure.

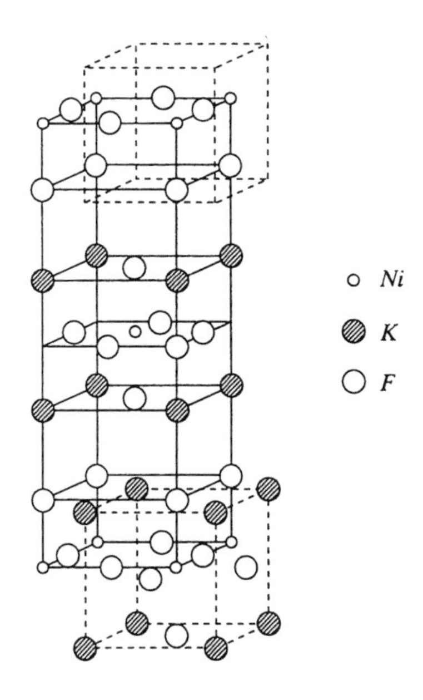

Fig. 1. The K_2NiF_4 structure.

Experimental Procedure

If a balance capable of weighing to ± 5 mg or better is available, weigh out approximately 0.600 g of NiF_2. Multiply the actual weight of NiF_2 by 1.01 and accurately weigh out this amount of KHF_2 (potassium hydrogen fluoride). If a balance good to ± 0.02 g is the best available, *very* carefully weigh out 0.60 g of NiF_2 and 0.60 g of KHF_2. Grind the NiF_2 and KHF_2 together in a clean mortar. If the mortar is not clean, grind about 0.2 g of KHF_2 in the mortar and rinse with water. Dry the mortar and pestle carefully before grinding the reactants together. Transfer the finely ground reactants from the mortar to a 20 mL nickel crucible. Place the cover on the crucible and fire at 700° – 750°C for half an hour to 45 minutes in a muffle furnace located in a hood. Remove the crucible from the furnace, allow it to cool and transfer the bulk of the green product from the crucible (without scraping the sides or bottom of the crucible too much) into a 250 mL beaker. Wash with two 10 mL portions of distilled water crushing the lumps with a stirring rod. Carefully decant the cloudy wash liquid. Add 100 mL of distilled water and bring to a boil for 3 to 5 minutes *only*. Decant the hot water and place the beaker with the damp solid in a drying oven. Transfer the homogeneous product into a sample bottle, label, and turn in.

Auxiliary Experiments

If the facilities are available, determine the X-ray powder pattern of the material you have prepared. Examine the powder pattern for evidence of impurities.

21-1002 JCPDS-ICDD Copyright (c) 1991 Quality: *

$KNiF_3$

Potassium Nickel Fluoride

Rad: CuKa1 Lambda: 1.54056 Filter: Mono. d-sp:
Cutoff: Int: Diffractometer I/Icor: 3.00
Ref: Natl. Bur. Stand. (U.S.) Monogr. 25, 7 42 (1969)

Sys: Cubic S.G.: Pm3m (221)
a: 4.0127 b: c: A: C:
A: B: C: Z: 1 mp:
Ref: Ibid.

Dx: 3.978 Dm: SS/FOM: F22=53(.018,23)

εα: nωβ: εγ: Sign: 2V:
Ref:

Color: Pale yellow-green
Pattern made at 25 C. The sample was prepared by adding hydrofluoric acid to a mixture of K2CO3 and NiCO3. The material was then heated to about 200 C. Perovskite type. Tungsten used as internal standard. PSC: cP5. Mwt: 154.79. Volume[CD]: 64.61.

d A	Int.	h k l
4.02	30	1 0 0
2.84	100	1 1 0
2.317	12	1 1 1
2.006	65	2 0 0
1.795	12	2 1 0
1.639	30	2 1 1
1.418	25	2 2 0
1.3376	4	3 0 0
1.2686	10	3 1 0
1.2096	2	3 1 1
1.1581	8	2 2 2
1.1129	2	3 2 0
1.0726	10	3 2 1
1.0032	4	4 0 0
0.9732	2	4 1 0
0.9457	6	3 3 0
0.9206	2	3 3 1
0.8973	8	4 2 0
0.8757	2	4 2 1
0.8554	4	3 3 2
0.8190	6	4 2 2
0.7870	4	5 1 0

Strong lines: 2.84/X 2.01/7 4.02/3 1.64/3 1.42/3 2.32/1 1.80/1 1.27/1

Fig. 2. PDF 21-1002 Potassium Trifluoronickelate(II)

Questions to be Considered

1. How is the perovskite structure related to the ReO_3 structure?

2. Recall the perovskite model and satisfy yourself that each potassium ion has 12 close-packed fluoride neighbors.

3. Explain in your own words the difference between the K_2NiF_4 structure and the perovskite structure.

4. What are the coordination numbers for each of the ions in the perovskite structure and in the K_2NiF_4 structure.

5. Note that this experiment is not a solid state reaction, in a strict sense, since KHF_2 decomposes at approximately 255°C to yield KF and gaseous HF. The presence of the gas (even though it does not enter directly in the reaction as written) is useful and probably necessary for several reasons. Can you give a reason for using KHF_2 instead of KF?

Helpful Background Literature

1. Wells, A. F. "Structural Inorganic Chemistry," 5th ed.; Clarendon Press: Oxford, 1984; p. 209, 602.

POTASSIUM TRIFLUOROMAGNESATE

Equipment Needed:	mortar and pestle nickel crucible and cover muffle furnace (600° – 700^oC)
Chemicals Needed:	0.3 g MgF_2 0.4 g KHF_2
Time Needed:	1.5 hours
Safety Notes:	Potassium hydrogen fluoride is corrosive and irritating to the skin and mucous membranes. Handle with care.

Purpose

This preparation illustrates a solid state reaction. The two reactants in the solid state are brought into intimate contact by grinding and/or pressing. The high temperature to which they are subjected allows for the diffusion of the ions creating a new solid structure.

Introduction

The synthesis of potassium trifluoromagnesate is intended to illustrate the structural interrelationships of two complex halides. Techniques used for solid state syntheses are also demonstrated.

Complex halides are those in which at least two metal atoms are combined with halogen atoms. A binary or simple halide might be generalized as M_aX_b. The most common complex halides are the ternary halides of the type $M_aM'_bX_c$. The stabilities of the ternary halides may vary from the very unstable to those which are quite stable. Potassium trichlorocuprate(II) is an unstable red compound. It readily decomposes to

$CuCl_2$ and KCl in cold water or moist air. $KMgF_3$, on the other hand, can survive boiling water. Although it decomposes slowly on extended boiling, practically no decomposition occurs in 10 to 15 minutes of boiling water treatment. There is a second potassium magnesium fluorine compound of the stoichiometry K_2MgF_4. This complex is of intermediate stability. It survives cold water treatment but decomposes completely to $KMgF_3$ if boiled in water for several minutes.

The syntheses of both $KMgF_3$ and K_2MgF_4 involve solid state reactions. The most common type of reaction studied in elementary chemistry involves reactions of solutes in a convenient solvent. Also familiar are examples of heterogeneous reactions in which interactions are observed between solid-gas, solid-liquid, liquid-liquid, and gas-liquid phases. Solid-solid reactions are generally unfamiliar. Yet in these days of "solid-state" technology, these reactions are of more than passing interest.

The general technique in solid state reactions is to heat the intimately mixed reactants at high temperatures. For some reactants, it is necessary to remove the partially reacted mixture, regrind, and refire several times. Problems are encountered in reactions with crucibles, volatilization of one of the reactants, removal of contaminants, and removal of excess starting materials. In this reaction potassium hydrogen fluoride (KHF_2 or KF.HF) is reacted with magnesium fluoride (MgF_2). A one to one molar ratio produces $KMgF_3$. A two to one molar ratio produces K_2MgF_4.

In order to achieve an intimate contact between two unlike solids, it is desirable that the particle size be small and that they are well mixed. Small samples are efficiently mixed and ground by using the "Wig-L-Bug." Containers of plastic, stainless steel, or agate are available for this mechanical mixer. Classically, samples are crushed and ground using a mortar and pestle. A mechanized mortar and pestle arrangement is also available.

A serious problem which is encountered is the selection of a crucible. A porcelain crucible is unsatisfactory. It is attacked by the reaction mixture and significant amounts of potassium aluminum silicate are found in the reaction product. Quartz or glass vessels are equally unsatisfactory. K_2SiF_6 or its decomposition products are formed and become contaminants. Nickel crucibles are reasonably priced and exhibit fair resistance to fluoride attack. It is essentially inert to MgF_2 but is slowly attacked by KHF_2. For this reason K_2MgF_4 can be made in nickel crucibles but generally shows a grey color due to the products formed by attack of the crucible.

Another problem is the purification of the product formed. If one reagent or the other is used in excess, it must be removed from the desired product. For this purpose selective solubility, volatility, or decomposition might be used. If none of these methods are applicable, it is necessary to weigh out exact stoichiometric amounts of the reactants and hope for a complete reaction. In this experiment an excess of MgF_2 in the product cannot be removed by dissolving in water. However, because it is a very fine powder, small amounts can be suspended in water and decanted off along with some of the fine particles of the product. An excess of KHF_2 produces some K_2MgF_4 along with $KMgF_3$. The purification of these two compounds is much cleaner since a short period in boiling water converts all the K_2MgF_4 into $KMgF_3$. This would seem to indicate that a large excess of KHF_2 should be used. This, however, is not desirable since increasing amounts of KHF_2 result in increased attack on the crucible. A weight ratio of KHF_2 : MgF_2 of 1.5 : 1 is the suggested upper limit. The closer one can achieve the molar equivalent weight ratio of 1.25 : 1, the better. The object is to get as close to the molar equivalent ratio while at the same time trying to insure that any error be on the side where the impurity is the easily decomposed one.

The decomposition of the solid K_2MgF_4 to the solid $KMgF_3$ in boiling water is an interesting reaction. $KMgF_3$ has the perovskite structure. The potassium and fluoride ions occupy positions of cubic close-packing with the potassium ions spaced out so that they are coordinated by twelve fluorides. The magnesium ions are coordinated by six fluorides. This structure is illustrated in the model experiments. The K_2MgF_4, on the other hand, crystallizes in the K_2NiF_4 structure. This structure is related to the perovskite structure in that there are interleaving KF layers with the spaced out "perovskite" layers being displaced or slid over.

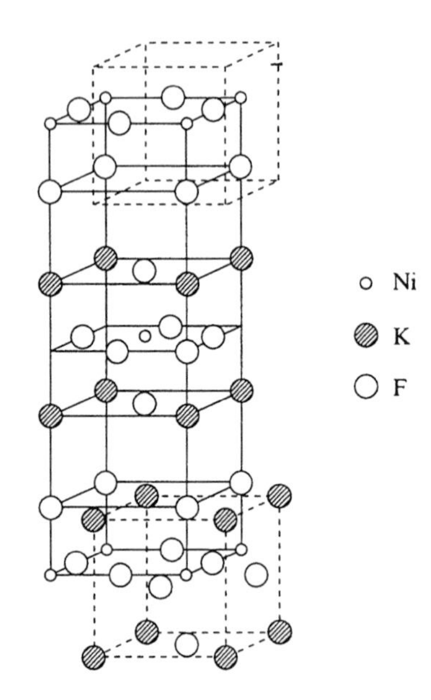

Fig. 1. The K_2NiF_4 Structure

How the K_2MgF_4 converts to the $KMgF_3$ structure lends itself to interesting speculation.

Experimental Procedure

If a balance capable of weighing to approximately 5 mg accuracy or better is available, carefully weigh out about 0.3 g of MgF_2. Multiply the actual weight of MgF_2 by 1.30 and accurately weigh out this amount of KHF_2 (Potassium hydrogen fluoride). If a balance which is only accurate to approximately 0.02 g is the best available (e.g. triple beam balance), carefully weigh out 0.3 g of MgF_2 and 0.4 g of KHF_2. Grind the MgF_2 and KHF_2 together in a clean mortar. If the mortar is not clean, grind about 0.2 g of KHF_2 in the mortar and rinse with water. Dry the mortar and pestle carefully before grinding the reactants together. Transfer the finely ground reactants from the mortar to a 20 mL nickel crucible. Place the cover on the crucible and fire at $600^o - 700^oC$ for half an hour in a muffle furnace placed in a hood. Remove the crucible from the furnace, allow to cool and transfer the bulk of the slightly off-white product to a 250 mL beaker without scraping the sides or bottom of the crucible too much. Wash with two 20 mL portions of distilled water crushing the lumps with a stirring rod. Decant off the cloudy wash liquid. Add 125 mL of distilled water and bring to a boil. Decant off the hot water and place the beaker with a damp solid in a drying oven. Place the product which is a white solid in a sample bottle, label and turn in. Determine the X-ray powder pattern if so assigned. Determine that the powder pattern is that of $KMgF_3$ from the d spacing listed in the ASTM card. Prove or disprove the detectable presence of MgF_2.

Auxiliary Experiments

1. X-Ray Powder Pattern

One of the methods used to characterize inorganic compounds is by X-ray diffraction. Single crystal X-ray diffraction data are used to determine the exact atomic positions within the crystal. X-Ray powder patterns are used to determine d spacings for the planes responsible for the reflection of the observed lines. (For a more detailed discussion of X-ray diffraction, refer to the appropriate section in Part I.) The X-ray powder pattern can be used to determine the type of unit cell (cubic, monoclinic, tetragonal, etc.) and the dimensions of that cell. It can also be used as a "fingerprint" to identify a compound or element. In order to determine whether or not $KMgF_3$ was actually prepared, determine the X-ray powder pattern of the product obtained. A copy of the PDF data for $KMgF_3$ is reproduced.

18-1033

Wavelength= 1.5405

KMgF3

Potassium Magnesium Fluoride

Rad.: CuKa1 λ: 1.5405 Filter: Ni Beta d-sp:

Cut off: Int.: Diffract. I/Icor.: 0.90

Ref: Natl. Bur. Stand. (U.S.) Monogr. 25, 6, 42 (1968)

Sys.: Cubic S.G.: Pm3m (221)

a: 3.9889(1) b: c: A: C:

α: β: γ: Z: 1 mp:

Ref: Ibid.

Dx: 3.150 Dm: SS/FOM: F_{19} = 81(.0102 , 23)

εα: ηωβ: 1.404 εγ: Sign: 2V:

Ref: Ibid.

Color: Colorless
Pattern made at 25 C. Made by adding HF to a slurry of K2 C O3 and Mg C O3 and evaporating to dryness. Pattern sharpened by heating sample to melting point.
Spectroscopic analysis: 0.001-0.01% each Al, Ca, Pt, Rb and Sr; 0.01-0.1% each Na, Pb and Si.
Perovskite type. Tungsten used as an internal stand. PSC: cP5. Mwt: 120.40. Volume[CD]: 63.47.

d(A)	Int	h	k	l
3.988	2	1	0	0
2.819	94	1	1	0
2.302	83	1	1	1
1.9943	100	2	0	0
1.7842	1	2	1	0
1.6284	24	2	1	1
1.4101	36	2	2	0
1.3298	<1	3	0	0
1.2614	6	3	1	0
1.2028	8	3	1	1
1.1516	8	2	2	2
1.0661	8	3	2	1
.9972	2	4	0	0
.9403	2	4	1	1
.91	2	3	3	1
.89	10	4	2	0
.8505	1	3	3	2
.8142	3	4	2	2
.7823	4	5	1	0

PCPDFWIN v. 1.31

Fig. 2. PDF 18-1033 for Potassium Trifluoromagnesate

2. Structural Models for $KMgF_3$ and K_2NiF_4

Refresh your memory of the ReO_3 structure prepared in *Session Two* of the models laboratory.

The oxygens are located by the corners of the octahedra and the rhenium by the centers of the octahedra in this ReO_3 structure.

The perovskite structure, such as $KMgF_3$, has the ReO_3 structure with the addition of a large positive ion in the center of the large hole formed by the eight octahedra. The fluorines are located at the corners of the octahedra and the small magnesium ions are located at the center of the octahedra.

> Confirm the 1 : 3 ratio of Mg : F from the structure. Indicate your argument for this stoichiometry to the instructor.
>
> Confirm the 1 : 1 ratio of K : Mg by looking at the structure. Indicate your argument for this stoichiometry to the instructor.

Take the Mo_4F_{20} structure provided for you in the laboratory and hold it one "bond" length above the top octahedral layer of the ReO_3 structure. Now translate this layer so that one of the bottom points of the Mo_4F_{20} is above the position of the potassium ion. Look at this model and diagram of the K_2NiF_4 structure. This should help you visualize the K_2NiF_4 structure which may not be evident from the two dimensional diagram even when supplemented with a verbal description.

> Confirm the 1 : 4 ratio of Ni : F by looking at the model. Explain your reasoning to the instructor.
>
> Confirm the 2 : 1 ratio of K : Ni by looking at the model. Once again explain your reasoning to the instructor.

Questions to be Considered

1. Hydrogen fluoride is a toxic gas. The recommended maximum atmospheric concentration for an 8 hour exposure is 3 parts of HF per million parts of air by volume. Assuming ideal gas behavior and a closed room, calculate the size of room needed if a 0.4 g sample of KHF_2 were being decomposed and the final concentration of HF is to be 3 ppm.

2. How is the perovskite structure related to the ReO_3 structure?

3. Recall the perovskite model and satisfy yourself that each potassium ion has 12 nearest fluoride neighbors.

4. Explain in your own words the difference between the K_2NiF_4 structure and the perovskite structure.

Helpful Background Literature

1. Wells, A. F. "Structural Inorganic Chemistry", 5th ed.; Clarendon Press: Oxford, 1984; pp 182, 267, 584 – 589, 602 – 603.

POTASSIUM IODATE AND CLOCK REACTION

Equipment needed:	balance with ± 5 mg sensitivity two 6 inch test tubes mortar and pestle Bunsen burner 10 mL graduated cylinder medicine dropper stopwatch one-hole rubber stopper rubber tubing
Chemicals needed:	1.5 g $KClO_3$ 1.75 g iodine 2.5 g KI 1 g KOH 20 mmoles of H_2SO_4 one drop concd. HNO_3 0.63 g Na_2SO_4 1 g soluble starch litmus paper small wad of cotton
Time needed:	3 hours
Safety notes:	Be sure to carry out the reaction of potassium chlorate and iodine in a hood in case some chlorine escapes the chlorine trap.

Purpose

This experiment illustrates the relative ease of oxidation of two halides. A "clock reaction" with potassium iodate is illustrated.

Introduction

Iodine can be oxidized to iodate by potassium chlorate. The chlorine in potassium chlorate is reduced to chlorine and the elemental iodine is oxidized to iodate. The examination of the reactants and the product would appear to indicate that iodine has been substituted for chlorine in potassium chlorate. In actuality, because iodine, with electrons further from the nucleus than chlorine, has a smaller electron affinity than chlorine, the redox reaction becomes energetically favorable. The reaction is found to go best in acid solution, possibly because of the intermediate formation of a little chloric acid.

The acid solution results in the formation of some $KH(IO_3)_2$. For this reason, the solution must be neutralized during the crystallization process in order to obtain pure KIO_3.

The reaction to form potassium iodate generates chlorine. Even though the experiment will only generate about 6 mmoles of chlorine, it is not a good practice to vent chlorine into the atmosphere. Chlorine can be trapped with KOH or with KI solution. Because of the possibility of suck back as the reaction test tube is sporadically heated, it is not desirable to bubble the evolved gases through a solution. A solid material which would absorb chlorine by a surface reaction is Ascarite. For many years Ascarite, which was alkali adsorbed on asbestos, was used to absorb carbon dioxide in the old microgravimetric analysis of carbon and hydrogen of organic compounds. However, asbestos is now declared to be undesirable and the need for Ascarite has disappeared with the newer gas chromatographic methods for carbon and hydrogen. A similar scheme might be used with a ceramic fibrous material, such as Fiberfrax, on which KOH might be adsorbed. Such a material would work well as a trapping material, but the preparation of small amounts of such a substance entails some hazards. Pellets of KOH might be used, but the surface area is relatively small compared to the volume of KOH. Since relatively large amounts of KOH in the pellet form will have to be used, disposal of the KOH becomes a consideration. In this experiment, the small amounts of chlorine generated is trapped with solid KI which is dispersed by using cotton in the trapping test tube.

A "clock reaction" is one which shows no observable reaction until a point is reached when a reaction product suddenly becomes visible. Iodate is reduced to iodine by sulfurous acid in a sequence of three consecutive reactions. First, iodate is reduced to iodide. Second, iodide, iodate, and hydrogen ion react to form iodine. Third, iodine oxidizes more sulfurous acid to sulfuric. The first stage is the slowest, the last the fastest. Consequently, no iodine appears until the sulfurous acid is all used up.

If starch is added to the reaction mixture, a dark blue-black starch iodine color will form when the iodine appears at the point when all of the sulfurous acid is used up. Depending on the concentrations, there will be a time period when nothing seems to be happening. When the sulfurous acid is all used up, the solution will suddenly turn to the dark starch-iodine color. The change is quite dramatic.

Experimental Procedure

Preparation of Potassium Iodate

Prepare, as a reaction tube, a 6 inch test tube with a one-hole rubber stopper into which a short glass tube is inserted. Prepare a second 6 inch test tube with a loose glass tube which reaches to the bottom of the test tube. Have a short piece of rubber tubing available which can be used to connect the reaction tube to the glass tube which goes to he bottom of the trapping test tube.

Prepare the trap by putting a bit of cotton at the bottom of the test tube so that it surrounds the glass tube. Grind 2.5 g of KI using a mortar and pestle. Add a bit of the KI to the test tube. Add a bit more cotton and add a bit more KI. By repeating this process several times, the KI will be "dispersed" in such a way that caking will be minimized and the KI will offer the largest area to the evolving chlorine. Because chlorine reacts with KI to form iodine, the trap will have a built-in indicator. The iodine color can be seen to form at the bottom of the tube and then slowly work its way to the top of the trap.

Add 1.5 g of $KClO_3$ and 1.75 g I_2 to the reaction test tube. Add 3 mL distilled water and one drop of concd. HNO_3. Close the reaction test tube with the stopper with the glass tube and connect this to the tube leading to the bottom of the trapping test tube prepared with KI and cotton with a short piece of rubber tubing. It is convenient to use a clamp to hold the reaction tube. The trap can be held with the same hand taking care not to pull the glass tube out away from the bottom of the test tube.

Gently heat the reaction test tube over a Bunsen burner. Some of the iodine will initially vaporize and condense in the upper part of the test tube. Continued gentle heating and some shaking will cause this iodine to also react. When the iodine has disappeared and the solution has a pale yellow color, remove the stopper from the reaction tube and heat the mixture until about half of the water is boiled off. At this point crystallization will have started. Cool and pour off the mother liquor. Dissolve the residual salt in 7.5 mL of hot water. Prepare a solution of 1 g of KOH in 10 mL of water. Add this base to the hot solution about 1 mL at a time until the solution is neutral or slightly basic to litmus. Filter the precipitate which forms on cooling using a coarse filter paper. Wash with about 2 mL of ice cold water. Place the filter paper and product on a watch glass and dry the product in a 110°C drying oven.

Weigh the product, calculate the yield, put in an sample bottle and label properly.

The "Clock Reaction"

A starch solution should be made up fresh for this laboratory by the teaching assistants. Heat 100 mL of water to boiling in a beaker. Add one g of soluble starch to this boiling water with stirring. Allow the starch solution to cool.

Take 3.33 mL of 6 M H_2SO_4 or its equivalent (20 mmole) and dilute to 500 mL. Add 0.63 g of Na_2SO_3 to the 500 mL of dilute sulfuric acid solution.

Weigh out 0.4 g of KIO_3 you have prepared. Dissolve in 200 mL of water.

Measure out 25 mL of the sulfurous acid solution prepared above into a 125 mL Ehrlenmeyer flask and add 0.5 to 1 mL of starch solution. (Keep in mind that starch slowly hydrolyzes in acid.) Measure out 25 mL of the KIO_3 solution. Add the KIO_3

solution to the sulfurous acid solution and mix by swirling the Ehrlenmeyer flask. Start the stopwatch or note the time on the second hand of your watch. Record the time it takes for the starch-iodine color to form. Change the sulfurous acid concentration by making 25 mL of various combinations of sulfurous acid solution and water. For example, try a solution of 15 mL of sulfurous acid solution and 10 mL of water with 25 mL of the KIO_3 solution. Note the time until the color forms for each different concentration of sulfurous acid.

Auxiliary Experiment

If the above reaction is carried with an excess of sulfite, free iodine periodically appears and disappears as a result of the following series of reactions.

$$IO_3^- + 3\,SO_3^{2-} \rightarrow I^- + 3\,SO_4^{2-}$$

$$5\,I^- + IO_3^- + 5\,H^+ \rightarrow 3\,I_2 + 3\,H_2O$$

$$3\,I_2 + 3\,SO_3^{2-} \rightarrow 6\,I^- + 6\,H^+ + 3\,SO_4^{2-}$$

Add an excess of sulfite to iodate to get the periodic on-off phenomenon.

Questions to be Considered

1. How else, besides changing the sulfurous acid concentration, might the time period for the appearance of the starch-iodine color be changed?

2. Given the accuracy of volumetric measurements used in this laboratory, estimate (or experimentally determine) the reproducibility of the time to color development in the clock reaction.

3. The reactions used in the clock reaction have been used on a large preparative scale to prepare iodine from iodate in Chili saltpeter. Write the half cell reactions for the three redox reactions shown in the auxiliary experiment.

Helpful Background Literature

1. Greenwood, N. N.; Earnshaw, A. "Chemistry of the Elements"; Pergamon Press: Oxford, 1984; p. 1012.

2. Chen, P. S. "Entertaining and Educational Chemical Demonstrations"; Chemical Elements Publishing Company: Camarillo, Calif., 1976; p. 25.

ETHYLENEDIAMINE COMPLEXES OF COBALT AND NICKEL

Equipment Needed: sintered glass funnel
hot plate
two 100 mL beakers
Mohr pipettes, 10 mL graduated
two test tubes

Chemicals Needed: 1.5 g $CoCl_2.6H_2O$
10 mL ethylenediamine solution (130 mL anh. ethylenediamine in 1 L water)
5 mL ethylenediamine solution (25 mL anh. ethylenediamine in 25 mL water)
2.1 mL 3M HCl
2 mL 30% H_2O_2
15 mL concd. HCl
50 mL ethanol
40 mL diethylether
1.3 g $NiCl_2.6H_2O$

Time Needed: 3 hours

Safety Notes: Do the evaporations in the hood.

Purpose

This preparation is representative of the formation of coordination compounds. The compounds formed illustrate the range of colors and stabilities encountered in coordination compounds.

* Developed by Prof. Ronald A. Krause, University of Connecticut.

Introduction

Of particular importance to the development of coordination chemistry are metal complexes of the type to be synthesized and characterized in this experiment. Prior to 1950, research in this area was almost exclusively concerned with the investigation of complexes of transition metal ions with such monodentate ligands as Cl^-, Br^-, I^-, NH_3, pyridine, CN^-, and $NO_2{}^-$, and bidentate ligands such as ethylenediamine ($H_2NCH_2CH_2NH_2$), oxalate ($^-O_2CCO_2{}^-$), glycinate ($H_2NHCH_2CO_2{}^-$), and $CO_3{}^{2-}$. These complexes still form the basis of a vast amount of research today despite the more recent discoveries of the ligand properties of -H^-, -CH_3, CO, $H_2C{=}CH_2$, and benzene to mention a few.

Coordination compounds of Co(III) and Cr(III) have been of particular interest because their complexes undergo ligand exchange very slowly as compared to complexes of many other transition metal ions. For example, $[Ni(NH_3)_6]^{2+}$ reacts virtually instantaneously with H_3O^+ to form $[Ni(OH_2)_6]^{2+}$. Under the same conditions the analogous reactions of $[Co(NH_3)_6]^{3+}$ occur very slowly. This difference in behavior of complexes of different metal ions has been accounted for by ligand field and molecular orbital theory.

In this experiment, a bidentate chelating ligand, ethylenediamine (en) will be utilized to prepare an inert cobalt(III) complex and a labile nickel(II) complex:

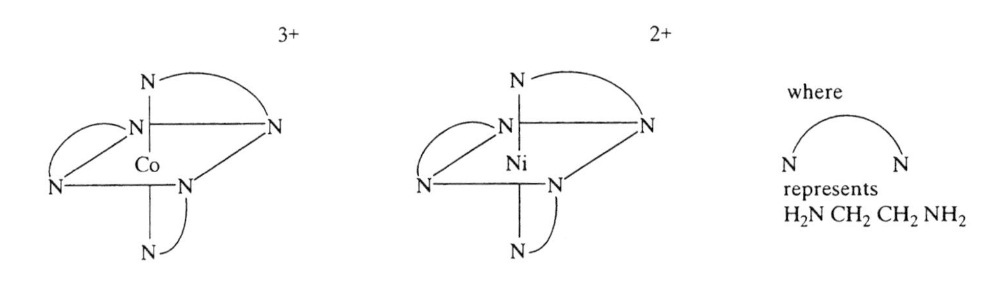

Fig. 1.

One important method of characterizing ionic substances is by determining the ability of their solutions to conduct an electric current. Those substances whose solutions have the highest conductivity consist of the greatest number of ions. Thus a one molar solution of $[Co(NH_3)_4.CO_3]NO_3$ will have a lower conductance than a solution of $[Co(NH_3)_5Cl]Cl_2$ of the same concentration. By measuring the conductivity of a solution of a compound, it is possible to determine if a formula unit of that compound consists of 2, 3, 4, or more ions. Although measurements will be done on water solutions of the complexes, the same information can frequently be obtained using organic solvents such as nitrobenzene or acetonitrile for ionic compounds which either are not very soluble in water or react with water.

Another powerful method for establishing the identity of a complex is infrared spectroscopy. This technique examines the frequencies of the vibrational modes of a molecule. Thus the infrared spectra of both of the preceding complexes exhibit absorptions at frequencies (commonly expressed in wave numbers, cm^{-1}, i.e., reciprocal wavelength $1/\lambda$) characteristic of stretching and bending modes of the NH_3 group. While the Co-N stretching modes are in principle also measurable, they sometimes occur at frequencies too low (below 650 cm^{-1}) to be observed in the standard infrared spectrophotometers. More specialized and expensive infrared instruments are, however, available for studying these vibrations. The complex $[Co(NH_3)_4CO_3]NO_3$ also exhibits absorptions characteristic of the carbonate group. Because of its coordination to the metal ion, the $(CO_3)^{2-}$ group has a somewhat different spectrum in this complex than that of the $(CO_3)^{2-}$ ion in Na_2CO_3. The spectrum should also contain absorption bands resulting from vibrational modes of the NO_3^- ion, quite similar to those observed in $NaNO_3$. In contrast the spectrum of $[Co(NH_3)_5Cl]Cl_2$ would be largely dominated by absorptions attributable to the NH_3 groups.

In general, the metal-chlorine stretching frequencies (below 400 cm^{-1}) cannot be observed on the usual infrared instruments; and, of course, the ionic Cl^- groups are not, in the solid state, strongly bonded to any other single atom and thus no absorptions are expected in the infrared spectrum to indicate their presence in the compound. While infrared spectroscopy and other instrumental methods of compound characterization are emphasized in this discussion, it should be stressed that a quantitative elemental analysis is an absolutely essential step in determining the composition and structure of a new compound.

The synthesis of $[Co(en)_3]Cl_3$ will be carried out according to the unbalanced equation:

$$[Co(H_2O)_6]^{2+} + (en) + H_2O_2 + H^+ \rightarrow [Co(en)_3]^{3+} + H_2O$$

while the synthesis of $[Ni(en)_3]^{2+}$ proceeds according to

$$[Ni(H_2O)_6]^{2+} + (en) \rightarrow [Ni(en)_3]^{2+} + H_2O$$

For the isolation of both complexes, advantage is taken of the low solubilities of the chlorides in aqueous alcohol.

Occasionally, the cobalt(III) complex will turn out to be green instead of orange-yellow. This is due to incomplete oxidation. If almost everyone in the class obtains a green compound, the indication will be that the 30% H_2O_2 may not have been at full strength due to partial decomposition. If only a few students obtain the green compound, the indication is that the procedure followed by those obtaining the green compound did not succeed in completely oxidizing the Co(II) to Co(III). If Co(II) is present, the complex ion $[CoCl_4]^{2-}$ is formed. In a high chloride concentration, such as in concd. HCl or in ethanol solution, the $[CoCl_4]^{2-}$ anion precipitates with the $[Co(en)_3]^{3+}$ cation.

$$2\,[Co(en)_3]^{3+} + 3\,[CoCl_4]^{2-} \rightarrow [Co(en)_3]_2[CoCl_4]_3$$

The $[Co(en)_3]^{3+}$ ion is yellow. The $[CoCl_4]^{2-}$ ion is blue. When these precipitate together, a green compound is formed.

The addition of water destroys the complex anion and the yellow color is observed. Addition of HCl or ethanol regenerates the complex anion and reforms the green compound.

Experimental Procedure

Cobalt(III) Complex

Dissolve 1.5 g $CoCl_2.6H_2O$ in 5 mL water. While the $CoCl_2.6H_2O$ is dissolving, measure out 10 mL of an ethylenediamine solution. (The solution is prepared by dissolving 130 mL of anhydrous ethylenediamine in one liter of water.) Cool the ethylenediamine solution in ice and slowly add 2.1 mL 3 M HCl. Add the $CoCl_2$ solution to the partially neutralized ethylenediamine and then slowly add 2 mL of 30% H_2O_2 with stirring. Stir until effervescence ceases, place the mixture on a hot plate and allow to boil gently. (At this point, you should proceed with the synthesis of the nickel complex.)

When the cobalt solution is evaporated to a volume of 6 to 10 mL, add an equal volume of concd. HCl and twice this amount of ethyl alcohol. Cool in ice, filter, press dry on the sintered glass funnel, and wash with two 10 mL portions of ethanol (95%) and two 10 mL portions of ether. Air dry, weigh, bottle, and label the orange-yellow product.

In your laboratory notebook experiment section report the volume of the solution after boiling down, the volume of HCl and alcohol added, the weight of complex obtained, and the percent yield.

Nickel(II) Complex

Dissolve 1.3 g $NiCl_2.6H_2O$ in 1.5 mL of warm water. With stirring, slowly add 5 mL of 50% ethylenediamine solution. Cool in ice for 15 minutes, add 10 mL ethanol (95%) and put back in the ice bath for 10 minutes. Filter, then wash with two 10 mL portions of ethanol and two 10 mL portions of ether. Air dry, weigh, bottle, and label the blue-purple product.

Comparison of Complexes

Dissolve a small amount (approximately 0.2 g of each) of $Co(en)_3Cl_3$ and $Ni(en)_3Cl_2$ in water in separate test tubes. Add dilute HCl to each. Record and explain your observations.

Auxiliary Experiment

Determine the ionic nature of the cobalt complex you have prepared by following the directions in the Conductivity of Ionic Solutions experiment found in Part III of this manual.

Questions to be Considered

1. Suggest names for the two compounds you have prepared.

2. Write balanced equations for the preparations of the two compounds.

3. How could you distinguish the two salts prepared in this exercise by conductivity measurements?

4. It is claimed that nine different compounds exist having the empirical formula $Co(NH_3)_3(NO_2)_3$. Write structural formulas for six, recognizing that the molecular formula may be twice the empirical formula in some of them.

5. Study the models of the coordinated cation and determine its symmetry. Is this cation an example of an optical isomer? Explain.

Helpful Background Literature

1. Krause, R. A.; Megargle, E. A. *J. Chem. Ed.*, 1976, *53*, 667.

2. Krause, R. A. *J. Chem. Ed.*, 1978, *55*, 453.

ETHYLENEDIAMINE COMPLEXES OF CHROMIUM

Equipment Needed:	drying oven (215°C and 100°C) large mortar and pestle steam bath, if available 8″ test tube air condenser hot plate surface thermometers for hot plate Buchner funnel watch glass
Chemicals Needed:	5 g chromium(III) sulfate hydrate 10 mL ethylenediamine (99%) 20 mL 1% solution of NH_4Cl 30 mL concd. HCl 40 mL ethanol 20 mL diethylether glass wool
Time Needed:	3 hours
Safety Notes:	Do the evaporations in the hood.

Purpose

This preparation is representative of the formation of coordination compounds. The compounds formed illustrate the range of colors and stabilities encountered in coordination compounds.

Introduction

The synthesis of $[Cr(en)_3]_2(SO_4)_3$ will be carried out according to the unbalanced equation:

$$Cr_2(SO_4)_3 \ + \ 6\,(NH_2)C_2H_4(NH_2) \ \rightarrow \ [Cr(en)_3]_2(SO_4)_3$$

The tris(ethylenediamine)chromium(III) sulfate prepared will be converted to the tris(ethylenediamine)chromium(III) chloride hydrate with concd. HCl. This product will then be heated to drive off an ethylenediamine molecule to give *cis*-dichlorobis(ethylenediamine)chromium(III) chloride. If the heating is too vigorous, the material will decompose beyond the desired product. It is important, therefore, to control the heating. The reaction can be written:

$$[Cr(en)_3]Cl_3.3.5\,H_2O \ + \ heat \ \rightarrow \ [Cr(en)_2Cl_2]Cl$$

The discussion which appears in the *Introduction* section to the experiment on Ethylenediamine Complexes of Cobalt and Nickel should be read prior to doing this experiment. Particular attention should be paid to the methods available for determining the structure and composition of the compounds prepared.

Experimental Procedure

Tris(ethylenediamine)chromium(III) Sulfate

Chromium(III) sulfate hydrate, a non-stoichiometric hydrate, is dehydrated by heating for several days in an oven at 100° to 110^oC. After three days, the material is removed from the oven and ground to a powder. It is then put back in the oven for one day for further drying. The chromium(III) sulfate has become anhydrous when the powder no longer dissolves in water.

The anhydrous chromium(III) sulfate will be prepared by the teaching assistants prior to this laboratory.

Five g of the anhydrous chromium(III) sulfate and 10 mL of ethylenediamine (99%) are placed in an eight inch test tube whose mouth is plugged lightly with a wad of glass wool. Select a beaker large enough so that the test tube leans over the side such that the steam will not be passing directly over the mouth of the test tube. Put water in the beaker so that the reaction mixture is covered. Gently boil the water for about an hour or until the sulfate loses its bright green color and its original powdery appearance. During the course of the reaction, shake the test tube to make sure that the unreacted sulfate comes into contact with the ethylenediamine. The product is a brown solid and there should be no liquid visible when the reaction is complete.

The product formed is then placed in a 100^oC oven overnight. The orange-yellow solid which results is then ground to a powder and washed with alcohol. After drying in air, the yield is about 9 g.

Tris(ethylenediamine)chromium(III) chloride hydrate

Make a solution of 1 mL of concd. HCl and 6 mL of water. While this is being warmed to 60° to 65°C on a hot plate equipped with a surface thermometer, weigh out 6 g of the tris(ethylenediamine) sulfate prepared above. Add the solid to the warm aqueous hydrochloric acid solution, and filter the resulting solution rapidly through a small Buchner funnel. Make a solution of 5 mL of concd.HCl and 8 mL of alcohol. Add the filtrate to this solution and cool in an ice bath with stirring. Filter the crude material which crystallizes as yellow needles.

cis-Dichlorobis(ethylenediamine)chromium(III) chloride

Recrystallize the $[Cr(en)_3]Cl_3 \cdot 3.5\ H_2O$ prepared in the section above from a 1 per cent aqueous solution of ammonium chloride. After the recrystallized salt has dried, carefully determine its weight. Since it is to be spread on a watch glass, you may wish to weigh the watch glass and then weigh the watch glass with the recrystallized sample. The solid is then spread in a thin layer and placed in a 210°C oven to carry out the thermal decomposition. Take care that the temperature of the oven does not exceed 215°C because more extensive decomposition occurs at higher temperatures. At temperatures below 200°C, the reaction is too slow. As the ethylenediamine evolution starts, the solid begins to darken and in about an hour will become red-violet. Follow the decomposition by determining the weight loss. Completion to the desired product involves a 30.6 per cent weight loss.

The product which is not entirely pure may be purified somewhat by washing with ice cold concd. HCl. If desired, recrystallization may be carried out by warming 4 mL of water for each gram of the product to 70°C and quickly dissolving the product in this warm water. The resulting solution should be filtered for any impurities and the filtrate cooled in an ice-salt mixture. Add one mL of ice-cold concd. HCl for each gram of the solid being recrystallized. The red-violet needles are filtered in a Buchner funnel and washed with alcohol and ether. The yield should be about 0.45 g for each gram of the material being recrystallized.

Place the product in a tared sample vial and determine the yield. The dried product is sensitive to light and to heat.

Auxiliary Experiment

Confirm how many of the chlorines in the molecule are coordinated and how many are ionic by doing the conductivity experiment.

If you have enough of the sulfate, determine the ionic character of the sulfate by the conductivity experiment.

Determine the number of unpaired electrons in the *cis*-dichlorobis(ethylenediamine)-chromium(III) chloride product by determining the magnetic susceptibility of the compound.

Questions to be Considered

1. Write the structures for both the *cis* and *trans* forms of dichlorobis(ethylenediamine)-chromium(III) chloride.

2. How many unpaired electrons would you expect for the product prepared?

3. How else besides conductivity might you determine how many of the chlorines in the final product are ionic?

4. How can infrared spectroscopy help in characterizing these compounds?

5. Does the final product have a plane or center of symmetry? What about the tris(ethylenediamine)chromium(III) sulfate?

Helpful Background Literature

1. Fernelius, W. C., "Inorganic Syntheses, Vol. II," McGraw-Hill: New York, 1946, pp. 196 – 202.

HEXAAMMINE COBALT(III) CHLORIDE

Equipment needed:	250 mL Buchner funnel safety bottle 125 mL filter flask one-hole rubber stopper 11 – 14 mm standard wall glass tubing aspirator pump watch glass
Chemicals needed:	4 g ammonium chloride 13 mL concd. ammonia 6 g $CoCl_2.6H_2O$ 0.13 g activated carbon or decolorizing charcoal 8 mL concd. HCl ice 20 mL alcohol litmus paper
Time needed:	3 hours
Safety notes:	Carry out the reactions with concd. HCl in the hood.

Purpose

This experiment illustrates the formation of a coordination compound of cobalt. It raises the question of whether or not the chlorine atom is coordinated to cobalt.

Introduction

The preparation of this compound is after that in Walton (1).

The cobalt 2+ ion is more stable than the cobalt 3+ ion for simple salts of cobalt. Only a few salts of Co(III) such as CoF_3 are known. However, complexation stabilizes the higher oxidation state, and a number of very stable octahedrally coordinated complexes of cobalt(III) are known.

The determination as to whether or not the chlorine atom is coordinated or ionic can be determined by gravimetric determination of the chloride precipitated with silver ions. Volumetric determination of the chloride with silver is difficult because the usual indicators do not work. Volumetric determinations have been done using mercury(II) nitrate. Because mercury(II) chloride is only slightly ionized, there are very few mercury(II) ions in solution as long as there are chloride ions present. The excess mercury(II) ions at the end point can be detected by using sodium nitroprusside as an indicator. The mercury(II) nitroprusside which forms from the excess mercury(II) ions is insoluble and separates as a white turbidity.

The cobalt(II) can be oxidized by hydrogen peroxide as illustrated in the chloropentamminecobalt(III) chloride experiment. It can also be oxidized by air when catalyzed by carbon as illustrated in this experiment. The carbon catalyst used in this preparation not only catalyzes the oxidation but also catalyzes the attainment of the equilibrium

$$[Co(NH_3)_5Cl]^{2+} + NH_3 \rightleftharpoons [Co(NH_3)_6]^{3+} + Cl^-$$

which under the conditions of the reaction is greatly in favor of the desired $[Co(NH_3)_6]^{3+}$.

Experimental Procedure

Preparation of Hexaamminecobalt(III) chloride

Mix 6 g of cobalt(II) chloride hexahydrate, $CoCl_2.6H_2O$, and 4 g of ammonium chloride with 5 mL of water and stir until most of the salts have dissolved; then add 13 mL of concd. ammonia solution and 0.13 g of activated carbon or decolorizing charcoal, preferably a good grade taken from a freshly opened bottle. Put the solution into a 125 mL Buchner flask which is fitted with a one-hole rubber stopper carrying a 13 to 14 mm o.d. standard wall tubing (10 mm bore) reaching down to the bottom of the flask. With the aspirator pump attached to the side arm of the flask, draw air through the solution at a brisk rate. There must be a check valve or large empty bottle between the aspirator pump and the reaction flask to prevent water sucking back from the aspirator in the event that the air inlet becomes clogged with crystals.

After an hour or an hour-and-a-half, the color of the solution in the flask should have changed from a reddish color to yellowish brown. The color may easily be observed by swirling the liquid around the sides of the flask while holding it up to a light. Pass air through the solution for 10 minutes after all the red color has gone from the liquid; then filter the solution on a Buchner funnel. The product, mixed with carbon, remains on the filter paper.

Stir the filter cake into 40 mL of water to which has been added 0.5 mL of concd. HCl. This should be enough to give the solution a slight acid reaction after all the solid has been added. Test the solution with litmus; if necessary, add a few more drops of hydrochloric acid. Then heat to 50° to 60°C and filter while hot. The desired product is in the solution. It may be precipitated by using the common-ion effect. Add 8 mL of concd. HCl to the hot solution and set it aside to cool slowly. When it has cooled to room temperature, set the beaker in crushed ice and cool to 0°C. Filter the fine yellow crystals on a Buchner funnel then wash, first with 12 mL of 60 per cent alcohol and then with 12 mL of 95 per cent alcohol. Using the water aspirator, suck as dry as possible; then dry on a watch glass in the oven at 80°C. The yield should be about 85%.

Questions to be Considered

1. What are some ways in which it can be shown that three chlorides are ionic?

Helpful Background Literature

1. Walton, H. F. "Inorganic Preparations," Prentice Hall: Englewood Cliffs, N. J., 1965, pp. 86 – 90.

2. Bjerrum, N. and McReynolds, J. P., *Inorganic Syntheses,* Vol. 2, p. 217.

3. Kolthoff, I. M. and Sandell, E. B. "Textbook of Quantitative Inorganic Analysis," Macmillan: New York, 1943, Chapter XXXV.

CHLOROPENTAAMMINECOBALT(III) CHLORIDE

NITROPENTAAMMINECOBALT(III) CHLORIDE

Equipment needed:	Magnetic stirrer-hot plate medicine dropper sintered glass funnel 250 mL Ehrlenmeyer flask ice bath Buchner funnel Whatman #1 filter paper surface thermometer
Chemicals needed:	5 g ammonium chloride 38 mL concd ammonia 10 g $CoCl_2.6H_2O$ 8 mL 30% H_2O_2 50 mL concd. HCl 25 mL 6M HCl 40 mL 2M HCl 5 g $NaNO_2$ ice 38 mL alcohol litmus paper
Time needed:	3 hours
Safety notes:	Carry out the reactions with concd. HCl in the hood. Wear plastic or rubber gloves when handling the hydrogen peroxide.

Purpose

This experiment illustrates the formation of several coordination compounds of cobalt. It raises the question of whether or not the chlorine atom is coordinated to cobalt or not, and the stability of the nitrito complex vs. the nitro complex with cobalt.

Introduction

The preparation of these compounds is after that in Jolly (1).

The cobalt 2+ ion is more stable than the cobalt 3+ ion for simple salts of cobalt. Only a few salts of Co(III) such as CoF_3 are known. However, complexation stabilizes the higher oxidation state, and a number of very stable octahedrally coordinated complexes of cobalt(III) are known.

The determination as to whether or not the chlorine atom is coordinated or ionic can be determined by gravimetric determination of the chloride precipitated with silver ions. Volumetric determination of the chloride with silver is difficult because the usual indicators do not work. Volumetric determinations have been done using mercury(II) nitrate. Because mercury(II) chloride is only slightly ionized, there are very few mercury(II) ions in solution as long as there are chloride ions present. The excess mercury(II) ions at the end point can be detected by using sodium nitroprusside as an indicator. The mercury(II) nitroprusside which forms with the excess mercury(II) ions is insoluble and separates as a white turbidity.

The equations for the preparation of $[Co(NH_3)_5Cl]Cl_2$ are written:

$$Co^{2+} + NH_4^+ + 1/2\ H_2O_2 \rightarrow [Co(NH_3)_5H_2O]^{3+}$$

$$[Co(NH_3)_5H_2O]^{3+} + 3\ Cl^- \rightarrow [Co(NH_3)_5Cl]Cl_2 + H_2O$$

The equations for the preparation of $[Co(NH_3)_5ONO]Cl_2$ and $[Co(NH_3)_5NO_2]Cl_2$ can be written as follows:

$$[Co(NH_3)_5Cl]^{2+} + H_2O \rightarrow [Co(NH_3)_5H_2O]^{3+} + Cl^-$$

$$[Co(NH_3)_5H_2O]^{3+} + NO_2^- \rightarrow [Co(NH_3)_5ONO]^{2+} + H_2O$$

$$[Co(NH_3)_5ONO]^{2+} \rightarrow [Co(NH_3)_5NO_2]^{2+}$$

Experimental Procedure

Preparation of Chloropentaamminecobalt(III) chloride

Make a solution of 5.0 g of ammonium chloride in 30 mL of concd. aqueous ammonia in a 250 mL Ehrlenmeyer flask. Place this flask on a magnetic stirrer-hot plate; and, while stirring, slowly add 10 g of finely powdered cobalt(II) chloride 6-hydrate. With continued stirring, add dropwise (use a medicine dropper) 8 mL of 30 per cent hydrogen peroxide. A brown slurry will form. When evidence of further reaction has ceased, by either color change or gas evolution, slowly add 30 mL of concd. HCl. At this point, while continuing the stirring, turn on the hot plate and adjust the temperature to about 85°C using a surface thermometer on the top of the hot plate. Heat at this temperature for 20 minutes. The mixture is then cooled to room temperature and the precipitated $[Co(NH_3)_5Cl]Cl_2$ is filtered using a medium porosity sintered glass funnel. Wash the purple crystals with several portions of ice water; the total wash should not exceed 20 mL. The crystals are then washed with 20 mL of cold 6M HCl and dried in an oven at 100°C for two hours. The yield is about 9 g of product.

Preparation of Nitritopentaamminecobalt(III) chloride

Start to heat a solution of 8 mL of concd. aqueous ammonia in 80 mL of water on the stirrer-hot plate used in the previous experiment. The surface temperature of the hot plate is not critical. While heating and stirring this solution, add 5.0 g of $[Co(NH_3)_5Cl]Cl_2$, or 6 g if the wet product is used. (If 5.0 g of the chloropentaamminecobalt(III) chloride was not obtained in the previous experiment, adjust the reactants to the amount you obtained.) Continue heating and stirring until the colored product dissolves. If a dark brown to black precipitate of cobalt oxide forms, filter it off. Cool the filtrate which should be a clear solution to about 10°C. Add 2 M HCl slowly while keeping the solution cold until it is just neutral to litmus. Add 5.0 g of sodium nitrite followed by 5 mL of 6 M HCl. After the solution has been in an ice bath for an hour, filter the precipitated salmon pink crystals of $[Co(NH_3)_5ONO]Cl_2$ using Whatman #1 or equivalent paper and a Buchner funnel. Wash with 25 mL of ice water, wash with 25 mL of alcohol, and then allow it to dry on the lab bench for one hour. The yield is about 4 g. The product is not stable and will slowly isomerize to the nitro compound.

Preparation of Nitropentaamminecobalt(III) chloride

The nitritopentaamminecobalt(III) chloride obtained in the experiment above is isomerized to the nitro compound by heating. The nitrito compound prepared above can be utilized before it is dried. Bring 20 mL of water to a boil, add a few drops of aqueous ammonia, and add 2.0 g of the $[Co(NH_3)_5ONO]Cl_2$. As this solution cools, add 20 mL of concd. HCl. After cooling the solution, the $[Co(NH_3)_5NO_2]Cl_2$ will crystalize from the solution. Filter the product in a Buchner funnel, wash the product with 13 mL of alcohol, and allow it to dry in air for two hours.

Questions to be Considered

1. Write the Lewis structures of the nitro and nitrito ligands.

Helpful Background Literature

1. Jolly, W. L. "The Synthesis and Characterization of Inorganic Compounds," Prentice Hall: Englewood Cliffs, N. J. 1970, pp. 461 – 463.

2. Schlessinger, G. *Inorg. Syn.*, 1967, *9*, 160.

3. Jorgensen, S. M. *Z. anorg. Chem.*, 1894, *5*, 147; *Z. anorg. Chem.*, 1898, *17*, 455.

FERROCENE

Equipment needed:

19/22 1-neck 250 mL round-bottom flask
19/22 Claisen Adapter
19/22 gas inlet adapter or thermometer adapter
glass "T"
19/22 condenser
25 mL 19/22 round-bottom flask
heating mantle for 25 mL round-bottom flask
stirrer-hot plate
small 19/22 dropping funnel
Whatman #1 filter paper
Buchner funnel

If dimer of cyclopentadiene is to be cracked individually:

19/22 condenser for distillation set-up

Chemicals needed:

21.6 mL of 1,2 dimethoxyethane
9 g KOH
1.98 mL cyclopentadiene
2.34 g $FeCl_2.4\ H_2O$
9 mL dimethylsulfoxide
source of dry nitrogen
32.4 mL 6 M HCl
90 g ice

Time needed: 3 hours

Safety notes:

Dimethylsulfoxide is rapidly absorbed through the skin. Although it, in itself, is not highly toxic, it is capable carrying many solutes through the skin into the body. For this reason, rubber gloves should be worn when handling dimethylsulfoxide (DMSO). Gloves can be made from different materials. To be wearing a glove through which DMSO can penetrate is worse than not wearing gloves at all. The Materials Safety Data Sheet for DMSO recommends rubber gloves when handling this material. There is a possibility that

DMSO can cause eye damage. Be sure that safety glasses are worn.

Purpose

This experiment illustrates the formation of an organometallic compound in which the bonding is by way of a pi cloud. The iron is bonded to two cyclopentadienyl moities which are parallel and form a "sandwich" structure.

Introduction

Organometallic compounds can widely vary in bonding and reactivity. The classical organometallic compounds such as Grignard reagents and organolithium compounds are reactive and partially ionic in character. They are polarized with the metal being partially positive and the carbon partially negative. The carbon-lithium bond is estimated to have about 40% ionic character and the carbon-magnesium bond about 35% ionic character. In reactions, these compounds act as if the carbon is negatively charged, i.e., a carbanion, and act as nucleophiles.

Metal atoms can also exist in a coordination sphere. Iron in hemoglobin and magnesium in chlorophyll are examples wherein the metal is coordinated by a cyclic tetradentate ligand called porphyrin which contains four pyrrole-type rings. In hemoglobin, the iron coordinated with the porphyrin is able to coordinate with oxygen and also to a certain extent with carbon dioxide. Thus it acts as an oxygen carrier to enable metabolism in the cells to take place. It also acts to some extent to carry carbon dioxide out to the lungs where it is exhaled. The reason that carbon monoxide and cyanide are such strong toxins is that they also coordinate to the iron in hemoglobin. For example, carbon monoxide coordinates several hundred times more strongly than oxygen. If hemoglobin is largely coordinated with carbon monoxide, it will be unable to carry oxygen to the cells and in effect will have the same result as if homocide were to be carried by blocking off the air supply.

Another type of organometallic compound is represented by ferrocene. In ferrocene, the iron is sandwiched by two cyclopentadienyl anions. The bonding is between iron and the delocalized π-electron cloud of the cyclopentadienyl rings. The cyclopentadienyl rings are planar with the energy associated with the electron delocalization contributing to the energy needed for the compression of the 120° bond angle of the sp^2 hybridized carbon to the 108° required for a planar cyclopentyl ring system.

In general, ferrocene is made by reacting cyclopentadienenyl carbanion with iron(II) chloride. One approach is to react the weakly acidic cyclopentadiene with sodium metal to form the sodium salt of cyclopentadiene and hydrogen. This sodium salt is then reacted with iron(II). Another approach is to use a strong base such as NaOH or KOH to form an equilibrium concentration of the cyclopentadienyl carbanion. This equilibrium mixture is then reacted with the iron(II) ion. A third approach is to use the Grignard, C_5H_5MgBr and to react this with the Fe(II) ion. In this experiment, the cyclopentadiene is reacted with KOH and the equilibrium mixture reacted with $Fe(II)Cl_2$.

Cyclopentadiene forms a dimer by a Diels-Alder reaction. In order for it to form the alkali metal salt, it must be cracked to form the monomer. This is easily done by distilling the dimer. Since the reverse Diels-Alder can be carried out relatively easily, one might think that the equilibrium might be shifted to the monomer by reacting the dimer directly with the KOH. However, this does not happen.

At room temperatures, the freshly prepared monomer will remain useful for 45 minutes after cracking. If kept cold, it will last for three to four hours. If kept at dry ice temperature, the monomer is stable indefinitely.

The yield can be improved by using purified iron(II) chloride. The commercially available iron(II) chloride can be purified by a Soxhlet extraction with tetrahydrofuran (THF). The THF complexed iron(II) chloride ($FeCl_2$.2 THF loses THF in air) is then used to react with KOH. Iron(II) sulfide, if pure, gives a better yield than iron(II) chloride. The product formed from the iron(II) sulfide is easier to clean up than that formed from iron(II) chloride. Technical grade iron(II) sulfide does not give good yields of ferrocene.

Other than the recrystallization indicated in the experiment below, the ferrocene can be nicely cleaned by passing the crude product through an alumina column using hexane as the solvent. The reason recrystallization from methanol is suggested is the disposal problem of large amounts of hexane that would be generated by the whole class.

A word here on the activity of alumina might be appropriate. The alumina is rated to be activity grades I through V. Alumina activity can vary depending on whether it is acid washed or base washed, but one of the main causes of loss of activity is the absorption of water. If the activity of alumina is uncertain, it can be activated by heating to 360°C for five hours. The length of heating time and temperature are dependent on the amount of alumina to be activated and temperature/time variables. For example, a small amount of alumina can be activated by heating in a drying oven at 110°C for a sufficiently long time, e.g., overnight. The activity grade can be determined using Brockman and Schodder's method employing azo dyes. The azo dyes used are *p*-hydroxyazobenzene, *p*-aminoazobenzene, Sudan red, Sudan yellow, *p*-methoxyazobenzene, and azobenzene. They are listed in the order of most strongly absorbing to least strongly absorbing. The alumina is put in a test column 5 cm long and 1.5 cm in diameter. Two mg each of *p*-methoxyazobenzene and azobenzene are dissolved in 2 mL of benzene and placed on the top of the column. This is followed by 8 mL of light petroleum and 20 mL of a 4 : 1 mixture of light petroleum : benzene mixture. Activity I alumina will absorb azobenzene at the bottom of the column and *p*-methoxyazobenzene at the top of the column. Activity II alumina will absorb *p*-methoxyazobenzene at the bottom of the column and Sudan yellow at the top of the column. It will not absorb azobenzene. The other activity classes are determined accordingly.

Experimental Procedure

If the dicyclopentadiene is not cracked for you and stored in dry ice, it will have to be cracked prior to use. Place about 10 mL of the dicyclopentadiene in the 25 mL round-bottom flask and set up the glassware for distillation. (See Fig. 1.) Distill out the needed amount of cyclopentadiene. Use the freshly distilled cyclopentadiene within about 45 minutes of preparation.

Assemble a 250 mL 19/22 round-bottom flask with a Claisen adapter. Fit the curved side-arm of the Claisen adapter with an outlet adapter or a 19/22 joint with a 7 or 8 mm tube. (An adapter can be devised by inserting a glass tube into the thermometer adapter with a short rubber sleeve.) The rubber hose carrying the nitrogen can then be attached via either the gas inlet adapter or the thermometer adapter modified to enable the nitrogen tube to be attached. A glass "T" is inserted into the nitrogen line which leads to a tube dipping into oil which acts as a pressure relief valve. The experimental set-up is shown in Fig. 1.

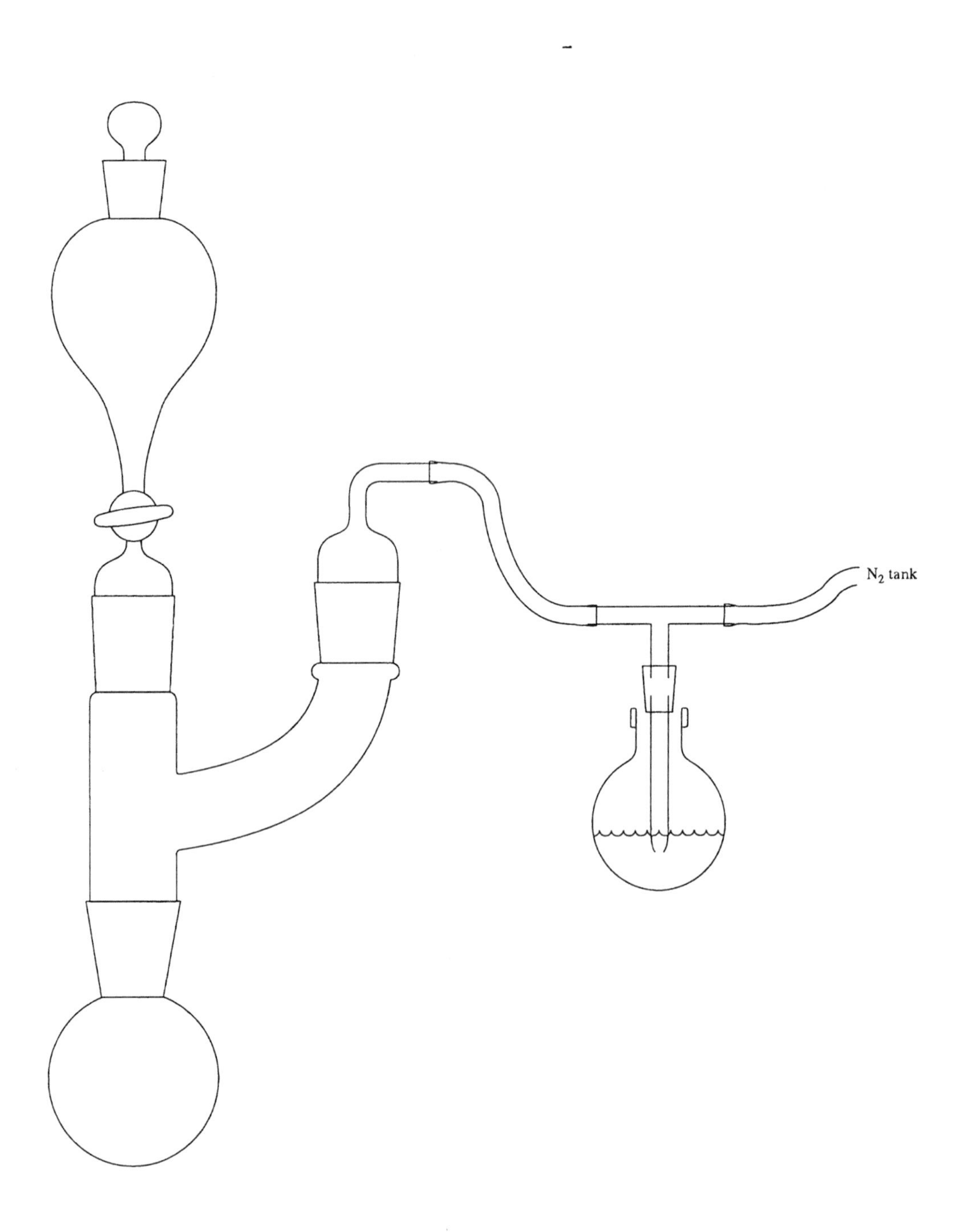

Fig. 1.

Place 22 mL of 1,2-dimethoxyethane (approx 0.2 moles) into the flask. Add 9 g KOH (approx. 0.16 moles). Stir this mixture while flushing with nitrogen. Add 2.0 mL of freshly distilled cyclopentadiene (0.024 moles) to this mixture. Continue flushing with a slow stream of nitrogen. Cyclopentadiene boils at 41.5°C so it is important that it react with the base before it is lost in a rapid nitrogen stream. Stir for 5 – 10 minutes until the reaction mixture turns black.

Place the 125 mL separatory funnel on the joint of the Claisen adapter directly above the flask. Have the stopcock open so that the separatory funnel is flushed with nitrogen. When most of the air in the separatory funnel is replaced with nitrogen, add a solution of $FeCl_2 4 H_2O$ in dimethylsulfoxide. This solution is made by dissolving 2.34 g of $FeCl_2.4 H_2O$ (0.012 moles) in 9 mL of DMSO. Immediately place a plug of cotton in the mouth of the separatory funnel to minimize the diffusion of air into the funnel. Stir the mixture in the flask vigorously for several minutes before starting the dropwise addition of the iron(II) chloride solution. After the addition of the iron(II) chloride solution, stir the reaction mixture for another 10 minutes.

At this point, the flow of nitrogen can be stopped. The reaction mixture is added to a mixture of 32.4 mL of 6 molar HCl and 90 g of ice. Stir the slurry for about 10 minutes and filter the precipitate using a Buchner funnel with a Whatman #1 filter paper. Wash the precipitate with four 4.5 mL portions of water. The product obtained is then dried and refluxed with a small amount, 10 mL in the 25 mL flask, of methanol. Any solid not dissolved in methanol is filtered out and the filtrate allowed to cool to room temperature. If the ferrocene does not crystallize out in a short time, it should be left until the next laboratory session.

A method of obtaining pure ferrocene is to pass a hexane solution through an alumina column. Reflux the crude ferrocene using 15 mL of hexane. Decant the hexane solution onto a column of alumina about 15 mm X 10 cm. Elute using hexane. Distill off the excess hexane and allow the ferrocene to crystallize. If the hexane is allowed to evaporate slowly, large crystals will form. The alumina column efficiently holds up the impurities of the synthesis. The disadvantage to this purification process is the relatively large amounts of hexane that will need to be disposed of.

Questions to be Considered

1. Draw a structure of ferrocene.

2. Determine the percentage yield you obtained in this experiment.

3. Why is hexane used for the chromotography instead of methanol dimethylsulfoxide?

4. Write the Diels-Alder reaction for the dimerization of cyclopentadiene.

5. Write the reactions for the three different procedures for producing ferrocene discussed above.

6. Name two other possible sources for the iron(II) ion which might be used in the ferrocene preparation.

Helpful Background Literature

1. Jolly, W. L. "Synthetic Inorganic Chemistry," Prentice-Hall: Englewood Cliffs, New Jersey, 1960, pp. 173 – 176.

2. Cotton, F. A.; Wilkinson, G.; Gaus, P. L. "Basic Inorganic Chemistry, Third ed.", John Wiley & Sons: New York, 1995, p. 689.

3. Shriver, D. F.; Atkins, P. W.; Langford, C. H. "Inorganic Chemistry," W. H. Freeman: New York, 1990, pp. 523 – 525.

4. Bochmann, M. "Organometallics 2, complexes with Transition Metal-Carbon π-Bonds," Oxford University Press: Oxford U. K., 1994, pp. 46 – 47.

PHASE DIAGRAM OF AN ALLOY

Equipment Needed:	5 cup furnaces or small tube furnaces 15 mL capacity crucible (e.g., Coors 60105) 6 or 7 mm borosilicate glass tube thermocouple readout devices thermocouples stopwatches
Chemicals Needed:	lead shot tin, mossy or granules carbon powder
Time Needed:	2 hours
Safety Notes:	Once an alloy of lead and tin is prepared and analyzed, it can be kept and used again. If it is to be disposed of, it should be done in a manner approved by your institution. Lead is a heavy metal.

Purpose

This experiment illustrates a method by which a portion of a simple phase diagram can be determined. It also serves as an introduction to the concept of the phases which can exist in simple binary alloys.

Introduction

Pure metals exhibit structures which are examples of sphere-packing. The three structures which are common are cubic close-packing, hexagonal close-packing, and body-centered cubic. Copper, silver, gold, palladium, platinum, rhodium, and iridium exhibit cubic close-packing. Beryllium, magnesium, technetium, ruthenium, and osmium represent metals in the hexagonal close-packed arrangement. Those with body-centered cubic structure are potassium, rubidium, cesium, barium, vanadium, niobium, tantalum, chromium and tungsten. Other metals exhibit several different structures. For example, iron at room temperature is body-centered cubic and is called α-iron. At 910°C, the α-iron undergoes a phase transition to γ-iron which is cubic close-packed. At 1390°C the γ-iron undergoes another phase transition to the δ-iron which is body-centered cubic but with different lattice parameters than for the body-centered cubic α-iron. Iron exists as the δ-iron from 1390°C to its melting point at 1535°C. Calcium is cubic close-packed at room temperatures. At 448°C, calcium undergoes transformation to the body-centered cubic structure. There are a number of reports that calcium exhibits a hexagonal close-packed structure at intermediate temperatures, but this phase seems to be catalyzed by hydrogen impurities. The fact that many of the metals undergo these phase transitions indicates that the energy needed for transformation from one type of packing to another is not large. Models indicate that only small shifts of atoms are needed to transform one type of packing to another.

The two metals studied in this experiment are lead and tin. Lead is cubic close-packed. Tin, on the other hand, shows two important polymorphs. α-tin, also called gray tin, is stable between 18°C and -130°C. It is non-metallic and has the diamond structure. β-tin, or white tin, is stable above 18°C and behaves as a metal. Each tin atom in white tin consists of four nearest neighbors in a flattened tetrahedra with two other tin atoms above and below so that each tin atom exists within a distorted octahedra of its neighboring tin atoms. The unit cell is tetragonal with a = 5.831 and c = 3.181 angstroms. The conversion from the α form to the β form and vice versa is slow. However, at times, tin disease (the conversion of metallic β-tin to grey tin) has undesirable consequences. For example, the lead/tin organ pipes in some European cathedrals have exhibited tin disease brought on by the cold winters.

When two or more metals are mixed, a number of added considerations need to be made. Two of the more important are relative sizes of the two atoms and the electronic structure of the metals.

Importance of Size

Similar metals form solid solutions. Examples are potassium-rubidium, silver-gold, and molybdenum-tungsten. It is not surprising that two metal atoms of similar size and electronic structure can substitute for one another in sphere-packing. If two elements are very different in size, then an interstitial solid solution can be formed. In titanium carbide and titanium nitride, all of the octahedral holes in cubic close-packed titaniums are filled with carbon or nitrogen. In titanium hydride, TiH_2, all of the tetrahedral holes are filled with hydrogen. In ZrH, one half of the tetrahedral holes are filled with hydrogen in cubic close-packed zirconium. This then is analogous to the zinc blende structure.

Questions might be raised about the relative size of spheres if the elements were to be considered ionic. However, they are not. If the difference in electronegativities of the elements are large so that ionic nature is introduced, e.g., metals and fluorides, then interstitial compounds do not form. For interstitial compounds, such as carbides, the six octahedral bonds from the non-metal atom are visualized as being 1/2 or 2/3 bonds with equivalence being achieved by resonance. Bonding, however, must be considerable in order to account for the hardness and high melting point of interstitial compounds. At the same time, they retain the metallic properties of luster and electrical conductivity.

A system of technological importance is iron and carbon. Due to phase changes that pure iron undergoes, the system is complex. However, parts of it involve the formation of interstitial carbon-iron systems.

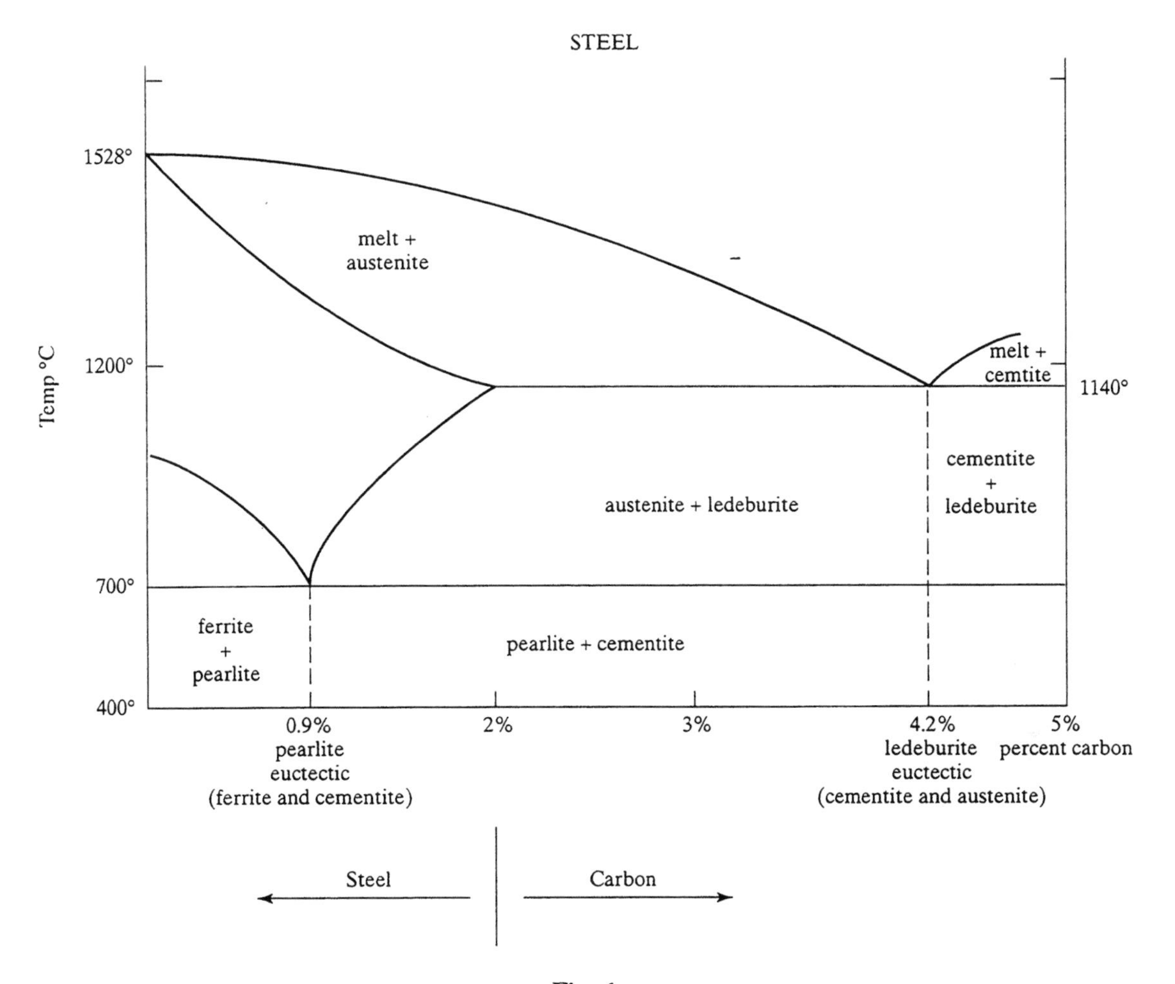

Fig. 1.

In the phase diagram for the carbon-iron system, it is indicated that iron with 2% or less carbon is considered to be steel and when the carbon content is greater than 2%, cast iron is the product. Austenite is the name given to a solid solution of carbon in γ-iron. In austenite, the carbon occupies part of the octahedral holes in a cubic close-packed iron structure. When the carbon content is 6.68%, the compound cementite, Fe_3C, is formed. Ledeburite is the name given to the eutectic formed by cementite and austenite and contains 4.2% carbon. Ferrite is the name given to nearly pure α-iron which contains about 0.06% of carbon forming an interstitial solid solution. A eutectic formed by ferrite and cementite is called pearlite and contains 0.9% carbon.

Importance of Electronic Structure

An example of the influence of the electronic structure is illustrated by the formation of electron compounds (1, p. 1310). In Table I, the formation of alloy phases governed by the electron to atom ratio are listed.

Table I

Relation between electron : atom ratio and crystal structure

Electron : atom ratio 3 : 2		Electron : atom ratio 21 : 13	Electron : atom ratio 7 : 4
β body-centered	β-Mn cubic structure	$\bar{\gamma}$ brass structure	ε close-packed hexagonal structure
CuBe	Ag_3Al	Cu_5Zn_8	$CuZn_3$
CuZn	Au_3Al	Cu_9Al_4	Cu_3Sn
Cu_3Al	Cu_5Si	Fe_5Zn_{21}	$AgZn_3$
Cu_5Sn	$CoZn_3$	Ni_5Cd_{21}	Ag_5Al_3
CoAl		$Cu_{31}Sn_8$	Au_3Sn
		$Na_{31}Pb_8$	

Hume-Rothery pointed out that the formulae for these electron compounds could be explained by dividing the sum of the valence electrons by the atoms involved. For example:

CuBe

Cu = 1 valence electron
Be = 2 valence electrons
CuBe = 2 atoms
3/2 = electron to atom ratio of 3 : 2

Cu_5Sn

Cu = 1 valence electron x 5 = 5
Sn = 4 valence electrons = 4
Cu_5Sn = 9 valence electrons
Cu_5Sn = 6 atoms
9/6 = electron to atom ratio 3 : 2

Ag_3Al

Ag = 1 valence electron x 3 = 3
Al = 3 valence electrons = 3
Ag_3Al = 6 valence electrons
Ag_3Al = 4 atoms
6/4 = electron to atom ratio 3 : 2

Cu_5Zn_8

Cu = 1 valence electron x 5 = 5
Zn = 2 valence electrons x 8 = 16
Cu_5Zn_8 = 21 valence electrons
Cu_5Zn_8 = 13 atoms
21/13 = electron to atom ratio 21 : 13

Ag_5Al_3

Ag = 1 valence electron x 5 = 5
Al = 3 valence electrons x 3 = 9
Ag_5Al_3 = 14 electrons
Ag_5Al_3 = 8 atoms
14/8 = electron to atom ratio 7 : 4

Other Systems

There are other alloy systems which show other types of phase changes. For example, in β-brass, the copper and zinc atoms occupy certain of the positions in a body-centered cubic lattice in a random way.

Phase diagrams for some systems can be quite complex. Bronze, an alloy formed from copper and tin, has a complex phase diagram. Bronze is an ancient alloy, the Bronze Age historically preceeded the Iron Age. The ancients obviously did not have the phase diagram, but they undoubtedly knew about some of the phases from experience obtained in formulating the alloy with different ratios of copper and tin to obtain alloys with different physical properties. Brass, an alloy of coppper and zinc, also has a complex phase diagram. Brass is valued as an easily machinable, slowly corroding metal. The phase diagrams for bronze and brass are shown.

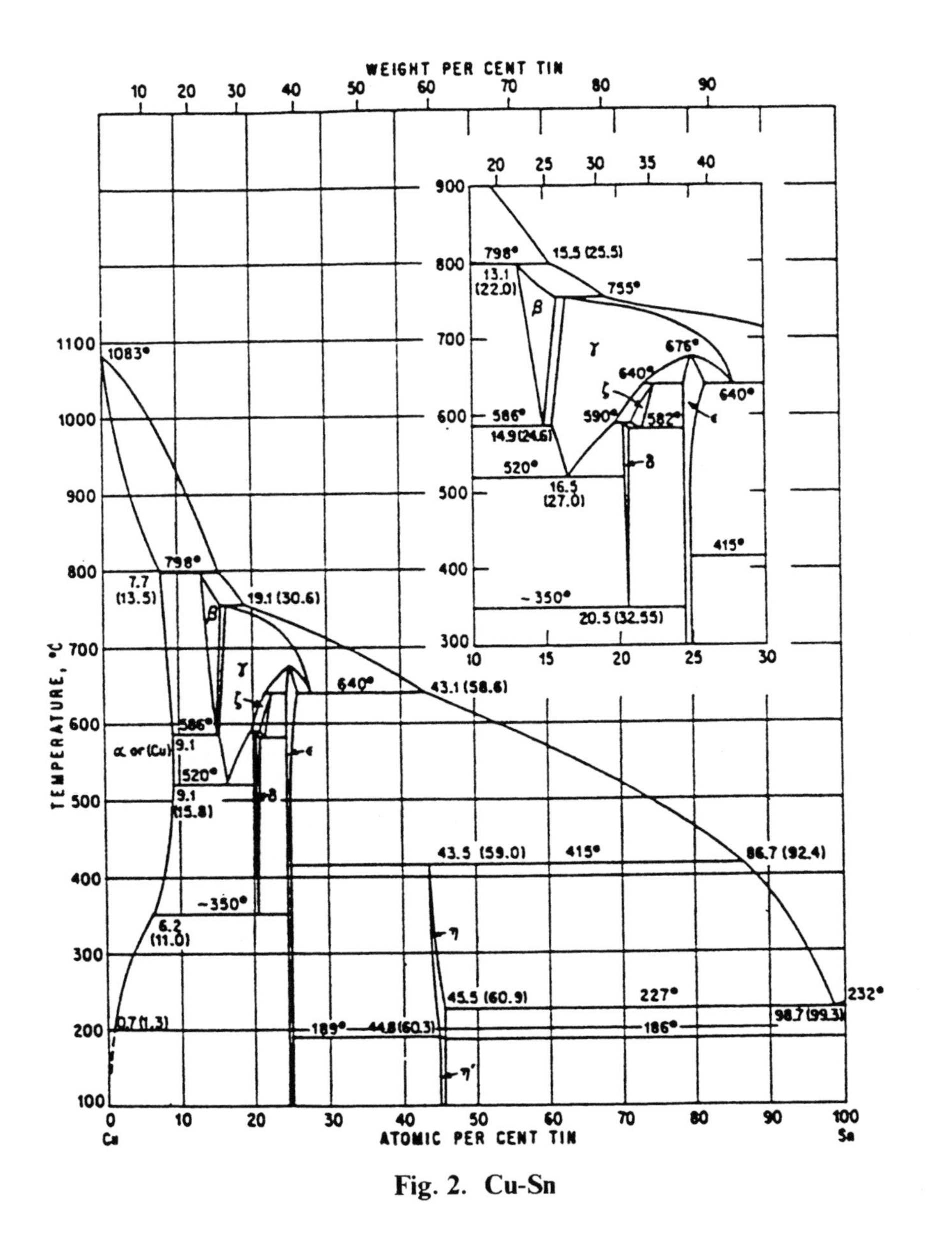

Fig. 2. Cu-Sn

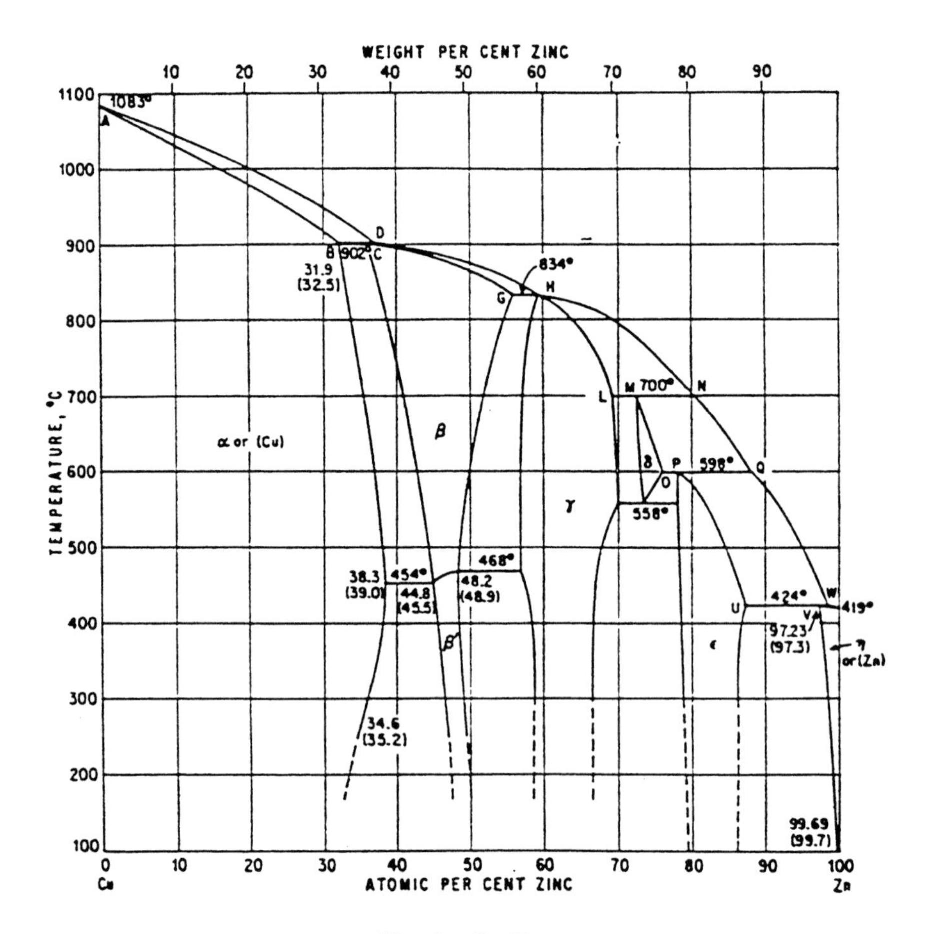

Fig. 3. Cu-Zn

Phase diagrams indicate liquidus and solidus curves. The liquidus curves indicate the temperature and composition when a solid phase is in equilibrium with the liquid phase. The liquidus curve will be determined in the experimental procedure described below. In effect, the liquidus curve can be derived from ideal solution theory. The curve descending from the melting point of lead is the freezing point of lead depressed with varying amounts of tin. The curve descending from the melting point of tin is the freezing point of tin depressed with varying amounts of lead. At the eutectic point, the phase rule dictates that the composition cannot vary. Remember that the phase rule states that $F = C - P + 2$, that is, the degrees of freedom equal the number of components minus the number of phases plus 2. Anywhere on the liquidus curve, the component is 2 (lead and tin), the phase is 2 (solid and liquid), and the degrees of freedom are equal to $2 - 2 + 2 = 2$. Thus, there are two degrees of freedom, the composition or the temperature and pressure. At the eutectic, there are two solid solution phases and the liquid phase with $F = 2 - 3 + 2 = 1$, or one degree of freedom, that of pressure. Since the experiments are usually carried out at atmospheric pressure, there are no degrees of freedom observed experimentally. Thus, if a liquid metal with the composition of the eutectic is cooled down, the composition of the solid which forms cannot change.

For the lead-tin alloy phase diagram (Fig. 4.), the solidus curve gives information on how much lead can be dissolved in solid tin and how much tin can be dissolved in solid lead. The solidus curve is determined by measurements of physical properties such as resistivities, tensile strength, initial melting, and microscopic analysis.

Note the liquidus and solidus curves in the phase diagram of the lead-tin system shown below.

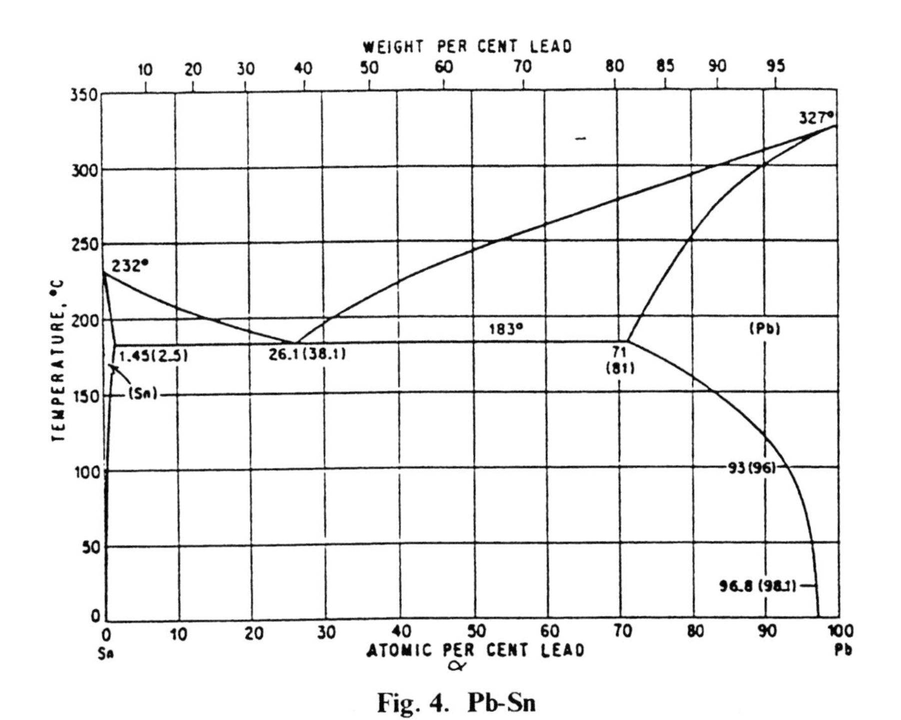

Fig. 4. Pb-Sn

Experimental Procedure

At least five alloy samples with different weight percentages of lead and tin in 15 mL porcelain crucibles will be available in the laboratory. These alloy samples will be heated in the cup furnaces or in the tube furnaces which are placed endwise with a plug so they can be used, in effect, as cup furnaces. An appropriate borosilicate glass tube with the end sealed flat can be used as a plug.

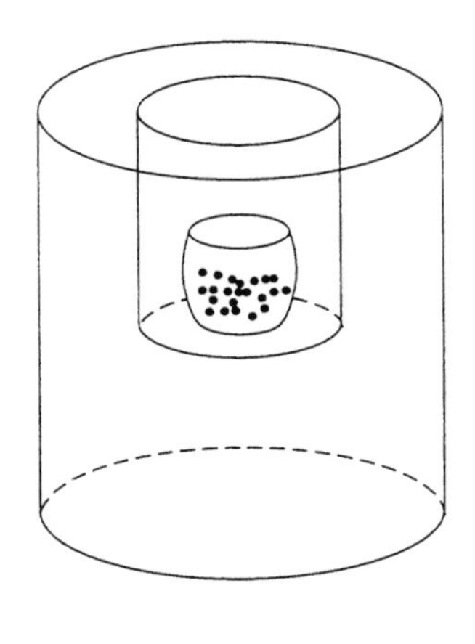

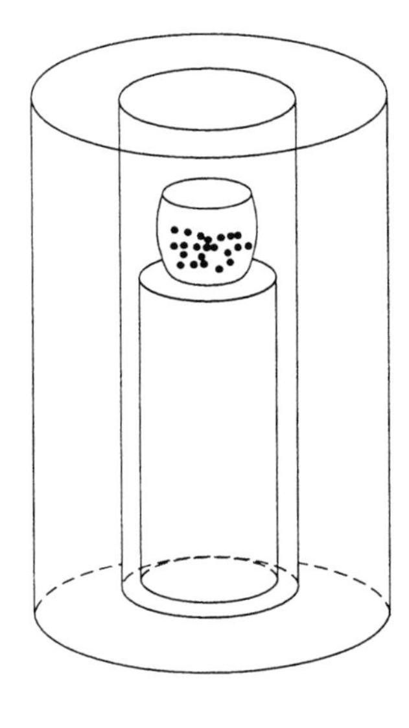

Fig. 5.

A 6 or 7 mm borosilicate glass tube with the tip pulled into a thin point is installed with a thermocouple. The thermocouple junction which is in the thin point is dipped into the molten alloy. By encasing the thermocouple junction in thin glass, electrical isolation of the thermocouple from the molten metal can be effected with a minimum of thermal insulation.

Students in groups of two or three should determine the cooling curve of at least two alloy samples.

In order to prevent oxidation of the alloy, lightly cover the surface of the alloy with carbon powder. If the alloy has been prepared for you, there will already be some carbon added. Check to see that there is black powder in the sample. The instructor will instruct the class on the variac setting needed in order to get a furnace temperature of 350°C or somewhat above. Heat the sample to this temperature and make sure that the sample is molten. Carefully stir the molten sample using the thermocouple protective tube as a stirring rod. The furnace can then be turned off and the temperature determined as a function of time. A reading every 30 seconds will be sufficient with a reading every minute being adequate when the temperature change vs. time elapsed is more or less constant. At the beginning, the liquid alloy will start to cool. At the point when a solid phase starts to separate, an inflection in the cooling curve will be detected. As the solid phase continues to separate, the rate of cooling will be observed to be different from that of the cooling of the liquid phase. When the eutectic point is reached, the temperature will remain constant until all of the eutectic separates as a solid. After this, the cooling curve indicates the cooling of the solid alloy.

When the different groups in the class have finished determining the cooling curves of the alloys, the groups having done the same alloy system should compare the temperature obtained when the solid phase first starts to separate. This will give some idea as to the reproducibility of the data. The data for all of the groups should then be shared and plots of the temperature-composition curve be determined. Unless one of the alloy samples happens to be exactly at the eutectic composition, the experimental data obtained will give the eutectic temperature, but not the composition of the eutectic. With enough points on the temperature-composition curve, the composition of the eutectic can be obtained by extrapolation. If, however, one of the alloy samples has the exact eutectic composition, then the cooling curve will indicate only the eutectic inflection, and the composition of the alloy with this cooling characteristic is that of the eutectic.

If a strip chart recorder or an x-y recorder is available, this can be connected to the thermocouple and the data will be collected automatically.

Questions to be Considered

1. How do the experimental data obtained in the laboratory compare to that recorded in the literature? (Refer to Fig. 4.)

2. In the phase diagram shown for lead-tin (Fig. 4.), indicate the phases present in each of the areas in the phase diagram.

3. On the phase diagram for lead-tin, identify the liquidus curve and the solidus curves.

4. It turns out that an alloy of a given composition is not easily prepared by weighing the required amounts of lead and tin. The surface effects, probably due to an oxide layer, prevent easy formation of a homogeneous melt. The composition of a melt is,

therefore, best determined by analysis. Suggest how one might analytically determine the composition of a lead-tin alloy.

5. Austenite is represented as a face-centered unit cell with carbons occupying some of the positions along the cell edge. Rationalize this unit cell structure with the statement that austenite is γ-iron with some of the octahedral holes occupied by carbon.

6. When austenite is quenched by rapid cooling, some of the structure is converted to martensite. Martensite is a body-centered structure of iron atoms with some of the face positions occupied by carbons. Because this cell becomes tetragonal, an internal strain is placed on the material thereby hardening it. Draw the unit cell of martensite indicating the possible positions of carbon atoms.

Helpful Background Literature

1. Wells, A. F. "Structural Inorganic Chemistry", 5th ed., Clarendon Press: Oxford, 1984, pp. 1274 – 1323.

2. Hansen, M.; Anderko, K. "Constitution of Binary Alloys", McGraw-Hill: New York, 1958.

3. Massalski, T. B.; Okamoto, H.; Subramanian, P. R.; Kacprzak, L "Binary Alloy Phase Diagrams" 2nd ed. The Materials Information Society: Materials Park, OH, 1990.

4. Villars, P.; Prince, A.; Okamoto, H. "Handbook of Ternary Alloy Phase Diagrams" ASM: Materials Park, OH, 1990.

METAL COMPLEX USED FOR DNA CLEAVAGE

Equipment needed:

Part A
heating mantle for 100 mL round-bottom flask
magnetic stirrer or stirrer-hot plate
stirring bar
19/22 organic glassware kit
Buchner funnel
Pasteur pipette
3 mL disposable syringe
reflux condenser

Part B
250 mL Ehrlenmeyer flask
magnetic stirrer or stirrer-hot plate
stirring bar
Pasteur pipette
Buchner funnel
Whatman #50 filter paper
pH meter or pH paper

Part C
Spectronic 20
3 mL disposable syringe
4 inch test tube
stopwatch

Chemicals needed:

Part A
3.5 g $Cu(NO_3)_2 \cdot 2.5\ H_2O$
0.75 g glycine
2.1 mL triethylamine
0.36 mL nitroethane
0.75 mL 35% formaldehyde
40 mL methyl alcohol

Part B
3.27 g zinc dust
15 mL concd. HCL
3.5 g $Cu(NO_3)_2 \cdot 2.5\ H_2O$

Part C
rhodamine B
Sodium ascorbate
Hydrogen peroxide, 3%

Time needed:

Part A is 3 hours
Part B is 1.5 hours + time to crystallize the product
Part C is 45 minutes for preparation and
1/2 hour for spectra

Safety notes:

Be careful of fire hazards when heating organic liquids. Avoid breathing the zinc dust. Do all reactions in a hood. Be careful with Rhodamine B. It easily dyes skin, benchtop, clothing, and most things with which it comes into contact.

Purpose

This experiment indicates one way in which inorganic chemistry contributes to biological chemistry. The synthesis portion illustrates the formation of a complex with an amino acid as a ligand. The glycine complex formed can cause nicking of DNA in vitro by producing hydroxy radicals. The assay of hydroxy radical production by rhodamine B degradation is illustrated.

Introduction

Damage to DNA can cause cell death as well as pathology in living organisms. Thus the study of DNA damage is necessary in assessing the role of repair mechanisms as well as in the search for chemotherapeutic agents.

A natural product that causes DNA scission is a group of compounds called bleomycins. Bleomycins can have as one part of the molecule a complex of iron or copper. It causes DNA double-strand cleavage in four steps: (1) binding and induction of damage at intact DNA surface, (2) binding at the new damaged site, (3) reactivation, and (4) complementary strand scission.

Because of the effectiveness of bleomycins in causing double strand cleavage, attempts have been made to find simpler synthetic analogs to accomplish the same type of cleavage. A number have been reported. One of these reports, (1), shows that a copper amino acid complex can produce nonrandom double-strand cleavage of DNA. This compound is ((2S,8R)-5-amino-2,8-dibenzyl-5-methyl-3,7-diazanonanedioato) copper(II), shown below as compound 1. It is a phenylalanine complex of copper(II) modified with a *tert*-butylamine group forming a bridge between the two alpha amino groups of the amino acid ligands. With the amine group converted to an ammonium ion, this compound can electrostatically interact with the negatively charged DNA backbone. It is thought that the hydrophobic benzyl groups interact with the hydrophobic interior of DNA. In addition to the binding, this complex can generate hydroxyl radicals which damage DNA by hydrogen abstraction from the sugar moities.

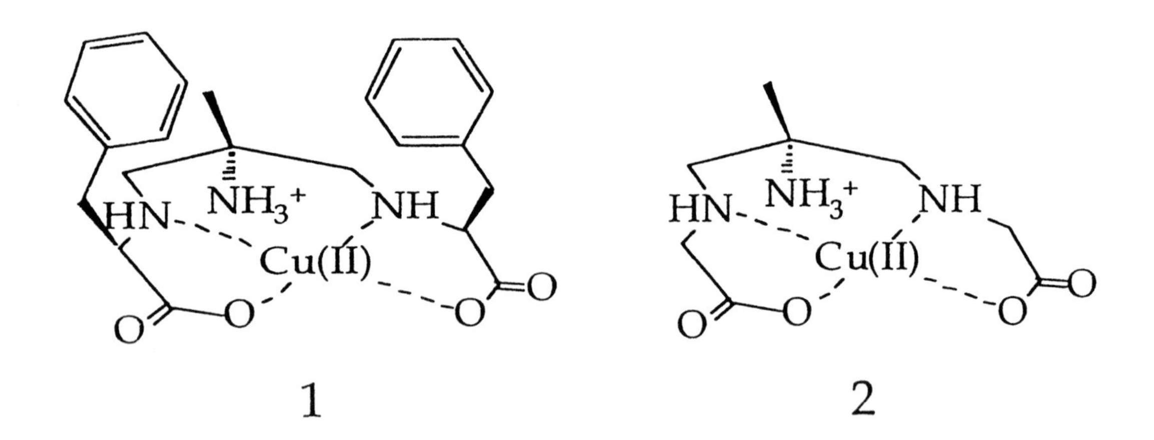

Fig. 1. Structures of 1 and 2 (note an apical coordinated H_2O has been omitted for clarity

The glycine complexed analog (compound 2) of the phenylalanine copper complex is to be synthesized in this laboratory primarily because glycine is less expensive than phenylalanine. It also serves as a model compound to the phenylalanine complex in that it too can generate hydroxyl radicals and the comparison of the two complexes with DNA can elucidate aspects of the cleavage mechanism. For example, unlike the phenylalanine complex, the glycine complex is not able to effect double-strand cleavage of the DNA. The fact that it can only do single-strand cleavage is ascribed to the lack of hydrophobic groups, found in the phenylalanine complex, which was pointed out to form additional interactions with the hydrophobic interior of DNA.

The interest in double-strand cleavage is that it is probably more difficult to repair than single-strand cleavage. In single-strand cleavage, the complementary sequence information is still present. In double-strand cleavage, the complementary base pair blueprint is lacking and the strands are no longer proximate to each other. Natural products such as bleomycins and synthetic cleavage agents such as the phenylalanine complexed copper are thus able to damage DNA more profoundly than material which only causes single-strand cleavage.

There are at least two possible mechanisms for the generation of hydroxyl radicals by the copper complexes. One is the Fenton mechanism and the other is the Haber-Weiss mechanism. In the Fenton mechanism, the Cu(II) is first reduced to Cu(I) which reacts with hydrogen peroxide in a redox reaction to produce the hydroxy radical.

$$Cu^{II}(L) + e^{-} \rightarrow Cu^{I}(L)$$

$$Cu^{I}(L) + H_2O_2 \rightarrow Cu^{II}(L) + OH^{-} + \bullet OH$$

In the Haber-Weiss pathway, Cu(I) reacts with dioxygen to form the superoxide anion and Cu(II). The superoxide anion then reacts with hydrogen peroxide to form the hydroxyl radical.

$$Cu^{I}(L) + O_2 \rightleftarrows Cu^{II}(L) + O_2^-$$

$$O_2^- + H_2O_2 \rightarrow O_2 + OH^- + \bullet OH$$

Both the Fenton and Haber-Weiss mechanisms involve the Cu(II)/Cu(I) couple. However, the Haber-Weiss chemistry involves both molecular oxygen and the superoxide anion; the Fenton mechanism requires neither. Studies carried out (1) have shown that the Haber-Weiss mechanism is involved in the generation of $\bullet OH$ using the phenylalanine complexed reagent.

Since the above two mechanisms both involve reactions with hydrogen peroxide, some rationale is needed for the formation of this compound. There is evidence that hydrogen peroxide is formed in living systems because of the existence of enzymes such as catalases that decompose peroxides. In a simplistic way, a possible mechanism for the formation of hydrogen peroxide would initially involve a redox process of a metal in a complex that would lead to the formation of superoxide ion from dioxygen. (See the first reaction in the Haber-Weiss mechanism.) The superoxide can then dismute to form dioxygen and hydrogen peroxide.

$$2\,O_2^- + 2\,H^+ \rightleftarrows O_2 + H_2O_2$$

Because hydroxyl radicals react rapidly with the dye, Rhodamine B, the diminishing of the strong absorption peak at 553 nm of Rhodamine B can be used to measure the formation of hydroxyl radicals. As the hydroxyl radical forms, it will react with Rhodamine B. Thus the decrease in the Rhodamine B concentration is a measure of the hydroxyl radical formation.

Ascorbic acid, or vitamin C, is easily oxidized leading to its use as an antioxidant. In addition to converting copper(II) to copper(I), ascorbate can also react to accelerate the reaction of the superoxide ion to hydrogen peroxide.

$$ascH_2 + O_2^- \rightleftarrows asc^- + H_2O_2$$

It should be noted that the above reaction and the metal complexes discussed in this experiment are not the only entities capable of generating hydrogen peroxide in a biological system. A review article by Halliwell and Gutteridge (8) outline a number of ways in which a redox reaction can take place to generate hydrogen peroxide.

Experimental Procedure

A. Nitro Substituted Glycine Complex of Copper

The compound 5-nitro-5-methyl-3,7-diazanonanedioato)copper(II) is prepared in the first step. Add 1.16 g (5 mmol) of $Cu(NO_3)_2 \cdot 2.5 H_2O$ to a 100 mL round-bottom flask. To the copper nitrate solid, add 0.75 g (10 mmol) glycine in 30 mL methyl alcohol. Attach a reflux condenser to the flask and heat until reflux starts. Using a disposable syringe or a Mohr pipet, add 2.1 mL (15 mmol) of triethylamine through the reflux condenser. A light purple color will form. To this solution, add 0.36 mL (5 mmol) nitroethane and reflux for another five minutes. To the hot solution, slowly add using a Pasteur pipette a solution of formaldehyde in methyl alcohol. This solution is made by mixing 0.75 mL (10 mmol) of 35% formaldehyde in 5 mL of methyl alcohol. At this point, the refluxing is reinitiated and continued for two hours. The solid product which forms is filtered and dried.

B. Reduction of the Nitro Group to the Amine

A 250 mL Ehrlenmeyer flask is charged with 10 mL (10% V/V) HCl and 15 mL water. Zinc dust, (3.27 g, 50 mmol) is added and the copper complex prepared in the first part of this experiment is added (1.56 g, 5 mmol). A Pasteur pipette can be used to wash down the sides of the flask. Gently swirl the flask to get a complete "wetting" of the zinc dust and the complex. This reaction mixture is then heated in a water bath at $60^o - 65^oC$ for 30 minutes. The solution is filtered to remove the Cu^o, and following this the pH is raised to about 12 using concd. NaOH solution. Use pH paper to monitor the pH. The $Zn(OH)_2$ which forms at this pH is then filtered. In order to filter the fine $Zn(OH)_2$ precipitate, a Whatman #50 filter paper is recommended. The pH is now lowered to 9 by addition of small amounts of concd. HCl. Copper is now reintroduced to reform the complex. 2.32 g (10 mmol) of $Cu(NO_3)_2 \cdot 2.5 H_2O$ is added, and the pH is lowered to about 3 using concd. HCl. At this point, the solution should be a rich purple/blue. If a fine microcrystalline precipitate does not form in a short time, put the solution away for several days. The precipitate forms slowly at times. The precipitate is then filtered and washed with methyl alcohol.

C. Formation and Detection of the Hydroxyl Radical

Make up a stock solution of Rhodamine B by dissolving 17.2 mg of the dye in a liter of water. Using a disposable syringe or a Mohr pipet, remove 1 mL of the Rhodamine B solution and place it in a 12 mm x 100 mm test tube used as a cell in the Spectronic 20.

The hydrochloride of the glycine complex contains two waters of hydration and has a molecular weight of 353. Dissolve 0.01059 grams of the amine complex from Part B in 50 mL of water. Place one mL of this solution in the Spectronic 20 test tube.

Dissolve 0.259 g sodium ascorbate in one liter of water. Take 1 mL of this solution and add to a clean 4 inch test tube.

Measure out 1.36 mL of 3% hydrogen peroxide and dilute to 1 liter. Take 1 mL of this dilute hydrogen peroxide and add to the sodium ascorbate solution in the 4 inch test tube. Add two mL of water and mix well. When this ascorbate-peroxide solution is added to the complex and rhodamine, the timing should be started. Stir the mixture in the Spectronic 20 test tube before starting to measure the absorption as a function of time. One technique for stirring the mixture is to hold a piece of polyethylene over the mouth of the test tube and, holding the polyethylene firmly in place, shaking the test tube by tilting it from end to end. The absorbance of the 553 nm peak will start falling immediately. Follow the generation of the hydroxyl radical by the degradation of the dye concentration. If time is available, continue the observation until the steady state is reached.

Questions to be Considered

1. Calculate the final concentration of each component in the Spectronic 20 cuvette in terms of μM.

2. Write the structure of uncomplexed glycine.

3. What is the geometry of the copper complex?

4. Write the formulas for the oxidized and reduced forms of sodium ascorbate.

5. A cuvette for a research grade spectrometer holds 1.4 mL instead of 6 mL. How would you prepare the solutions for this research grade spectrometer?

Helpful Background Literature

1. Pamatong, F. V.; Detmer, C. A., III; Bocarsly, J. R. *J. Am. Chem. Soc.*, 1996, *118*, 5339.

2. Detmer, C. A., III; Pamatong, F. V.; Bocarsly, J. R. *Inorg. Chem.*, 1996, *21*, 6292.

3. Barton, J. K. In *Bioinorganic Chemistry:* Bertini, I.; Gray, H. B.; Lippard, S. J.; Valentine, J. S., Eds.; University Science Press: Mill Valley, CA, 1994, p. 455.

4. Kazakov, S. A. in *Bioorganic Chemistry: Nucleic Acids;* Hecht, S. M. Ed.; Oxford University Press: New York, 1996, p. 244.

5. Comba, P.; Hambley, T. W.; Lawrance, G. A.; Martin, L. L.; Renold, P.; Varnagy, K. *J. Chem. Soc, Dalton Trans.*, 1991, 277.

6. Balla, J.; Bernhardt, P. V.; Buglyo, P.; Comba, P.; Hambley, T. W.; Schmidlin, R.; Stebler, S.; Varnagy, K. J. *J. Chem. Soc., Dalton Trans.*, 1993, 1143.

7. Comba, P.; Curtis, N. F.; Lawrance, G. A.; Sargeson, A. M.; Skelton, B. W.; White, A. H. *Inorg. Chem.*, 1986, *25*, 4260.

8. Halliwell, B.; Gutteridge, M. C. *Biochem. J.*, 1984, *219*, 1 – 14.

9. Beckman, J. S.; Koppenol, W. H. *Am. J. Physiol.* 271 (*Cell Physiol.* 40), 1996, C1424 – C1437.

SYNTHESIS OF HETEROGENEOUS CATALYSTS AND CATALYST SUPPORTS

Equipment Needed:
- magnetic stirrer-hot plate
- magnetic stirrer bar
- 15 cm diameter crystallizing dish
- Buchner funnel

Chemicals Needed:
- 146 mL sodium silicate
- 57 mL 4 M HCl
- 1 mL 0.1 M $AgNO_3$ solution
- 10 g CP grade $AlCl_3$
- 1 mL concd. NH_4OH
- pH paper
- 25 mL dilute ammonia solution (0.1 M)
- 1 g Al_2SO_4
- 100 mL 2% $Al_2(SO_4)_3.18\ H_2O$

Time Needed: 3 hours (not including thermal treatments)

Safety Notes: The chemicals used in this experiment are not highly toxic. Ammonia can be unpleasant. It should be used in a hood.

Purpose

Aluminosilicates are widespread in nature as rocks and minerals. Some of these naturally occurring minerals such as zeolites are useful as catalysts or catalyst supports. Synthetic aluminosilicates can be prepared as "designer" catalysts or substrates to many catalysts. This series of experiments illustrate some of the principles used in preparing synthetic aluminosilicates.

Introduction

Many heterogeneous catalyst systems are based on oxides of silicon, aluminum or mixtures of these two elements. Several reasons exist for the tremendous quantity of these types of systems. First of all, such oxides are very abundant and inexpensive. Second, they are very thermally stable materials. A variety of structural types are also available. Finally, the acid-base properties of silica, alumina, silica-alumina, and related systems are well known and can be exploited in the preparation of catalysts.

Silica is believed to contain Brönsted acid sites when it has been heated to temperatures above 150°C. Alumina on the other hand is considered to contain only Lewis acid sites, although this point is still under debate. Transition metal complexes can be ion-exchanged on these materials, tethered to their surfaces via covalent attachment of siloxane groups, or deposited by a variety of methods as discussed in the essay on crystal growth.

There are various forms of silica and alumina and therefore the starting materials and subsequent treatments will govern the phase or mixture of phases that are obtained. Depending on the application, one phase may be more desirable than another. For example, the most important alumina phases are γ-Al_2O_3 and η-Al_2O_3 since they have high surface areas and are thermally stable over a wide range of temperature. Both can be described as spinel phases, although the η-form is more distorted than the γ-form. A phase diagram of various forms of alumina is given below. Note that the resultant phases markedly depend on the precursor, typically bayerite (β-Al_2O_3 .$3H_2O$), boehmite (α-Al_2O_3 .H_2O), or gibbsite (α-Al_2O_3 .$3H_2O$).

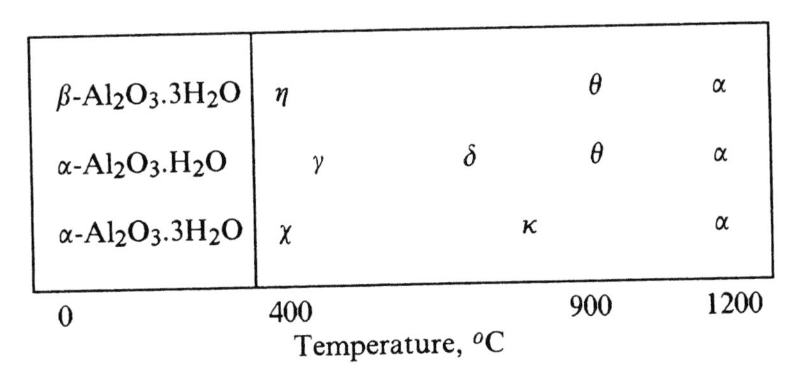

Fig. 1. Formation of Various Forms of Alumina

In this experiment, silica, alumina, and silica-alumina materials will be prepared. Characterization of these supports will be important to verify the identity of the phases that are formed. Some auxiliary experiments that can be done include deposition of transition metals onto these supports, calcination, and reduction of the transition metal to produce materials that are active hydrogenation and dehydrogenation catalysts.

Experimental Procedure

Silica

Prepare a 100 mL solution of water glass having a SiO_2 : Na_2O ratio of 3 : 2 with 200 mL H_2O. Prepare another solution of 50 mL 4 M HCl with 60 mL H_2O. Both of these solutions should be cooled to ice water temperature (O^{o}C) and then slowly add the water glass solution into the acid solution with vigorous stirring. Measure the pH right after both solutions are mixed. The sol that is formed from this mixture will harden to a gel over various periods of time depending on the pH. Before hardening, pour the sol into a flat container. After the hard gel is formed it can be cut into small pieces. The pieces should be treated with washings of 1M HCl so that the HCl solution just covers the gel. After this, wash the gel with distilled water and test the effluent with $AgNO_3$ solution (0.1 M) until no Cl^- is detected. Dry the gel to 150°C for various periods of time from 1 hour to 5 hours. The gel should be transparent and you should notice a correlation of heating time and the degree of transparency of the gel. This particular preparation yields microporous silica with pore sizes of 10 Å. Characterize the gel with X-ray powder diffraction and FTIR spectroscopy. If possible carry out surface area and pore size determinations as well as thermal analysis.

Alumina

Prepare a solution of 10 g CP grade $AlCl_3.6H_2O$ in 100 mL of distilled water. Prepare another aqueous solution of 1 mL concd. NH_4OH to a 14.5 mL total volume. The ammonia solution should be added to the aluminum chloride solution which is being stirred. The final pH should be somewhat basic, (around pH = 8) and should turn litmus paper or pH paper to a basic color. Filter the product in a Buchner funnel. Wash the product 5 times with 5 mL of dilute ammonia solution, for example 0.1 M. Dry the product at 150°C. The final product can be obtained by drying for 12 hours at 550°C. This material should be characterized with X-ray powder diffraction, FTIR and surface area, thermal analyses, and pore size determinations.

Silica-Alumina

Prepare a solution of 46 mL water glass in 90 mL distilled water. Prepare another solution of 6.7 mL of 4 M HCl, 1 g Al_2SO_4, and 40 mL distilled water. Add the water glass solution into the acidified aluminum sulfate solution which is vigorously stirred. The sol should be poured as fast as possible after mixing into a flat tray. After 1 hour of aging cut the gel into small pieces. Age the gel for various periods, up to 2 days. Then wash the material in a Buchner funnel 3 times with a 2% $Al_2(SO_4)_3.18H_2O$ solution. Then wash with distilled water. Dry this material for 5 hours at 180°C, or longer until the gel is transparent. To activate the material heat it in air for 12 hours at 550°C. This material should be characterized with X-ray powder diffraction, thermal analyses, FTIR, and surface area and pore size determinations.

Auxiliary Experiments

Similar ion-exchange experiments to those discussed in the zeolite experiments can be done in order to introduce small amounts of cations onto these supports. Similar characterization experiments can also be done.

Some catalysis experiments can be done either with the oxide materials prepared above or after incorporating active species into these materials. Alumina can be used for the catalytic dehydration of ethanol. A flow reactor similar to the one described in the zeolite experiment could be used for these studies.

Several metals supported on alumina can be used for hydrogenation reactions of olefins like ethylene. Group VIII metals are most active. One specific example would be deposition of nickel on alumina. In fact, an interesting experiment would be to prepare nickel on silica, alumina, and on silica-alumina to compare differences in activity, selectivity, and stability of the catalyst. A 25% nickel on 75% alumina catalyst can be prepared by mixing 45 g $Al(NO_3)_3.9H_2O$ in 300 mL distilled water. Another solution of 20 g NaOH in 100 mL distilled water should be prepared. Both solutions should be cooled to ice water temperature and the basic solution should be added dropwise through a separatory funnel to a vigorously stirred aluminum nitrate solution. Another solution of $Ni(NO_3)_2.6H_2O$ in 60 mL distilled water and 4.5 mL concd. HNO_3 should be prepared and cooled to ice water temperature. The nickel solution should then be added to the sodium aluminate solution which is stirred vigorously for 1 hour. The product should be filtered in a sintered glass frit and should be washed 5 times with 10 mL portions of distilled water. The filtrate can be cut into small pieces and dried in at 105°C for 8 hours.

The solid should be ground in a mortar and pestle and a size fraction of 8 to 12 mesh should be used in catalysis reactions. In order to activate the catalyst, the Ni^{2+} ions need to be reduced by using a flow rate of 30 mL/min H_2 at 350°C for 8 hours. This catalyst should be active for the reduction of ethylene, toluene and similar hydrocarbons. Test the activity of this catalyst at temperatures of 200° to 350°C in a flow reactor similar to the one described in the zeolite experiment. Various flow areas and feeds can be used. A 10% C_2H_4 feed in Ar or He is suggested. Vary the flow rate of the feed and check the activity.

Questions to be Considered

1. What is the advantage of making a silica-alumina material over a mixture of silica and alumina?

2. Other mixed systems like chromia alumina can be prepared. How would you make this material?

3. How would you characterize the Ni alumina catalyst after reaction? What about some *in situ* methods of characterization?

Helpful Background Literature

1. Satterfield, C. N. "Heterogeneous Catalysis in Practice"; McGraw Hill: New York, 1980.

2. Gates, B. C. " Catalytic Chemistry"; John Wiley and Sons: New York, 1991.

3. "Environmental Catalysis"; Armor, J. N., Ed.; ACS Symposium Series 552, ACS: Washington, 1994.

4. "Perspectives in Catalysis"; Thomas, J. M.; Zamaraev, K. I., Eds.; Blackwell Scientific Publications: London, 1992.

PREPARATION OF A ZEOLITE, B/Al ZSM-11

Equipment Needed:	stainless steel autoclave 175°C oven 90°C drying oven 600°C furnace
Chemicals Needed:	25 g Ludox AS40 (DuPont, 40% SiO_2) 2.07 g boric acid 52.89 g of 55% tetra-n-butylammonium hydroxide tank of nitrogen
Time Needed:	3 hours + reaction time
Safety Notes:	Cool the autoclave before opening in a hood. An autoclave with a blowout valve is recommended.

Purpose

Zeolites with specific catalytic properties have been one of the most important catalysts for the petroleum industry. The syntheses of zeolites with specific catalytic properties has been, and will continue to be, an important contribution of inorganic chemists to industrial chemistry.

Introduction

Zeolites are crystalline aluminosilicate microporous materials which have channels on the order of molecular dimensions (1). Zeolite pores typically range from about 4 Å to about 8 Å although considerable recent research has been aimed at increasing this range. The framework structure of zeolites is anionic due to the substitution of Al^{3+} for Si^{4+}. Cations balance this anionic charge and these cations can be replaced by ion exchange reactions. Zeolites are hydrated materials that when heated can yield Brönsted sites. High surface areas on the order of 300 to 800 m^2/g are usually found for zeolites.

The cationic sites of zeolites can act as Lewis acid sites in catalytic reactions. In addition, Brönsted sites can also act as active sites depending on the reaction at hand. In certain cases, base sites are also believed to be present in zeolites, especially when the heavier alkali metal cations are incorporated.

One of the most common applications of zeolites is in the cracking of petroleum range hydrocarbons into gasoline (2). Modified large pore zeolites like zeolite Y are used in such reactions. Zeolites are also used in a variety for other applications including use in detergents as builders, for drying applications, as abrasives in toothpaste, and in other areas.

The structure of two zeolites are given below. The silicon and aluminum ions in the framework are in tetrahedral coordination. The channel size limits the size of molecules that can enter and leave the pores. This geometrical limitation is often referred to as a shape selective effect. Shape selectivity can be governed by the size of the reactant (if too large it cannot enter the pores), by the product (if too large it cannot leave the large internal pores), or transition state selectivity where the size of a transition state complex is limited by the size and shape of the central cavity.

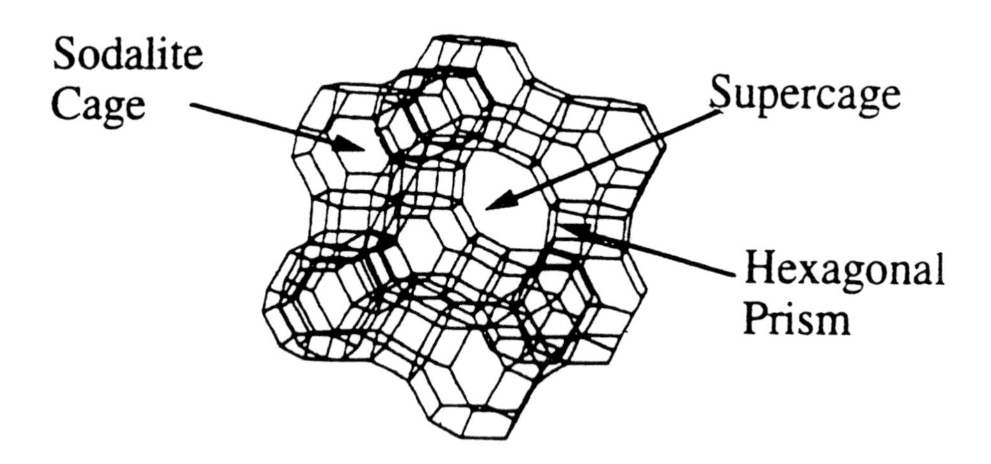

Fig. 1. Structure of NaX Zeolite. The structure extends in three dimensional space. The intersections of the lines are positions of silicon or aluminum. They are four-coordinate with oxygens on the line connecting the silicons and aluminums.

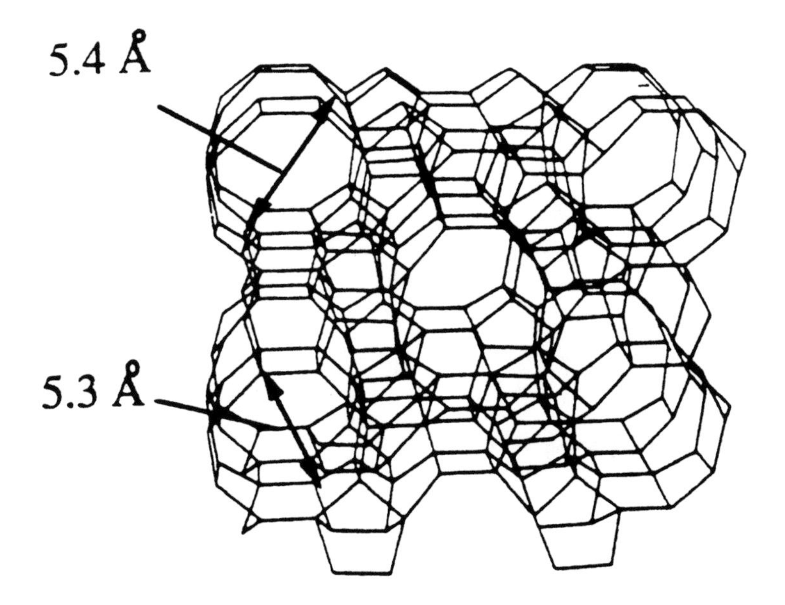

Fig. 2. Structure of ZSM-11 Zeolite. As with Fig. 1, this is a portion of an infinite three-dimensional structure. All of the intersections which are four-connected are locations of silicon or aluminum. The oxygens lie along the line connecting the silicons or aluminums.

Templates or structure directors are often used to control the size and shape of the pores of zeolites (3, 4). In some recent examples it has been possible to computer design such templates and then make the template and then the zeolite (5, 6). In this experiment, an intermediate pore (~ 5.5Å) zeolite, ZSM-11, will be doped with boron and aluminum cations. The product can be used to isomerize n-butenes to isobutylene (7) as described in the auxiliary experiment.

Experimental Procedure

Slowly add 25 g Ludox AS40 (DuPont, 40% SiO_2) while vigorously stirring to a solution which contains 2.07 g H_3BO_3, 52.89 g of 55% tetra-n-butylammonium hydroxide (TBAOH), and 189 mL H_2O. A gelatinous solution will form with a molar composition of 3.36[(TBA)2O], 1.0(B_2O_3), 10(SiO_2), and 680(H_2O). This solution should have a pH of 13.0. The mixture should then be transferred to a stainless steel autoclave. Heat the autoclave in an oven at 175°C for 2 days.

The autoclave should be cooled to room temperature and the white powder product recovered by washing with 3 volumes of 10 mL distilled deionized water. The sample should be dried at 90°C in air for 12 hours. In order to remove the organic TBA template, the dried sample should be calcined at 600°C under an nitrogen stream for 4 hours and then in air for 4 hours.

The above procedure is long, however it is relatively easy, inexpensive, and affords a catalytic material with unique properties. If time does not permit, or if an autoclave is not available, then a commercial catalyst like ferrierite which can be obtained from Tosoh, Atlanta, GA, can be purchased for the catalytic experiments.

The B/Al ZSM-11 can be characterized with X-ray powder diffraction, FTIR spectroscopy, scanning electron microscopy, surface area BET measurements, luminescence, chemical analysis via inductively coupled plasma methods or atomic absorption (7). Most of these methods can be used on the commercial ferrierite sample, although it does not emit so luminescence methods will not be useful. If available, ^{11}B magic angle spinning NMR experiments can be done to study the environment of the boron species.

Auxiliary Experiments

Ion Exchange of Commercial Zeolites

Many zeolites are commercially available from Universal Oil Petroleum, Alfa Ventron, and other companies. Purchase A, X and Y zeolites in powder form since pellets are often contaminated with iron. To ion-exchange such systems, take 100 mL of a 0.05 M aqueous solution of any cationic species and add 1 g of the zeolite powder (typically a sodium or proton form). Stir the solution for 1 to 2 hours and then filter and wash sparingly with distilled deionized water to remove surface cationic species. Of course the nature of the cationic species much be chosen carefully to avoid hydrolysis. Many transition metal and lanthanide ions can be readily incorporated. Some examples are given below. The anion of the salt used for exchange is often nitrate since it can readily be removed or burned off during thermal activation.

Cupric salts can be used for ion exchange in neutral pH solutions. The resultant cupric zeolites will be EPR active. Study the EPR spectra as various amounts of water are removed by dehydration. This should be done on a vacuum line so that the quartz EPR tubes can be sealed off in the absence of oxygen. More sophisticated experiments can be done to show that acid base reactions can occur in zeolites by introducing small amounts of base such as NH_3 or pyridine to the dehydrated zeolites where $CuL_4{}^{2+}$ complexes will form. These can also be characterized by EPR spectroscopy. Some other examples of paramagnetic ions that can be ion exchanged and studied with EPR include Co^{2+}, Ti^{3+}, Mn^{2+}, and VO^{2+} (although the latter may need to be done under a nitrogen atmosphere in, for example, a glove bag environment).

In the case of lanthanide ions, the resultant ion-exchanged zeolites are excellent luminescent materials and can also be used in Friedel Crafts catalytic reactions (8). Mixed valent Ce^{4+}/Ce^{3+} materials can be made by starting with either Ce^{3+} and oxidizing this in air or oxygen at temperatures above 200°C, or by exchange of Ce^{4+} and reduction with hydrogen. It is always important to characterize such systems after any type of activation or exchange since it is possible that the structure of the zeolite might change.

Catalytic Isomerization of n-Butenes with B/AlZSM-11 Zeolites

Take 0.04 to 0.4 g B/AlZSM-11 after sieving above 200 mesh but below 240 mesh and support it in a microreactor on a 250 mesh stainless steel screen with glass wool above and below the bed. Do not pack the bed tightly. The microreactor should have a volume of 1.5 cm^3 and should behave like an ideal mixed flow or continuously stirred tank reactor (CSTR) as determined by response experiments. Details concerning the performance of such a microreactor can be found elsewhere (9). A diagram of the reactor is given below.

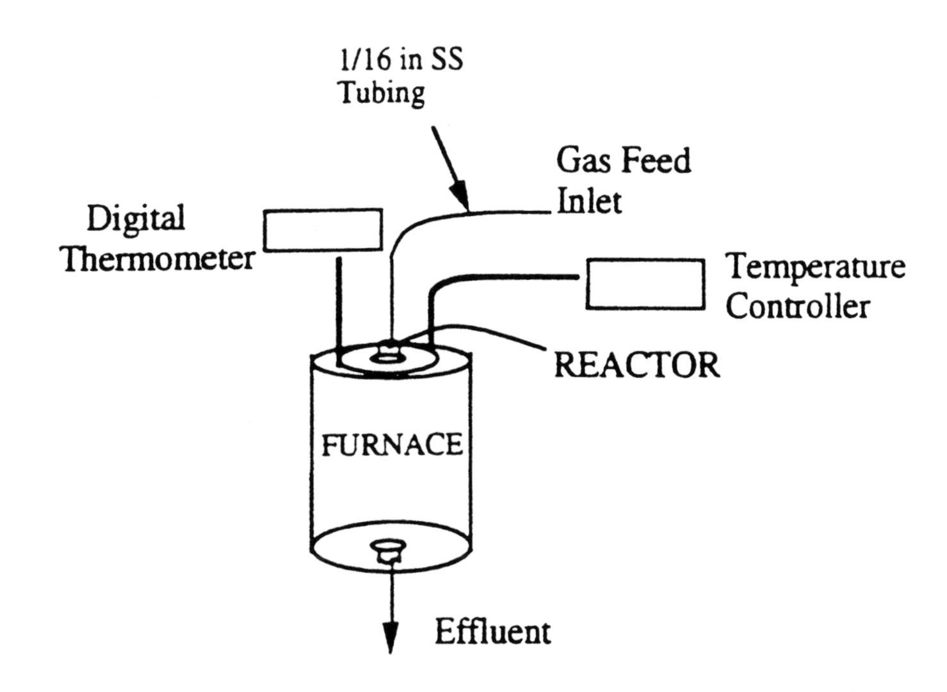

Fig. 3. Diagram of a Microreactor

The zeolite catalysts should be activated under an Ar purge 30 cc/min. The temperature should be elevated at 50°C intervals. The reactor should be operated at one atm and a bypass valve around the reactor into a gas chromatograph detector would be useful for calibration before each analysis. Flow rates should be measured under ambient conditions with bubble meters before and after the catalyst bed. Temperatures should be controlled with a K-type thermocouple and conformed if possible with a digital thermometer placed near the catalyst bed. A diagram of the flow reactor system is given on the following page. (See Fig. 4.)

A GasCHROM C8 column with a He carrier gas flow of 5 cc/min under isothermal conditions and a flame ionization detector should be used. A 10.0% 1-butene in Ar feed can be used. FID response factors can be used to calibrate the concentrations of products which are predominantly between C1 and C4. To determine the hydrogen produced during reaction a mass spectrometer will be needed. A reaction temperature near 520°C should be used, although one of the variables would be to vary this temperature to obtain the maximum activity. Changes in selectivity with reaction temperature should also be noted.

The isomerization of n-butenes to isobutylene is currently of interest since isobutylene is an important reagent in the synthesis of methyl tertiary butyl ether (MTBE) which is being used as an additive in gasoline. Try to find the optimum conditions for selective formation of isobutylene in these catalysis experiments.

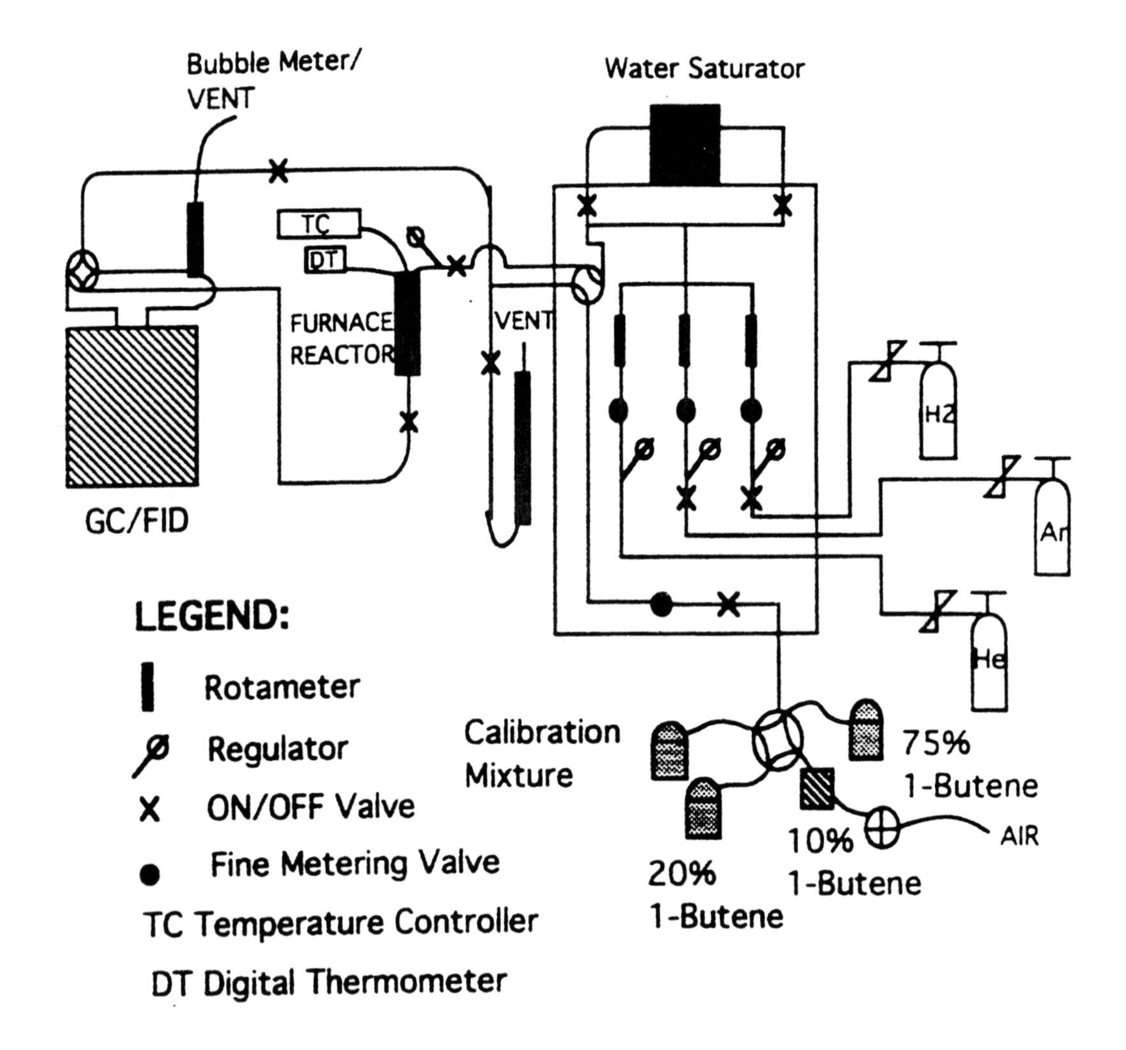

Fig. 4. Diagram of a Gas Flow Reactor

Questions to be Considered

1. There is currently considerable debate about use of the word template in zeolite synthesis. What is the function of a template and how might this term be inaccurate?

2. What other species besides B might be substituted into the framework of zeolites?

3. There is current interest in the mechanism of isomerization of butenes over these and other zeolite catalysts. What types of experiments would you do to try to distinguish the type of mechanism of isomerization?

Helpful Background Literature

1. Barrer, R. M. "Hydrothermal Chemistry of Zeolites"; Academic Press: London, 1982.

2. Occelli, M. L.; O'Connor, P. "Fluid Catalytic Cracking III"; ACS Symposium Series 571, ACS: Washington, 1994.

3. Davis, M. E. Design for Sieving; *Nature*, 1996, *382*, p. 583.

4. Zones, S. I.; Nakagawa, Y.; Yuen, L. T.; Harris, T. V. *J. Am. Chem. Soc.*, 1996, *118*, pp. 7558 – 7567.

5. Lewis, D. W.; Willcock, D. J.; Catlow, C. R. A.; Thomas, J. M.; Hutchings, G. J. *Nature*, 1996, *382*, 604 – 606.

6. Lewis, D. W.; Sankar, G.; Wyles, J.; Thomas, J. M.; Catlow, C. R. A.; Willcock, D. J. *Angew. Chem. Int. Ed.* (Engl.), 1997, *109* in press.

7. Simon, M. W.; Nam, S. S.; Xu, W.; Suib, S. L.; Edwards, J. C.; O'Young, C. L. *J. Phys. Chem.*, 1992, *96*, pp. 6381 – 6388.

8. Tanguay, J. M.; Suib, S. L. *Catal. Rev.-Sci. Eng.*, 1987, *29*, 1 – 40.

9. Bennett, C. O., Suib, S. L. *Catal. Today*, 1992, *15*, pp. 503 – 514.

MOLECULAR MODELING COMPUTATIONAL LABORATORY

Equipment needed:	Computer lab and necessary software
Chemicals needed:	None
Time needed:	3 hours
Safety notes:	None

Purpose

This experiment illustrates the actual use of computational methods to describe the molecular and atomic interactions of two of the compounds prepared in the chloropentaminecobalt chloride experiment.

Introduction

Computational chemistry is introduced in an essay which is found in Part I of this laboratory manual. However, in order to thoroughly understand the principles being discussed, it is important to actually experience a computation first hand. This is done by calculating the minimum energy for a particular topologically connected set of atoms. Finally energy relationships of the kinetic and the thermodynamic product are calculated.

Computational programs have been written for Macintosh computers and PCs, specifically IBM compatible computers. Chem3D by CambridgeSoft Corp., which is used in our laboratories, and MacSpartan by Wave Function Corp. are two of the programs

*Developed by Prof. Jeffrey Bocarsley and Prof. Carl David, University of Connecticut

available for the Macintosh computers. Versions for the PC are also available. In the rapidly changing software and hardware environment, it is difficult to make an exhaustive list of available tools that will remain definitive for any length of time. This experiment will have to be rewritten for each specific condition of place as well as time. The specific directions used in our laboratory are presented to serve as an example of the calculations made on two of the products prepared by our students.

Experimental Procedure

You will use Chem3D by CambridgeSoft Corp. for your molecular modeling. This software uses the standard Macintosh interface common to all Mac software. Your teaching assistants will guide you in the drawing of your molecule in Chem3D. As you conduct your calculations, remember the key points concerning the quality of your calculation:

A. The more chemically reasonable your input structure, the less likely you are of finding a local minimum, and the better your chances are of finding a global minimum. So, examine your input structure for bond lengths and angles which are extremely unreasonable, and change them so that they are chemically reasonable!

B. Make sure the software has correctly identified your atoms, since the parameters (k_{str}, k_{bend}, k_{vdw}, etc.) for one type of carbon (e.g., an alkane) are different from those of another (e.g., a carbonyl carbon) even though in reality they are chemically the same element. You can find atom definitions simply by placing the cursor tool on an atom and waiting for the information to appear. You will have to name atoms you input with the correct names. In order to distinguish between the various types of atoms in the structure, we have used the following convention:

- the first part of the name is the atomic symbol of the element
- the next part of the name is the chemical group which the atom is part of (where applicable)
- the final part of the name is the letter M if the atom is directly bonded to the metal and L if it part of a ligand but not bonded to the metal.

Some examples:

1. A proton on an ammonia ligand is designated: H amine L
   Explanation: this atom is a hydrogen ('H') that is part of an amine which is a ligand ('L'); the H is not directly bonded to the metal.
2. The nitrogen of the amine ligand: N amine M
3. The oxygen in the nitro ligand: O nitro L
4. The oxygen of the nitrite ligand not bonded to the metal: O nitrito L
5. The oxygen of the nitrite ligand bonded to the metal: O nitrito M
6. The nitrogen of the nitro ligand bonded to the metal: N nitro M
7. The nitrogen of the nitrite ligand: N nitrito L
8. The nitrogen of the amine ligand *trans* to the nitro group: N amine M trans
   (This has a special designation because it has a longer bond to the cobalt than the other amine ligands.)

Your task will be to input the structures of nitritopentaaminecobalt(III) chloride and nitropentaaminecobalt(III) chloride and to calculate the respective energies of these linkage isomers.

The steps for the calculation are as follows:

1. Launch Chem3D Pro in the CS ChemOffice folder. Note the toolbar on the left. From the top of the toolbar down, the tools are: pointer, rotation, add single bond, double bond, triple bond, zero order bond, and erase. The pointer is used to select atoms and to translate them. The rotation tool allows real-time rotation of your molecule. The bond tools are used to create new bonds.

2. Under the File menu, find the Chem 215 folder and double-click. There are four documents: 'cono2' (for nitropentaaminecobalt(III) chloride) and 'cono2 template', and 'coono' (for nitritopentaaminecobalt(III) chloride) and 'coono template.' You will use each of the template files to construct your molecules and calculate the energies. When you open the files, a dialog box will appear telling you that the files are locked; just continue ('OK') and the template file will open. (The files are locked so that you can't accidentally overwrite them.) Start with nitritopentaaminecobalt(III) chloride using the file 'coono template'.

3. Check the atom definitions of the atoms in the template by putting the arrow tool on the atom for a few seconds. You should check that the central atom is cobalt, and that the ligand atoms are nitrogens bonded to cobalt. Using the single bond drawing tool on the palette at the left, build your molecule. When you try to make your first change in the structure, you will again see a dialog box which tells you that the file is locked; again continue ('OK') and the system will allow you to proceed in an identical untitled file. Add in the correct numbers of hydrogens on the amine ligands. Note that when you add an atom, it is automatically assigned as carbon (the default atomic type). After you add in all the atoms for hydrogen, select all the new atoms by shift-clicking on each atom with the arrow tool. Now, in the upper left of the window, there is an unlabeled input field. Type the correct definition of the new atoms ('H amine L') while they are all selected, followed by <enter>. The system should assign the correct name to each hydrogen. You should check the geometry of the groups in which you have input atoms. If the geometry is poor, fix it. You can move atoms by selecting them with the arrow tool, and then holding the mouse button down on the atom and sliding the mouse. When you are done adjusting the geometry, deselect any selected atoms by single clicking on the background. Now complete the nitrite ligand by using the same approach.

4. Deselect any selected atoms by single clicking off the structure. Under the Analyze menu, select "Minimize Energy." This performs a molecular mechanics minimization of your structure. Make sure that you have checked off the box "Display Every Iteration" and no others. In the Message box at the bottom of the screen, the energy of your input structure and its constituent components will appear, followed by the total energy of the structure during minimization. When your minimization starts, the Message box will display the following type of message:

Note: 1,3 Van der Waals interactions are used to compute the energy of angles around atoms with more than 4 coordinate bonds.
Warning: Some parameters are guessed (Quality = 1).

Stretch:	**1030.5126**
Bend:	**6.9371**
Stretch-Bend:	**7.3536**
Torsion:	**0.0003**
Non-1,4 VDW:	**-0.4278**
1,4 VDW:	**-1.1269**
Charge/Charge:	**0.0000**
Total:	**1043.2488**

Iteration 0	**Steric Energy**	**1043.249**	**RMS Gradient**	**171.198**		
Iteration 1	**Steric Energy**	**537.100**	**RMS Gradient**	**82.180**	**RMS Move**	**0.0589**
Iteration 2	**Steric Energy**	**463.226**	**RMS Gradient**	**87.475**	**RMS Move**	**0.0589**
Iteration 3	**Steric Energy**	**42.661**	**RMS Gradient**	**12.445**	**RMS Move**	**0.1215**

Iteration 4	Steric Energy	31.954	RMS Gradient	13.109	RMS Move	0.0300
Iteration 5	Steric Energy	28.051	RMS Gradient	8.066	RMS Move	0.0154
Iteration 6	Steric Energy	25.571	RMS Gradient	8.825	RMS Move	0.0113
			*			
			*			
			*			
Iteration 73	Steric Energy	5.100	RMS Gradient	0.046	RMS Move	0.0000

Minimization terminated normally because the gradient norm is less than the minimum gradient norm

Stretch:	**0.8401**
Bend:	**6.3138**
Stretch-Bend:	**-0.0166**
Torsion:	**0.0002**
Non-1,4 VDW:	**-0.4920**
1,4 VDW:	**-1.5453**
Charge/Charge:	**0.0000**
Total:	**5.1002**

The software calculates the energy for the structure based on your input geometry, and begins to minimize it. Note the high energy of the input structure (1043 kcal mol^{-1} in the case above). For every iteration, the total strain energy ("Steric Energy") and the RMS ("Root Mean Square") Gradient are reported. The gradient may be thought of as the local slope of the potential energy surface; the software will look for the lowest slope it can find, which should be at the bottom of a valley. The RMS Move is a measure of the geometry movement the software makes in attempting to find the downward direction. Your minimization may take hundreds (or even thousands) of steps, depending on the quality of your input structure. When the minimization converges, the termination notice appears along with the energy terms and the total energy. Your minimized structure will look something like:

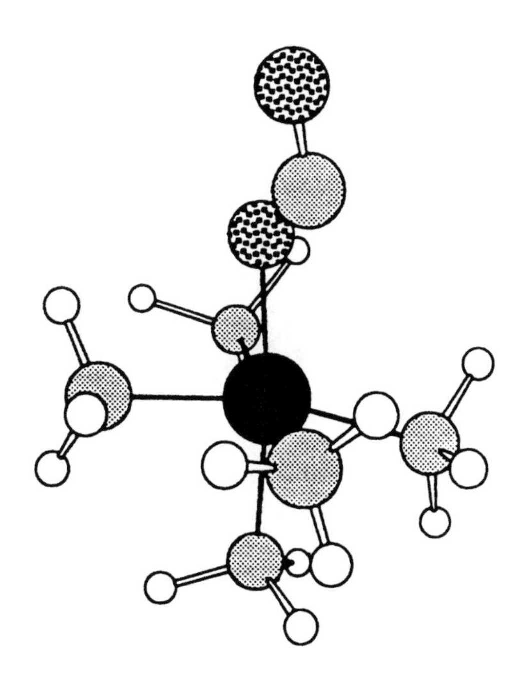

Fig. 1.

The software uses the term 'steric energy' for what we have called strain energy. The units are kcal mol^{-1}, not kJ mol^{-1}. Reasonable values at convergence (the bottom of the valley) are below 10 kcal mol^{-1}. Anything higher should be examined very critically, because you have most likely found a local and not a global minimum. Pay special attention to your amine groups.

5. Record in your laboratory notebook the total strain energy for the system you have calculated, along with the components of the sum. Also record the following bond lengths and angles of the calculated structure:
   - all bond lengths and angles in the nitrite (or nitro) group
   - the Co – nitrite (or nitro) bond length and the Co-O-N angle
   - the Co – amine nitrogen lengths

6. Try to assess the effect of your input structure on the final calculated strain energy by moving some of the atoms and resubmitting the calculation. For example, select an entire amine ligand and elongate the Co – N distance, and resubmit. Or, change bond lengths or angles in the nitrite group and resubmit the calculation. Finally, check the effect of moving a few hydrogens on the starting structure (you should look at the structure and think of an intelligent way of doing this). You may input a horribly distorted complex just to see the effect on the final energy. Copy the total and component energies into your notebook and the bond lengths and angles you are monitoring. Try to assess what starting structure gives the lowest minimized energy.

7. The file "coono" (or "cono2") contains the experimentally determined structure of the metal complex (non-hydrogen atoms only). Open this file and write down in your notebook the same bond lengths and angles you have been following in your calculated structures.

8. Redo the entire process for nitropentaaminecobalt(III) chloride using the file "cono2 template". You may receive several error messages at the start of the minimization with a dialog box informing you of impending trouble; ignore these messages. Your structure will look like:

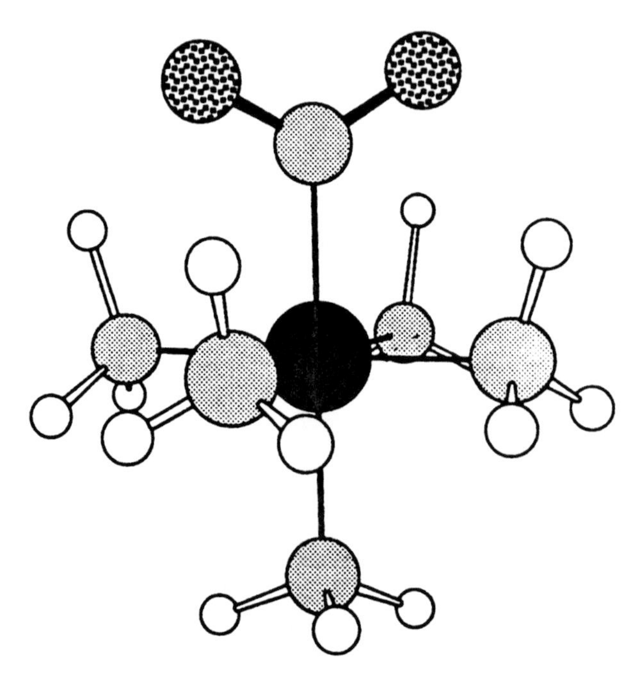

Fig. 2.

Chemical Significance

Kinetic vs. Thermodynamic Reaction Products

In the classic picture of how chemical reactions occur, a set of reactants that have sufficient energy (activation energy, E_A) interact to form a high energy transition state which can decompose either to re-form the reactants or to form products. In some cases, the transition state can form an intermediate, which may be thought of as a short-lived product of the first step of the reaction, that has sufficient energy to continue reacting to form final products. It happens that sometimes, an intermediate can continue to react to form one of two *different* transition states which each lead to *different* products. In this case, two different products will be found at the end of reaction, and the energy relationships between the various species involved control how much of each product forms.

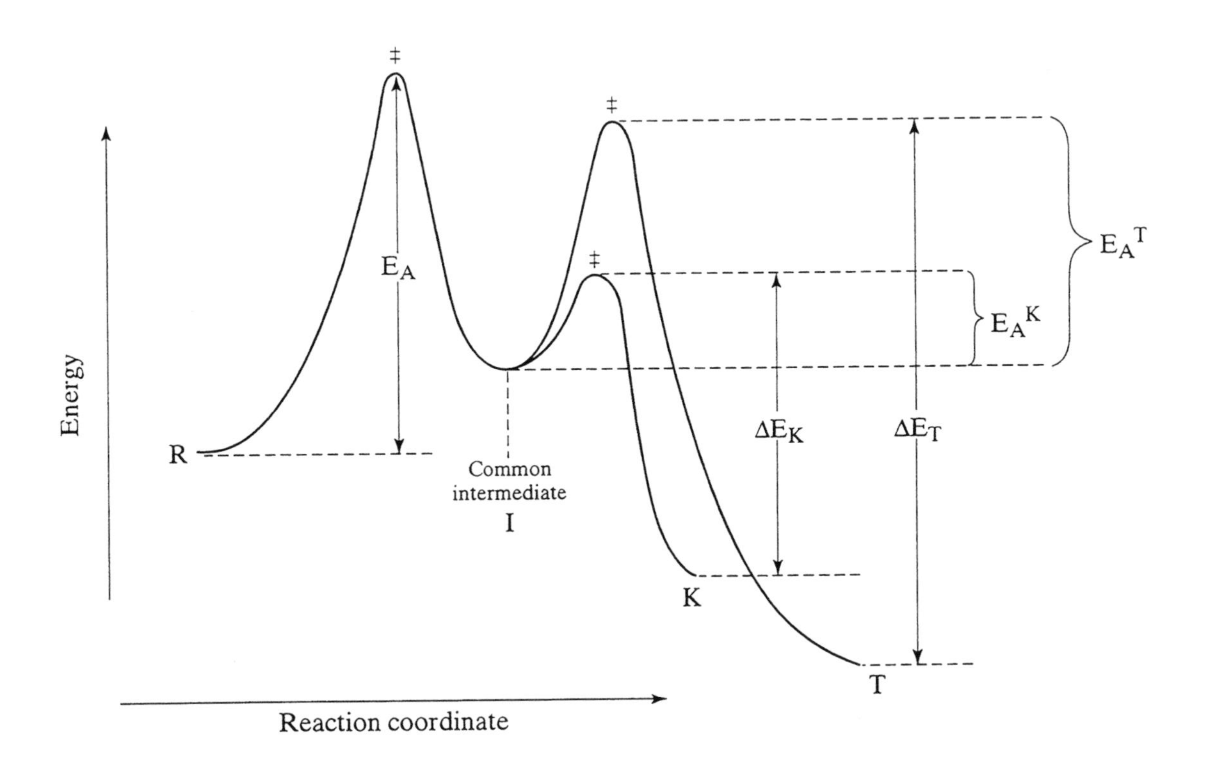

Fig. 3. Energy along reaction coordinate for a reaction with two products and a common intermediate.

Now, suppose we call our reactants R and our two products K and T. Let us further suppose that when reactants R acquire the activation energy E_A a (common) intermediate I forms which can react over one of two subsequent activation barriers either to form product K (activation barrier $E_A{}^K$) or to form T (activation barrier $E_A{}^T$). Also suppose that the two second activation energies are different with $E_A{}^T > E_A{}^K$, and that the back-reaction energy barrier between the common intermediate I and K (ΔE_K) is less than the energy barrier between the common intermediate I and T ($\Delta E_T > \Delta E_K$). These energy relationships are shown in Fig. 3. Since $E_A{}^T > E_A{}^K$, it requires less activation energy to form K than T, and at a given moderate temperature, where there is enough energy for K to form readily but not T, K will form more rapidly than T. This occurs even though the product K is actually *higher* in energy than product T. K is therefore referred to as the *kinetic product.*

Now consider further: if we run the reaction at a temperature such that there is roughly ΔE_K energy available but not ΔE_T, then the back-reaction from K to the common intermediate will have enough energy to occur but not the back-reaction from T to the common intermediate. Note that $\Delta E_K > E_A{}^T$. This means that when we have ΔE_K worth of energy available, so that the common intermediate I re-forms from the back-reaction of K, we also have enough energy available to overcome the activation energy for formation of T. Therefore, under these conditions, any K that forms has a chance of back-reacting to form the common intermediate I, and then forward-reacting to form the product T. But, any T that forms does not have sufficient energy to back-react to form the common intermediate I. The product T will therefore accumulate at the expense of the K product. Note this is not an equilibrium situation, since T can't back-react.

Now, for most reactions, the values of ΔE_K and ΔE_T are not known, so product K cannot be selectively driven to T as described above. However, consider the situation if we run the reaction at a sufficiently high temperature so that at least ΔE_T energy is present. Then all reaction pathways can be accessed (both T and K can back-react to form the common intermediate I). All reactions are therefore reversible, and the product distribution reflects the chemical equilibrium between products and reactants, which depends on the relative thermodynamic stabilities of the final products, K and T. Since T is lower in energy than K, it will dominate. T is therefore termed the *thermodynamic product.*

Questions to be Considered

1. Consider the practical consequences of the discussion above. The kinetic product should be favored under milder reaction conditions, while the thermodynamic product should be favored under more vigorous conditions. In your syntheses of nitritopentaamine cobalt(III) chloride and nitropentaaminecobalt(III) chloride, which reaction requires mild conditions and which requires vigorous conditions? Which product, kinetic or thermodynamic, should have a lower energy?

2. What can you conclude about the influence of the starting structure on the final minimized energy?

3. Compare your best (lowest energy) calculated structure to the experimental molecular structures using the bond lengths and angles you collected. How well did the calculation mirror the experimental determination? For which bond lengths and angle is the computational result correct? For which bond lengths and angle is the computational result lacking?

4. Which of the two isomers is the thermodynamic product and which is the kinetic product? How is this reflected in your best calculated energies for each structure?

Helpful Background Literature

1. See Computational Chemistry section in Part I of this laboratory manual.

Part III CHARACTERIZATION TECHNIQUES

CONDUCTIVITY OF IONIC SOLUTIONS

Equipment Needed:	conductivity bridge conductivity cell
Chemicals Needed:	0.01 N KCl 0.001 N solution coordination compound
Time Needed:	1.5 hours
Safety Notes:	Take care not to spill mercury if it is being used to make electrical connections in the conductivity cells.

Purpose

This experiment considers the theory and the practice of using conductivity measurements to characterize coordination compounds.

Introduction

Conductivity finds extensive analytical applications beyond conductometric titrations. A few of the other uses are to check the quality of water, to determine the moisture content of soil and to prepare solutions of known concentrations.

The principal use of conductivity measurements in inorganic chemistry is to determine the nature of the electrolytes formed by inorganic salts. For example, in the table below, three of the solvate isomers of chromium chloride exhibit these conductivity values.

Table I

Compound	Type Salt	Concentration	Conductivity
$[Cr(H_2O)_6]Cl_3$	$(A^{3+})(X^{1-})_3$	0.0079 M	353.1
$[CrCl(H_2O)_5]Cl_2.H_2O$	$(A^{2+})(X^{1-})_2$	0.01 M	208
$[CrCl_2(H_2O)_4]Cl.2H_2O$	$(A^{1+})(X^{1-})$	0.008 M	103.1

Clearly the tri-univalent salt has a higher conductivity than the uni-univalent salt.

If some correlation can be found for conductivity measurements and ion types, conductivity determinations will be useful in characterizing inorganic compounds.

Before looking for a correlation, it will be useful to consider how conductivity measurements are made. Because conductivity is the reciprocal of resistance, conductivities are determined by measuring the resistance in ohms of a solution in a bridge circuit.

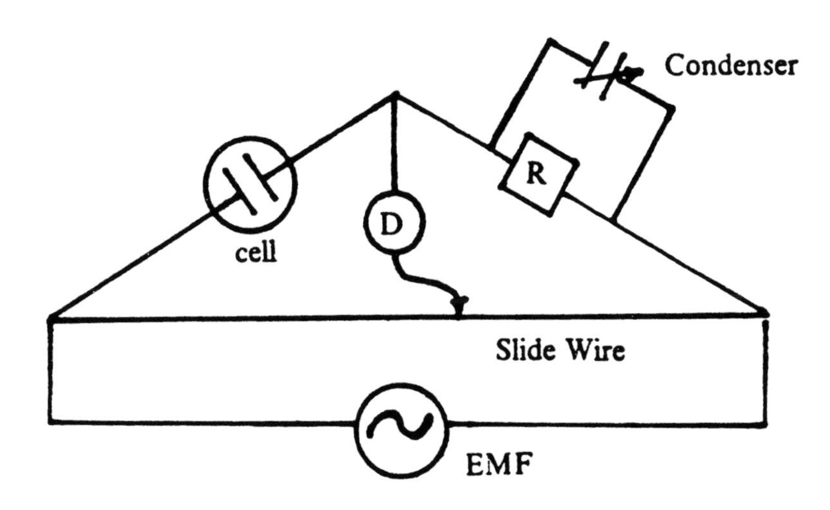

Fig. 1

If direct current is used, errors are created by the gases liberated by electrolysis at the electrodes. The source labeled EMF is, therefore, an alternating voltage of 60 cps or higher. By proper adjustment on the slide wire the bridge can be balanced and a null obtained on the detector (D). The older instruments used a telephone earpiece, a galvanometer type null detector, or a "magic eye" tube. The newer instruments use an electron ray tube null detector or give the conductance-resistance directly using digital readouts. In most commercial conductivity bridges the resistance (R) is a series of resistors mounted on a step switch so that the range of the slide wire can be extended. Provisions are also provided for connecting an external condenser in parallel to the resistance. This condenser is not needed unless the resistance of the solution is very low. In this case the capacity and self inductance of the circuit will cause the alternating current in the two arms of the bridge to be out of phase. This will not change the position of balance, but may make the determination of the exact position difficult. If a telephone earpiece is used, a sharp minimum in hum will not be obtained. With the "magic eye" which indicates a null when the dark segment reaches a maximum, the dark segment will close down and may approach zero at the maxima for extreme cases of phase shifts. The use of an external condenser ranging in value from 0.01 mfd to 10 mfd will make the balance position easier to determine. The higher capacitance condensers are used with lower resistance solutions.

The value obtained by achieving a balance on the conductivity bridge is a reading of resistance. If the value of a simple resistor is being determined, the job is finished. However, in measuring resistances of solutions, it is immediately evident that the value depends on the concentration of the solution, size of electrodes, the distance separating the electrodes, etc. For this reason, several convenient definitions have been set forth in order to enable the assignment of values characteristic of the solute and solvent.

Specific Resistance

The specific resistance is the resistance in ohms of a specimen one cm in length and one square cm in cross section. If

r = specific resistance
ℓ = length in cm
a = cross section in square cm
R = measured resistance

Then

$$R = r\frac{\ell}{a} \text{ ohms}$$

Specific Conductance

The specific conductance is the reciprocal of the specific resistance. If specific conductance is K, then by definition

$$K = \frac{1}{r} \quad \text{or} \quad r = \frac{1}{K}$$

Substituting in the expression for specific resistance,

$$R = \frac{\ell}{(K)(a)} \text{ ohms}$$

then rearranging,

$$K = \frac{\ell}{(a)(R)} \text{ ohms}^{-1}\text{ cm}^{-1}$$

Equivalent Conductance

This term, usually represented by Λ, expresses the conductivity of all ions produced by one gram equivalent of an electrolyte. If

$$N = \frac{\text{gram equivalent}}{\text{liter}}$$

K = specific conductance

$$\Lambda = 1000\,\frac{K}{N} \text{ ohms}^{-1}\text{ cm}^{2}\text{ equivalent}^{-1}$$

The gram equivalent is used in the classical sense. Thus, equivalents of $NaCl$ = FW/1, $BaCl_2$ = FW/2, $LaCl_3$ = FW/3 etc.

Equivalent Conductance of Separate Ion

This is identical to equivalent conductance except that Λ is a measure of the conductivity of just the anion or cation. N in this case is the number of gram equivalents of the particular ion per liter. A gram equivalent of Cl^- = (atomic weight)/1, Ca^{2+} = (atomic weight)/2, $[Fe(CN)_6]^{4-}$ = (formula weight of ion)/4, etc.

Molar Conductance

This term, which will be represented by μ, is similar to the equivalent conductance except that M, the number of gram moles per liter is used instead of equivalents per liter.

$$\mu = 1000 \frac{K}{M} \text{ ohms}^{-1} \text{ cm}^2 \text{ mol}^{-1}$$

Effect of Variables

In examining conductivity data, it becomes immediately evident that there are several other variables which must be specified. One of these is temperature.

Table II

The Equivalent Conductance of the Separate Ions

Ion	0°C	18°C	25°C	75°C	100°C	156°C
K^+	40.4	64.6	74.5	159	206	317
Na^+	26	43.5	50.9	116	155	249
$(NH_4)^+$	40.2	64.5	74.5	159	207	319
Ag^+	32.9	54.3	63.5	143	188	299
1/2 Ba^{2+}	33	55	65	149	200	322
1/2 Ca^{2+}	30	51	60	142	191	312
1/3 La^{3+}	35	61	72	173	235	388
Cl^-	41.1	65.5	75.5	160	207	318
$(NO_3)^-$	40.4	61.7	70.6	140	178	263
Ac^-	20.3	34.6	40.8	96	130	211
1/2 $(SO_4)^{2-}$	41	68	79	177	234	370
1/2 $(C_2O_4)^{2-}$	39	63	73	163	213	336
1/4 $[Fe(CN)_6]^{4-}$	58	95	111	244	321	
1/3 $[Fe(CN)_6]^{3-}$			100.9			
1/3 $[Co(NH_3)_6]^{3+}$			101.9			
1/6 $[Co_2(trien)_3]^{6+}$			68.7			
1/4 $[Ni_2(trien)_3]^{4+}$			52.5			
H^+	240	314	350	565	684	777
OH^-	105	172	192	360	439	592

For example, an examination of Table II shows that there is a marked increase in conductivity with increase in temperature. Near room temperature the change is about 2.5% per degree for many of the electrolytes. This necessitates accurate temperature control while making conductivity measurements. It is also important to note the temperature associated with a reported conductivity value. If rough comparisons are to be made of conductivity values reported at two different temperatures, the following formula can be used.

$$R_{25} = R_t(1 + 0.025\Delta t)$$

Where

R_{25} = Specific resistance at 25°C
R_t = Specific resistance at temperature t°C
Δt = Difference between 25°C and t°C

For accurate work, the variation of resistance with temperature must be determined independently for each salt studied.

Another important variable shown in conductivity data is that of concentration. This is tabulated in various ways. For example, in the *Handbook of Chemistry and Physics* (3) the concentrations are given in gram equivalents per 1000 cm^3. For $AgNO_3$ at 25°C, the following values are given.

Table III

Conc.:	Inf.Dil.	0.0005	0.001	0.005	0.01	0.02	0.05	0.1
Λ	133.36	131.36	130.51	127.20	124.76	121.41	115.24	109.14

In Gmelin, the concentrations are given as V where V is the number of liters of solvent in which one mole or one equivalent of solute is dissolved. The concentration in g moles or g equiv is then 1/V. For $Cs_2[Co(NO_2)_3C_2O_4NH_3]$ at 25°C the following values are given.

Table IV

V	2000	1000	500	250
μ	265	258	236	225

The reason for the drop in equivalent or molar conductance with concentration is the increased interionic attraction at the higher concentrations. This phenomena has been explained by Debye and Hückel and by Onsager. Discussions of the Onsager equation can be found in most elementary physical chemistry texts and all electrochemistry texts.

Because of the effect of the interionic attraction, conductivity data is usually determined for a series of concentrations. On plotting the molar conductance vs. concentration, a limiting molar conductance can be determined. Some sources list this limiting conductance as that at infinite dilution. This, of course, is not quite accurate since a solution at infinite dilution is the pure solvent which has quite a different conductance from that of the solution.

A variable which is not immediately evident from an examination of tables is that of solvent. Most of the tabulated data is for water. Comparisons cannot be made directly from data obtained in one solvent with that obtained in another. Work has been done and conductivity data reported in other solvents such as glycols, liquid ammonia, amines, bromine trifluoride, hydrogen fluoride, etc. The solvent is assumed to be water for the purposes of this experiment.

Now that the terms are defined and the variables understood, one can cast about for a correlation between molar conductance and ionization of a salt. On examining the equivalent conductance at 25°C in Table II, it will be observed that the values range from 40 for acetate ion to 111 for the ferrocyanide ion. The exceptional values for the hydrogen and hydroxyl ions are omitted from consideration. If one assumes that the average value for the equivalent conductance of an ion is 60, then the molar conductance will be 120 for a uni-univalent electrolyte (A^+X^-), 240 for electrolytes of type $A^{2+}(X^{1-})_2$ or $(A^{1+})_2X^{2-}$, 360 for electrolytes of type $A^{3+}(X^{1-})_3$ or $(A^{1+})_3X^{3-}$ etc. Table V shows the experimental data which can be compared to the estimated values.

Table V

Limiting Molar Conductances at 25°C
Expected (2 x 60) = 120

Salt	μ_o	Salt	μ_o
$AgNO_3$	133.36	$KReO_4$	128.20
KBr	151.9	LiCl	115.03
KCl	149.86	$LiClO_4$	105.98
$KClO_4$	140.04	NH_4Cl	149.7
$KHCO_3$	118	NaCl	126.45
KI	150.38	$NaClO_4$	117.48
KNO_3	144.96	NaI	126.9
		NaOAc	91.0

Expected (4 x 60) = 240

Salt	μ_o
$BaCl_2$	279.96
$CaCl_2$	171.68
$CuSO_4$	267.2
$MgCl_2$	258.8
Na_2SO_4	259.8
$SrCl_2$	271.6
$ZnSO_4$	265.6

Expected (6 x 60) = 360

Salt	μ_o
$K_3Fe(CN)_6$	523.5
$LaCl_3$	437.4

It can be seen that, although many salts fall in the right ballpark, some do not. Some arbitrarily selected electrolytes whose molar conductivities at 9.85 x 10^{-4} molar and 25°C would give misleading results are listed.

Table VI

Salts with Low Values

Salt	μ
$[CoOH(H_2O)(NH_3)_4](NO_3)_2$	131.8
$[CoCrO_4(NH_3)_4]NO_3.H_2O$	56.79
$[Co_4(OH)_6(NH_3)_{12}](NO_3)_6$	354.2

Salts with High Values

Salt	μ
$[CoClO_4(NH_3)_5](ClO_4)_2$	330
$[CoBrO_3(NH_3)_5](BrO_3)_2$	369
$K_4Fe(CN)_6$	1476

However, formulation of the solvate isomers for chromium chloride mentioned at the beginning of this experiment would have been assigned correctly based on conductivities. The conductivities for the nitroammine cobalt compounds are consistent. The molar conductivities for 0.001 molar solutions at 25°C are shown.

Table VII

Nitroammine Cobalt Compounds

Salt	μ
$[Co(NH_3)_6]Cl_3$	461
$[Co(NH_3)_5NO_2]Cl_2$	263
$[Co(NH_3)_4(NO_2)_2]Cl$	105
$Co(NH_3)_3(NO_2)_3$	1.6
$K[Co(NH_3)_2(NO_2)_4]$	106
$K_3[Co(CN)_6]$	459

The values for the AX_3 and A_3X compounds are higher than the 6 x 60 = 360 value but entirely consistent with the $LaCl_3$ experimental value. Because the 3,3 and 4,4 salts quite consistently show conductivities higher than the 6 x 60 = 360 and 8 x 60 = 480 values, some sources list 430 as an arbitrary average value for 3,3 electrolytes and 550 as an arbitrary average value for 4,4 electrolytes.

Even when the values are well out of the estimated range, a series of measurements will point out the source of the deviation. For example, the following sequence of measurements were made on diphenyliodonium dicyanobis(dimethylglyoximato)cobalt(III). The conductivity values quoted are the limiting values of equivalent conductances at 20°C.

Table VIII

Limiting Molar Conductances at 20°C

Iodonium Salt	Λ_o, in mho cm^{-1}
$\phi_2I^+[Co(CN)_2(DMG)_2]^-$	43
$\phi_2I^+Cl^-$	97
The equivalent ion conductance for Cl^- from tables is	70
By difference, equivalent ion conductance for ϕ_2I^+ is	27
$K^+[Co(CN)_2(DMG)_2]^-$	87
The equivalent ion conductance for K^+ from the tables is	68
By difference, equivalent ion conductance for $[Co(CN)_2(DMG)_2]^-$	19
The equivalent conductance of the salt obtained from the individual equivalent ion conductance is 27 + 19 =	46

This confirms the low value 43 and indicates that both the anion and cation contribute to the low equivalent conductance.

Experimental Procedure

Before measurements are made on an unknown solution, the cell constant will have to be determined. Because the dimensions ℓ (distance between electrodes) and a (area of electrode) are difficult to measure accurately, the quantity ℓ/a is determined by measuring the conductance of a standard KCl solution. From the definition of specific conductance,

$$K = \frac{\ell}{aR}$$

If ℓ/a is the cell constant k,

$$K = \frac{k}{R}$$

Rearranging

$$k = KR$$

K, the specific conductance of KCl for various concentrations and various temperatures, is listed in handbooks and other references. The data for 0.01 N KCl prepared by dissolving 0.7459 g pure dry KCl in enough water to make 1 liter at 18°C is shown in Table IX.

Table IX

Specific Conductance of 0.01 M KCl

Temperature (^oC)	Spec. Cond. (Mho/cm)	Temperature (^oC)	Spec. Cond. (Mho/cm)
15	.001147	23	.001359
16	.001173	24	.001386
17	.001199	25	.001413
18	.001225	26	.001441
19	.001251	27	.001468
20	.001278	28	.001496
21	.001305	29	.001524
22	.001332	30	.001552

From the value of K obtained from Table IX and the measured value R, the cell constant k can be determined.

Because the conductivity of water can vary markedly with impurities, the next step is to clean the cell thoroughly with distilled water and measure the conductivity of distilled water. Pure water is reported to have a specific conductivity, $K = 5 \times 10^{-8}$ mho cm^{-1} at 25°C. When water of this purity is exposed to air, it reaches an equilibrium specific conductivity, $K = 0.8 \times 10^{-6}$ mho cm^{-1}. This is mainly due to CO_2 from the air dissolving in the water. Ordinary distilled water usually exhibits conductivities higher than this. Besides CO_2, there is often NH_3 or NH_4Cl and organic matter dissolved in it. If the specific conductivity of the distilled water is high, then there are two choices. Either a correction will have to be made in the conductivities of the solutions measured or the water used will have to be purified by distilling from an alkaline permanganate solution in an all glass system. The latter, of course, is to be preferred.

Clean the cell once again and place a 10^{-3} molar solution of the salt to be measured in the cell. Use an external bath to hold the temperature of the cell constant.

In general, solutions with high resistance should be measured in cells with large electrodes close together and solutions with low resistance in cells with electrodes further apart. For electrolytes with specific resistance less than 250 ohms, a cell with a constant of 10 or more is recommended. For electrolytes with specific resistance 250 to 200,000 ohms, a cell with a constant of 1 should be used. For electrolytes of higher specific resistance, cells with a constant of 0.1 is best.

Observe whether there are changes in conductivity with time. Most complex ions undergo aquation. Some undergo aquation reactions so rapidly that the species measured is always the equilibrium species. Of the first row transition series, Cr^{3+} and Co^{3+} form complexes whose reactions are generally slow. Even so, the chromium chloride values quoted in the first part of this experiment are values extrapolated back to zero time. A change of conductivity with time can be noted for these ions due to the formation of an equilibrium aquated form.

Sample Calculation:

Measured resistance of 0.01 M KCl solution at 20°C — 1175 Ω

Specific conductivity of 0.01 M KCl at 20°C from Table IX — 0.001278

Calculating for the cell constant

$$k = KR$$

$$k = (0.001278)(1175)$$

$$k = 1.501$$

Measured resistance with water at 20°C — 299,000 Ω

Calculating for specific conductance of water

$$K = \frac{k}{R}$$

$$K = \frac{1.501}{299{,}000}$$

$$K = 5.02 \times 10^{-6}$$

Measured resistance of 9.85×10^{-4} molar solution of X — 3750 Ω

Calculating for specific conductance of solution X

$$K = \frac{k}{R}$$

$$K = \frac{1.501}{3750}$$

$$K = 4.00 \times 10^{-4}$$

Because the conductivity of water is two orders of magnitude or less, no corrections need to be made.

Calculating for molar conductance,

$$\mu = 1000 \frac{K}{M} \text{ ohm}^{-1} \text{ cm}^2$$

$$\mu = 1000 \frac{4.00 \times 10^{-4}}{9.85 \times 10^{-4}}$$

$$\mu = 406.5 \text{ ohm}^{-1} \text{ cm}^2 \qquad (\text{ohm}^{-1} = \text{mho})$$

Questions to be Considered

1. Conductivity water is prepared by distillation from alkaline permanganate. What is the role of the permanganate?

2. Why are platinum electrodes preferred for conductivity cells?

Helpful Background Literature

1. Dodd, R. E.; Robinson, P. L. "Experimental Inorganic Chemistry", Elsevier: Amsterdam, 1954; pp. 377 – 379.

2. Kolthoff, I. M.; Sandell, E. B.; Meehan, E. J.; Bruckenstein, S. "Quantitative Chemical Analysis", 4th ed.; The Macmillan Co.: Riverside, N.J., 1969; pp. 951 – 953.

3. Lagowski, J. J. "Modern Inorganic Chemistry"; Marcel Dekker: New York, 1973; pp. 193 – 194.

4. Huheey, J. E. "Inorganic Chemistry: Principles of Structure and Reactivity", 3rd ed.; Harper & Row: New York, 1983; pp. 361 – 362.

5. Lide, D. R. editor "CRC Handbook of Chemistry and Physics", CRC Press: Boca Raton, 1992.

MEASUREMENTS OF MAGNETIC SUSCEPTIBILITY

Equipment Needed:	magnetic susceptibility balance (Gouy) susceptibility tube (Gouy) agate mortar and pestle
Chemicals Needed:	mercury(II) tetrathiocyanatocobaltate nickel(II) chloride
Time Needed:	1.5 hours
Safety Notes:	If mercury(II) tetrathicyanatocobaltate is used to calibrate the magnetic susceptibility balance, take care to dispose of it properly. It is a compound containing a heavy metal, mercury.

Purpose

The magnetic susceptibility measurement using the Gouy method is used to determine the number of unpaired electrons in a coordination compound.

Introduction

Although a number of different magnetic characteristics are observed, most of the compounds encountered in the laboratory are diamagnetic or paramagnetic.

In diamagnetic compounds, all the electrons are paired. The external magnetic field induces a field in the completed electron shells which is opposite to that of the applied field. This induced field causes a diamagnetic substance to be weakly repelled from the applied field. Because the diamagnetic susceptibility is a ratio of the induced field to the applied field, the susceptibility does not depend on the applied field. It does depend on the size of the atoms and the total number of electrons. Since all atoms have some completed shells, all compounds exhibit a diamagnetic effect. The diamagnetic effect is experimentally observable by weighing a substance in the absence of and in the presence of a magnetic field. The repulsion effect will be noted by a slight loss in weight of the diamagnetic compound in the presence of an inhomogeneous magnetic field.

Paramagnetism is observed when there is one or more unpaired electrons in a molecule. The unpaired electrons, whether they are d electrons in a coordination compound, organic free radicals, or unpaired p electrons as in O_2 or NO, align themselves in the same direction as the field and are attracted to the field. Substances with paramagnetic properties are, therefore, pulled into a field. When a paramagnetic substance is weighed in the presence of an inhomogeneous magnetic field, it will appear to be heavier than without the field. Unlike diamagnetism, paramagnetic effects are affected by temperature. This is because thermal agitation will destroy some of the alignment of the magnetic dipoles. Paramagnetic effects of a molecule are easy to measure in the presence of the diamagnetic effects from the completed shells because the paramagnetic effect is 10 to 1000 times that of the diamagnetic effect. Even though the paramagnetic effect is much larger than the diamagnetic effect, a paramagnetic characterization of a substance requires corrections for the diamagnetic effects.

From the discussion in Part I on magnetic susceptibility, one starts with

$$\chi = \frac{C\ell(F - \delta)}{wt} + \frac{\kappa_o \, vol}{wt} \qquad [1]$$

Where:

χ is the magnetic susceptibility

C is a constant if the same tube is used for all measurements (tube constant)

ℓ is the length (or height) of the sample in the tube

F is the force exerted on the material (in mg)

δ is the diamagnetic correction (in mg) for the tube material. For compensating tubes $\delta = 0$

wt is the weight of the sample expressed in grams

κ_o is the volume susceptibility of the air displaced by the sample. It is equal to $+0.029 \times 10^{-6}$ cgs units/cc. This corrects for the paramagnetic properties of oxygen in air. For a nitrogen atmosphere, κ_o can be ignored.

vol is the volume of the sample expressed in cc.

The entire second term can be ignored if the sample is small. Assuming the sample to be small and a compensating Gouy tube is being used,

$$\chi = \frac{C\ell(F)}{wt} \qquad [2]$$

If the same length (height) of sample is used for both the standard and unknown, $C\ell$ can be considered a constant that can be determined by use of a standard material with a known susceptibility. Once $C\ell$ is determined, a susceptibility can be determined for any other material.

The gram susceptibility, χ, determined for the substance in question is then converted to the atomic susceptibility, χ_A, by multiplying the mass susceptibility by the atomic weight of the metal atom. When the atomic susceptibility is corrected for the diamagnetic components of ligands and associated ions, then χ'_A is obtained. These corrections are shown in Table I of the essay on magnetic susceptibility in Part I. Also, as shown in this essay, the number of unpaired electrons for first row transition elements can be calculated using:

$$n = -1 + \sqrt{1 + 8.0656\chi'_A T} \qquad [3]$$

Experimental Procedure

A compensating Gouy tube is one which is long enough so that both ends of the tube extend out of the effective magnetic field. The tube is constructed so that there is a glass membrane dividing the upper and lower sections of the tube. See Fig. 1 for the Gouy setup in the essay on magnetic susceptibility.

Use a compensating tube 5 to 6 mm in diameter and 25 to 30 cm long. One half of the tube will be the compensating portion below the glass membrane. The other half of the tube will be the portion containing the sample above the membrane. A sample chamber of 12.5 to 15 cm will be adequate for a sample height of 10 cm as described below. In weighing the compensating Gouy tube, the glass membrane should be placed at the center of the magnet poles.

Observe the temperature in the weighing chamber and enter it in your laboratory notebook.

Weigh the empty Gouy tube in and out of the magnet field and confirm that it is a compensating tube.

Fill the compensating Gouy tube to a height of 10 cm with mercury(II) tetrathiocyanatocobaltate. The magnetic susceptibility for this compound is 16.44×10^{-6} at 20°C. The most likely source of error in using this substance is in failing to pack the Gouy tube properly. Always use the following technique for packing solids in Gouy tubes. The sample is first powdered using an agate mortar and pestle. Add about 1 cm of the powdered sample to the Gouy tube. Tap 50 to 60 times on a wooden block. This may be done by dropping the Gouy tube through a short length of a larger tube onto a wooden block. Add another cm and repeat. If static electricity develops, wipe the outside of the tube with a piece of cloth moistened with ethyl alcohol. Continue the procedure until the Gouy tube is filled to a height of 10 cm.

When a liquid is used for calibrating purposes, packing is not a problem. However, the volume of the miniscus may need to be considered. The liquids commonly used are water and nickel(II) chloride solution. The susceptibility of water at 20°C is $\chi = -0.720 \times 10^{-6}$. Near 20°C the change in χ per degree is 0.12 per cent. The susceptibility of nickel(II) chloride solution is

$$10^6\chi = 10{,}030\frac{y}{100T} - 0.72(1 - \frac{y}{100})$$

where y is the weight per cent of nickel(II) chloride in the solution. The nickel concentration is determined by analysis.

Weigh the filled Gouy tube in and out of the magnetic field. The weight difference, in milligrams, gives the value for F. The weight of the filled Gouy tube minus the weight of the empty Gouy tube gives the weight of the sample.

Calculate the constant $C\ell$.

Fill the Gouy tube to a height of 10 cm with a complex transition metal compound prepared in the laboratory and determine its χ. χ_A, the atomic susceptibility, is the gram susceptibility, χ, times the atomic weight. The χ_A is then corrected for the diamagnetic components of the ligands to get χ'_A. From the value of the χ'_A and the temperature measured in the sample compartment, calculate the value of n for the compound using equation [3] above.

Questions to be Considered

1. What is the definition for μ_{eff}?

2. What phenomena does the Curie Law address?

3. What is ferromagnetism?

4. Tabulate in your notebook some experimental and calculated values of the effective magnetic moment (2).

Helpful Background Literature

1. See the essay on Magnetic Susceptibility.

2. Jolly, W. L. "The Synthesis and Characterization of Inorganic Compounds," Prentice-Hall, Inc.: Englewood Cliffs, N.J., 1970, pp. 369 – 382.

ABSORPTION SPECTRA OF TRANSITION METAL COMPLEXES

Equipment needed:	Absorption spectrometer Reflectance attachment for above
Chemicals needed:	Coordination compounds (prepared in previous laboratory experiments.)
Time needed:	3 hours
Safety notes:	Take care to dispose of the samples properly.

Purpose

The object of this experiment is to characterize the optical properties of inorganic solids that have been prepared in previous experiments. The student will learn how to use Tanabe-Sugano diagrams, to interpret absorption spectra, to collect diffuse reflectance spectra and will compare the optical properties of pure and impure samples. Some of the compounds that were made during the semester that can be studied are $Co(en)_3{}^{3+}$, $Ni(en)_3{}^{2+}$, $Cu(thiourea)_3Cl$, and $KCuCl_3$. Note that all these compounds are colored and should have absorptions in the visible region of the spectrum.

Introduction

A. Absorption

Absorption of light by molecules is dependent on the specific molecular orbitals that are available in that molecule. As a simple example, most molecules in their ground states have two paired electrons in the highest occupied molecular orbital (HOMO). On absorption of light, electrons from the HOMO can be raised to the lowest unoccupied molecular orbital, LUMO. This process is shown in Fig. 1.

LUMO

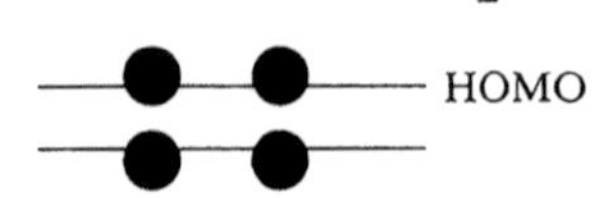

Fig. 1. MO Diagrams for Ground and Excited States.

One of the important selection rules in absorption spectroscopy is that the spin multiplicity is given by the formula:

$$2S + 1, \text{ where } S = s_1 + s_2 + s_3 + \ldots s_n \qquad [1]$$

The total spin quantum number, S, is the sum of the individual spins. In the above situation, $S = s_1 + s_2 = 1/2 + -1/2 = 0$, and, therefore, the spin multiplicity $2S + 1$ is equal to 1. Both the ground and excited states are therefore, called singlet states. The selection rule for optical absorptions is that the spin multiplicity must not change from the ground state to the excited state. Such transitions are called spin-allowed transitions.

The other important selection rule depends on the symmetries of the orbitals that are involved in the transition. An electric dipole mechanism is the only important mechanism for light absorption. The electric dipole moment operator, r, needs to be considered in the following way. The transistion moment integral,

$$\int_{-\infty}^{+\infty} \psi_a \, r \, \psi_b d\tau \qquad [2]$$

must have a finite value for a transition to occur between two energy states a and b. According to Laporte's Rule, only transitions between even (g) and uneven (u) states are allowed. In order to tell whether such transitions between states a and b are LaPorte-allowed or LaPorte-forbidden, group theory must be used.

For example, if ψ_a and ψ_b transforms as Γ_a and Γ_b of the point group to which the molecule belongs, and if r transforms as Γ_r of that group, then the integral will be nonzero only if the reducible representation ($\Gamma_a \times \Gamma_r \times \Gamma_b$) contains the totally symmetric representation of that point group (A_1, A_{1g}, etc.). One can easily show that the d-d electronic transitions of transition metals are LaPorte-forbidden.

The intensity of a band is related to the oscillator strength, f. The oscillator strength is given by the relationship:

$$f = 4.32 \times 10^{-9} \int_{\nu_2}^{\nu_1} E \, d\nu \qquad [3]$$

where v is the frequency and E is the extinction coefficient. The oscillator strength is also related to the transition moment integral, P_e, by:

$$f = 1.06 \times 10^{11} \, v \, P_e^{\,2} \qquad [4]$$

B. Types of Electronic Transitions

A comparison of types of electronic transitions, oscillator strengths and extinction coefficients is given in Table I.

Table I

Types of Electronic Transition	f	E
1. Spin-forbidden, LaPorte-forbidden	10^{-7}	10^{-1}
2. Spin-allowed, LaPorte-forbidden	10^{-5}	10
3. Spin-allowed, Laporte-forbidden, with d-p mixing.	10^{-3}	10^{2}
4. Spin-allowed, Laporte-forbidden, with vibronic coupling.	10^{-2}	10^{3}
5. Spin-allowed, LaPorte-allowed	10^{-1}	10^{4}

Spin-forbidden and Laporte-forbidden transitions are very weak and often difficult to observe. Spin-allowed and Laporte-forbidden transitions are more intense and are the typical electronic transitions observed for transition metal complexes. When orbitals can mix, such as d and p orbitals in a tetrahedral field (which has no center of symmetry and, therefore, the orbitals are not described as even (g) or uneven (u)), the intensity is even larger. When vibrational wave functions can mix with electronic wave functions, this leads to vibronic coupling and even more intense transitions. In order to tell whether vibronic coupling is possible, the vibrational modes of the ground and excited states are multiplied into the direct product described earlier. Again, if the reducible representation contains the totally symmetic representations of that point group, then the intensity of that transition can be enhanced due to vibronic coupling. The most intense transitions are spin-allowed and LaPorte-allowed transitions. Charge-transfer bands are of this type.

Charge-transfer can be from the metal to the ligand, from the ligand to the metal or they can be in the ligand itself (intraligand). Charge-transfer bands commonly occur in the UV region of the spectrum but can tail off into the visible region.

C. Predicting the Number of Bands in the Absorption Experiment

The number of bands that we expect to observe in a UV-visible experiment for a transitional metal complex is related to the symmetry of the molecule. For an octahedral system the crystal field splitting diagram is as follows:

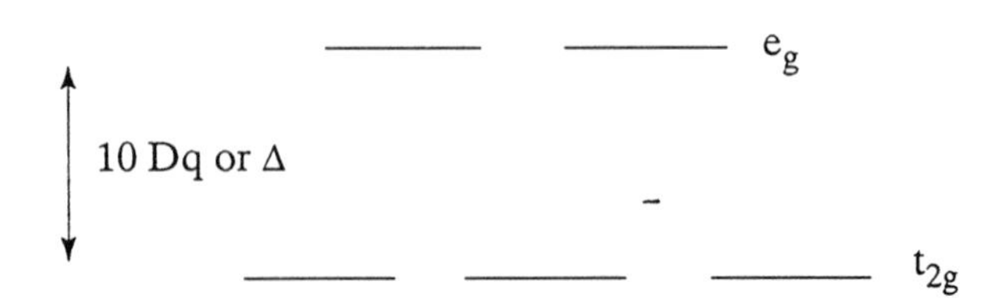

Fig. 2. Crystal Field Splitting for Octahedral System.

The low energy orbitals (xz, yz, xy) are labelled as t_{2g} orbitals according to group theory procedures. They are even (g) type orbitals. The z^2 and $x^2 - y^2$ orbitals are of higher energy and are labeled e_g. Spin-allowed transitions are of higher intensity than spin-forbidden transitions and only the former type will be discussed.

Let us take, for example, a d^1 system such as $Ti(H_2O)_6{}^{3+}$. The crystal field splitting diagram with the one d electron in the ground state is given below:

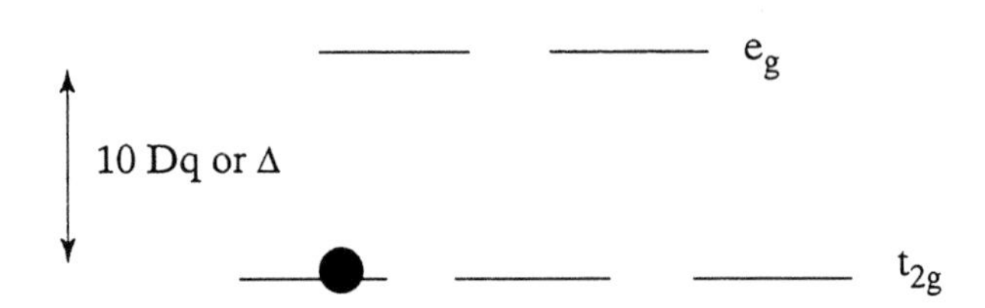

Fig. 3. Crystal Field Splitting for d^1 System.

The energy between the t_{2g} and e_g sets of orbitals is referred to as 10 Dq or Δ. The ground state of this system is labelled $^2T_2$ which is of the general form $^{2S+1}x_q$. We have already discussed the 2s+1 spin-multiplicity term. The symbol x refers to the number of ways that one d electron can go into the low energy (ground state) t_{2g} orbital. It turns out that it can go into any of the 3 (xz, yz, xy) t_{2g} orbitals and, therefore, is given the symbol T. Other symbols for x are A for 1 way and E for 2 ways. The subscript q is obtained from group theory and is not critical to this discussion.

When the single d electron of $Ti(H_2O)_6{}^{3+}$ is excited in an absorption experiment, the resultant crystal field diagram looks like the following:

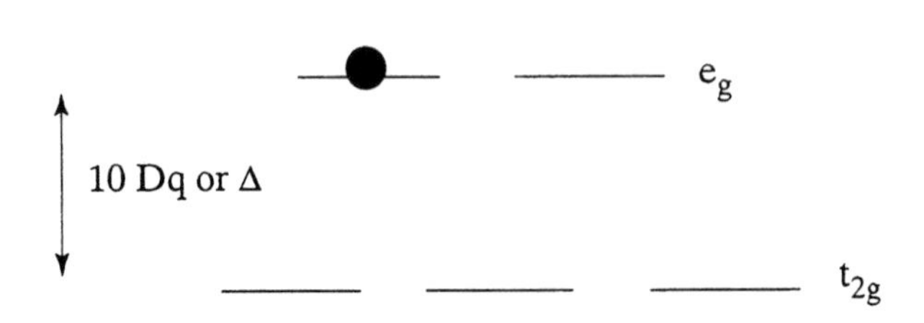

Fig. 4. Single d Electron in Excited State.

The energy of this absorption is given by the quantity 10 Dq and the $^{2S+1}x_q$ label for the excited state is now ^2E. There can be excited-state configurations that involve 2 electrons being excited. These are of higher energy and, again, to be allowed must result in no change of spin-multiplicity.

The relative energies of all the possible excited-state configurations could be generated using the above principles for a variety of crystal fields and electron occupancies. Instead, we will focus on octahedral systems and in particular on energy level diagrams that have already been established for ground- and excited-state configurations. Such diagrams are called Orgel or Tanabe-Sugano diagrams depending on how they are constructed. We will limit our discussion to Tanabe-Sugano diagrams.

D. Tanabe-Sugano Diagrams

Tanabe-Sugano diagrams have been constructed for octahedral systems of various d electron populations from optical spectra of known complexes. A Tanabe-Sugano diagram for an octahedral d^7 system is shown below:

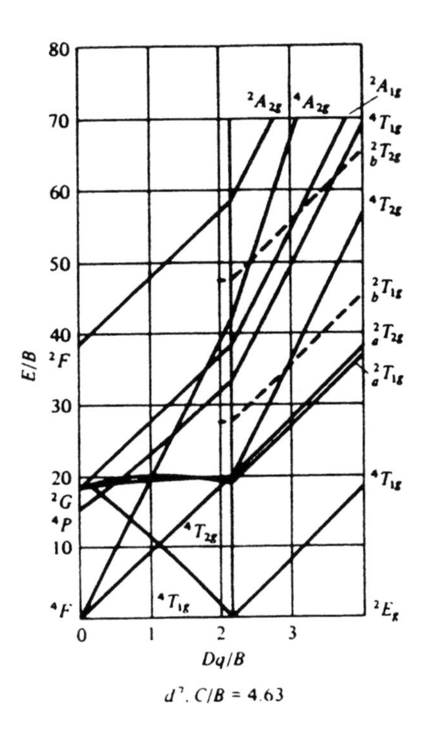

Fig. 5. Tanabe-Sugano Diagram for Octahedral d^7 System.

There are several features of this diagram that are crucial to the understanding of how these diagrams are used to predict the number of bands in an optical spectrum of an octahedral complex. First of all, the y-axis represents a unit of energy. It is the relative energy between the various orbitals of the molecule. The symbols on the y-axis refer to ground and excited states for a *free* ion or atom. These states have nothing to do with the nature of the ligands that are bound to the metal. Therefore, we can see that the ground state for a d^7 *free* ion or atom is $^4F_2$ and that the excited states (higher in energy) are ^4P, ^2G, and ^2F. These can be derived from Russell-Saunders coupling in combination with Hund's Rules.

The second feature of the diagram that should be noticed is the x-axis. The x-axis is related to the relative separation in energy of the ground and excited states. This will mainly depend on the nature of the ligands around the metal. Notice the label Dq on the x-axis and remember that in the octahedral diagram this is related to the separation between the low energy (t_{2g}) and the high energy (e_g) orbitals.

A very important concept dealing with the relative separation of these energy levels is that if the separation is low then electrons can more easily go into the high energy orbitals (e_g). The converse is also true. The two different crystal field diagrams for a d^7 system are given below:

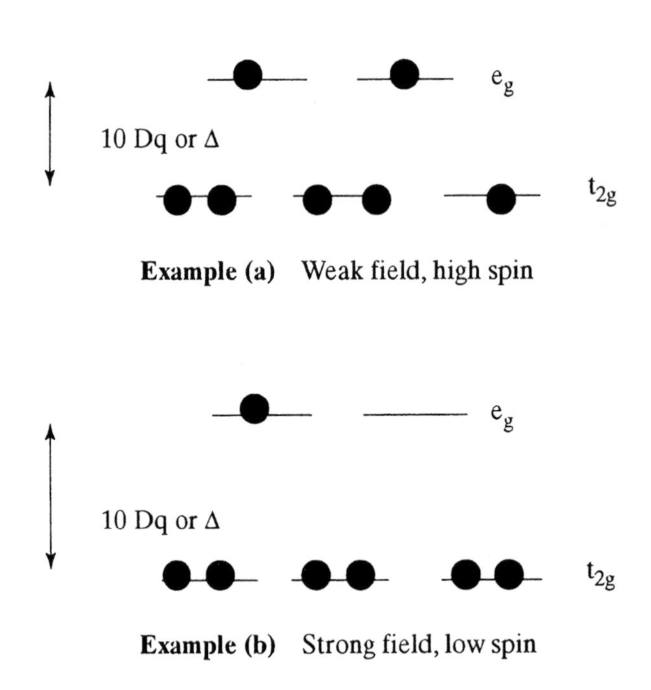

Fig. 6. Crystal Field Diagrams for d^7 System.

In Example A, the energy separation 10 Dq is small. Here electrons would rather enter the e_g orbitals than pair up in the t_{2g} orbitals. For example, the 4th and 5th electron would go into the e_g orbitals rather than the t_{2g} orbitals. This situation is called high spin or weak field and results in 3 unpaired electrons and a T_{1g} ground state.

In Example B, the first 6 electrons go into the t_{2g} orbitals and pair up because 10 Dq is very large. This results in a system with 1 unpaired electron and a ^2E_g ground state. This situation is called low spin or strong field.

If we now return to the d^7 octahedral Tanabe-Sugano diagram and study the x-axis, we observe that on the left hand side of the vertical line at 2.0 Dq/β that the ground state (lowest horizontal line) is labelled $^4T_{1g}$ as shown above for the high spin or weak field Example A. To the right hand side of the vertical line we see the ground state of the d^7 system is labelled ^2E_g. Therefore, high spin or weak field systems occur on the left hand side of such diagrams at low values of Dq/β and low spin or strong field systems have large values of Dq/β.

Finally, if we want to predict the total number of transitions, we look for the total number of excited states that have the same spin multiplicities as the ground state. For example, for a low spin d^7 system (assuming a value of $Dq/\beta = 4$) the first excited state is $^2T_{1g}$ since the spin multiplicity $2S+1$ is equal to 2 for the ground state (^2E_g). The $^2T_{2g}$ state is the second excited state and so on. These transitions listed in order of increasing energy are $^2E_g \rightarrow {}^2T_{1g}$, $^2E_g \rightarrow {}^2T_{2g}$. Transitions to the $^2T_{1g}$, $^2T_{2g}$, and $^2A_{1g}$ excited states will be at higher energies.

Experimental Procedure

Prepare aqueous solutions of $Co(en)_3{}^{3+}$ and $Ni(en)_3{}^{2+}$ and load these into cuvettes for the absorption experiment. The exact concentration of the solute depends on the extinction coefficient for that compound. For these transition metal complexes and for $KCuCl_3$, about 50 mg in 20 mL of water should be close to the right concentration. If reflectance attachments are available for analyses of solids, then load solid salts into the cuvette.

Turn on the spectrometer, insert the samples, and collect absorption spectra for the solids and solutions according to the procedures outlined by your instructor. A plot of intensity versus wavelength will be collected for each sample. Run these experiments between 800 nm and 200 nm.

Other compounds that you have made this semester such as $Cu(thiourea)_3Cl$ and $KCuCl_3$ should also be analyzed.

Some precautions:

1. Do not open the sample compartment during a scan.
2. Run a background scan first with an empty cuvette.
3. Load your sample into a cuvette in the reflectance apparatus.

Predict the number of UV-visible bands for the ethylenediamine complexes assuming an octahedral geometry. Based on your observations, are these complexes high or low spin? Try to label the ground and excited states for the transitions that you observe. Are there differences between the solution and solid spectra? If so, what are these due to?

Questions to be Considered

1. Do you expect intraligand charge-transfer bands to be dependent on the nature of the transition metal? Why?

2. Using Tanabe-Sugano diagrams predict the number of transitions for the following species:

(a). $Mn(H_2O)_6{}^{2+}$

(b). $Cu(H_2O)_6{}^{2+}$

(c). $Fe(CN)_6{}^{3-}$

3. Compare and contrast the optical absorption spectra for the Ni^{2+} and Co^{3+} ethylenediamine complexes as solids and as complexes in aqueous solution.

4. Show via group theory that d-d transitions are LaPorte-forbidden. Hint: The electric dipole moment operator transforms like a translational mode.

Helpful Background Literature

1. Huheey, J. "Inorganic Chemistry"; Harper and Row: New York, 1983.

2. Purcell, K. F.; Kotz, J. C. "Inorganic Chemistry"; W. B. Saunders: Ft. Worth, 1977.

3. Day, M. C.; Selbin, J. "Theoretical Inorganic Chemistry"; Van Nostrand Reinhold Co.: New York, 1969.

X-RAY POWDER DIFFRACTION OF PILLARED CLAYS

Equipment Needed:	X-ray powder diffraction instrument
Chemicals Needed:	Sample of Bentolite L Sample of pillared clay (prepared in a previous laboratory experiment)
Time Needed:	3 hours
Safety Notes:	Do not expose your eyes to the X-ray beam, as it can cause blindness. When using ionizing radiation, a film badge should be worn.

Purpose

It is desired to measure the extent of pillaring, or for that matter, whether pillaring took place at all. This experiment is designed to expose the student to the technique of obtaining X-ray powder diffraction data and interpreting the results obtained.

Introduction

The goal in the preparation of the pillared clay was to change the structure of a two-dimensional infinite-network inorganic compound in the third dimension using rather mild conditions. In order to find out to what extent the separation of the layers was accomplished, it is necessary to experimentally determine the cell parameters of the solid crystalline material by X-ray powder diffraction. Before starting this experiment, the essay on X-ray diffraction in Part I of this manual should be read.

Experimental Procedure

Collect the X-ray powder diffraction data for the precursor bentonite and the pillared sample. The exact directions for collecting this data is omitted here because it depends on the equipment available. The availability of modern instrumentation will make this data collection easier than some of the older techniques such as the use of Debye-Scherrer cameras with film.

It is important that data are collected over the same range of 2θ values for each material so that a valid comparison can be made. For a pillared clay system, it is important to collect data at low values of 2θ, on the order of about 3° 2θ. If possible, start the X-ray scan at about 2° 2θ. Bentonite has peaks as far out as 40° 2θ so it important to scan at least out to this angle.

Once the X-ray data are collected, make plots of each of the diffraction patterns and get a copy of the raw data (if possible) which give a listing of 2θ values, d-spacings, and intensities. If only the 2θ values are available, use Bragg's law to calculate the d values for each peak. Make sure that you know the correct wavelength of radiation which was used in the diffractometer; this wavelength can change depending on the type of X-radiation (X-ray anode) that is used.

Once the powder diffraction data are obtained, it is important to compare the X-ray data for the bentonite and the pillared bentonite sample. Pillaring occurs along the c direction of these materials. The first reflection of clay materials like the bentonite sample is generally an (00ℓ) reflection. This means that this plane is infinite along the a and b directions. Determine the d-spacing for this reflection.

If pillaring of bentonite is successful, then the pillaring agent such as aluminum chlorhydrol will intercalate between the layers of the bentonite and the material will expand. There are several applications of pillared type materials because the expansion causes an increase in surface area and open space in the material. This can lead to enhanced adsorption of the pillared material and an accordion-like appearance of the particles.

Once the bentonite expands, the surface will increase substantially. In addition, the X-ray d-spacings that have a component along c (all of the reflections where ℓ does not equal zero) will change due to this expansion. The planes that will change the most are the (00ℓ) planes since they have no a or b component, neither of which should influence the expansion. It is very simple then to calculate the amount of expansion of the bentonite by subtracting the d-spacing value of the same (00ℓ) reflection of the bentonite sample from the d-spacing value of the same (00ℓ) reflection of the pillared bentonite material.

It is important to look at all of the other diffraction peaks and see how these peaks change from the bentonite sample to the pillared bentonite sample. Make a list of each of the X-ray peaks in terms of d-spacings and note the change in d-spacing after intercalation of the pillar.

Questions to be Considered

1. What value of an increase in d-spacing did you obtain? Is this a reasonable value?

2. Did you observe the same expansion for all of the diffraction lines? Why or why not?

3. Are there any reflections that did not shift after intercalation of the pillar? If so, why is this the case?

4. Compare the widths and intensities of the peaks for the pillared and precursor materials. Are there any noticeable differences? Why?

Helpful Background Literature

1. Wells, A. F., "Structural Inorganic Chemistry," 5th ed., Clarendon Press: Oxford 1984.

2. Occelli, M. L.; Tindwa, R. M., *Clays Clay Min.*, 1983, *31*, 107.

3. Suib, S. L.; Carrado, K. A.; Skoularikis, N. D.; Coughlin, R. W., *Inorg. Chem.*, 1986, *25*, 4217.

4. Ocelli, M. L.; Robson, H. S. "Expanded Clays and Other Microporous Solids"; Van Nostrand Reinhold: New York, 1992.

LUMINESCENCE SPECTROSCOPY

Equipment needed:	Luminescence spectrometer pellet press
Chemicals needed:	Copper doped phosphor (prepared in a previous laboratory experiment.)
Time needed:	3 hours
Safety notes:	Though minimal, heed should be taken for shock hazards whenever electrical equipment is being used.

Purpose

The purpose of this experiment is to study the luminescence of inorganic compounds. The luminescence excitation, emission, and lifetime data will be collected for luminescent compounds that have been prepared this semester.

Introduction

When a molecule absorbs light, electrons can enter excited states. When a ground-state singlet is excited, an excited-state singlet can result as shown below:

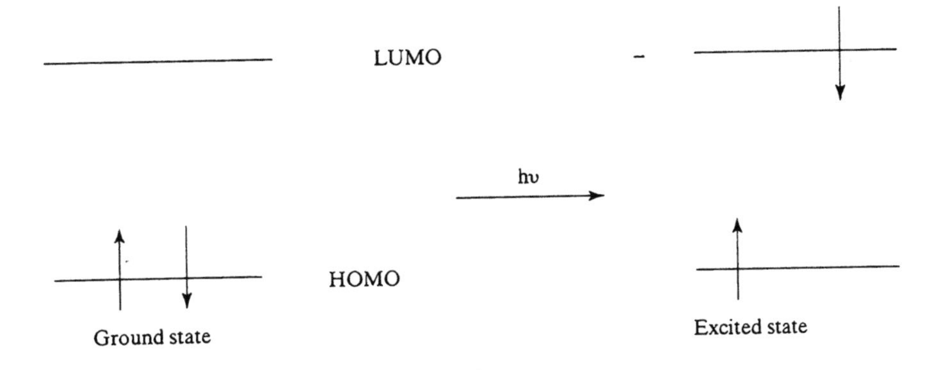

Fig. 1

The absorption or excitation experiment can monitor this process.

The excited state has an electron in a high energy state. Such electrons can do many things depending on the symmetry of the system and the location of molecular orbitals. The electron can return to the ground state giving off heat to the surroundings. This is shown in Fig. 2 as process 1. Another possibility is that light energy can be given off as the electron returns to the ground state. This is a radiative transition and is labelled as process 2.

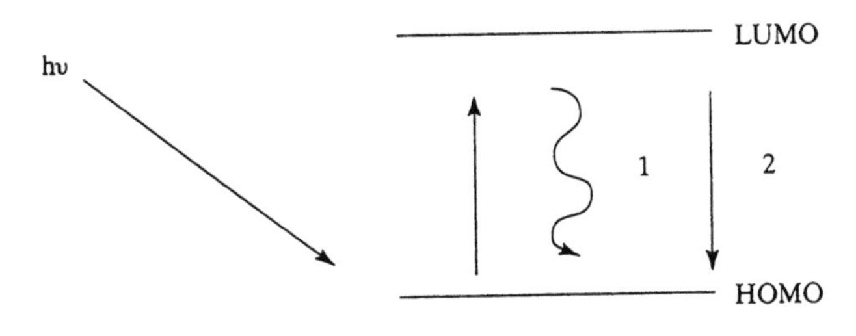

Fig. 2

The radiative process is known as fluorescence and occurs on a relatively fast timescale usually between microseconds and nanoseconds. Organic and inorganic molecules that have molecular orbitals that are sufficiently separated in energy can fluoresce. This fluorescence usually occurs in the ultraviolet or visible region of the spectrum.

If the electron in the excited state can undergo a flip in spin and reach a lower energy excited state (a triplet state) then nonradiative and radiative transitions can again occur. These processes are shown in Fig. 3.

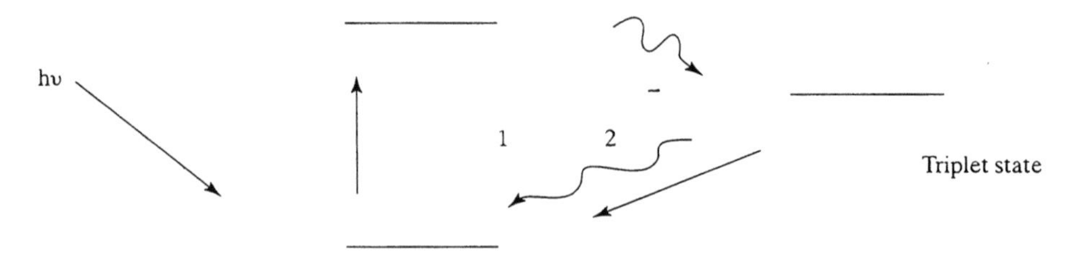

Figure 3

The transition between excited states of different spin is called intersystem crossing (process 3) and is forbidden. Forbidden transitions between the excited-state triplet and the ground state are either nonradiative (process 4) or radiative (process 5).

The radiative process involves spin flips and takes a considerable amount of time, typically longer than a microsecond. For this reason, process 5 (known as phosphorescence) occurs between seconds and microseconds. Phosphorescence and fluorescence together are known as luminescence. Such definitions can become blurry when spin-orbit coupling is large.

Another thing that can happen with the excited-state energy is a change in the chemical components of the systems. Bonds can be broken and made as a result of this excited state energy. These processes are known as photochemical processes.

A diagram linking these different processes is given below:

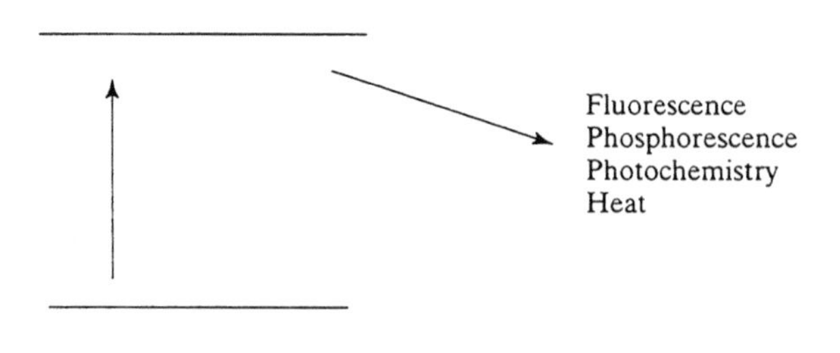

Fig. 4

Types of Luminescent Inorganic Compounds

The simplest inorganic luminescent species are isolated atoms. Certain atoms like Ne have excited energy levels that allow light to be emitted during radiative decay. This process forms the basis of neon lights.

Metals do not luminesce because they are quenched or "grounded." Usually fluorescent molecules near the surface of metals are quenched (do not emit) because of this grounding.

Semiconductors on the other hand can luminesce if the energy separation of the conduction and valence bands is in the ultraviolet or visible region of the spectrum. If the separation is too low (< 1 eV) or too high (> 6 eV) the material will not luminesce provided that it is relatively pure. Donors and acceptors in the band gap can modify the efficiency of the luminescence process and can modify the energy (frequency or wavelenth) of emission.

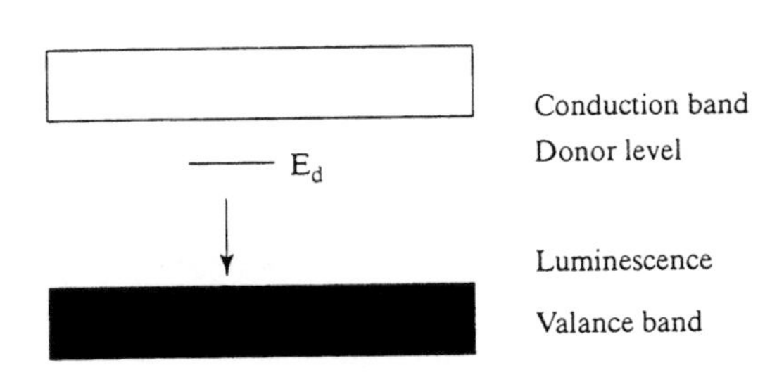

Fig. 5

Pure insulators ($E_g > 6$ eV) would not luminesce. However, just like the doping of semiconductors, insulators can be doped so that luminescence can be observed. Ruby is Cr^{3+} doped alumina (a great insulator) and luminescence due to the energy levels of Cr^{3+} which are between the large band gap of the insulator.

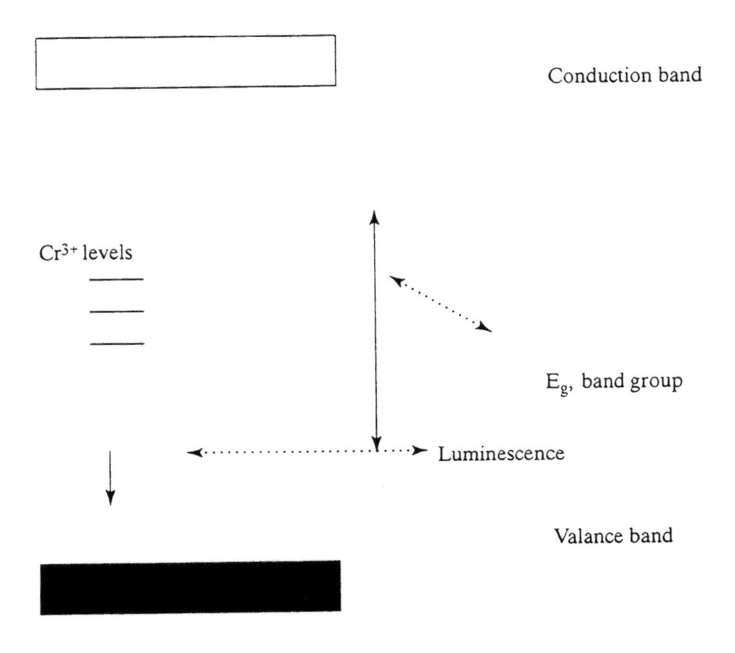

Fig. 6

Experimental Procedure

Collect excitation and emission spectra for the Cu^{2+} doped ZnS you made this semester. Excite the sample at 366 nm.

Questions to be Considered

1. Why is the sample excited at 366 nm?

2. Is there a difference between the excitation and emission spectra? Are they related?

3. Spectra for two different preparations of copper doped zinc sulfide were collected. The intensity of one material is 100 times more intense than the other which is a weak emitter. What do you conclude about the two samples?

4. The other attached spectra (Appendix I) are those of uranyl acetate dihydrate. The $UO_2{}^{2+}$ ion luminesces. Note that there is a major difference between the solid state and solution spectra. Why? What is the structure of the solid state spectrum due to? Why does it not occur in the solution spectrum?

5. Shown in Appendix II is a plot of the luminescence lifetime decay of the above uranyl salt on a zeolite support. The plot is of the natural log of the luminescence intensity versus time. Give a kinetic expression to explain this process. Define all terms. Interpret the chi-squared value and the sum of the squares of the residuals.

6. List three commercial applications of luminescence.

Helpful Background Literature

1. Goldberg, M. C., Ed. "Luminescence Applications"; ACS Symposium Series 383, American Chemical Society: Washington, D.C., 1989.

2. Rabek, J. F., Ed. "Photochemistry and Photophysics Volume 11"; CRC Press: Boca Raton, 1990.

3. Krasovitskii, B. M.; Bolotin, B. M. "Organic Luminescent Materials"; VCH: Weinheim, 1988.

4. Skoog, D. A.; West, D. M.; Holler, F. J. "Fundamentals of Analytical Chemistry" 5th ed.; Saunders College Publishing: New York, 1988.

Appendix I

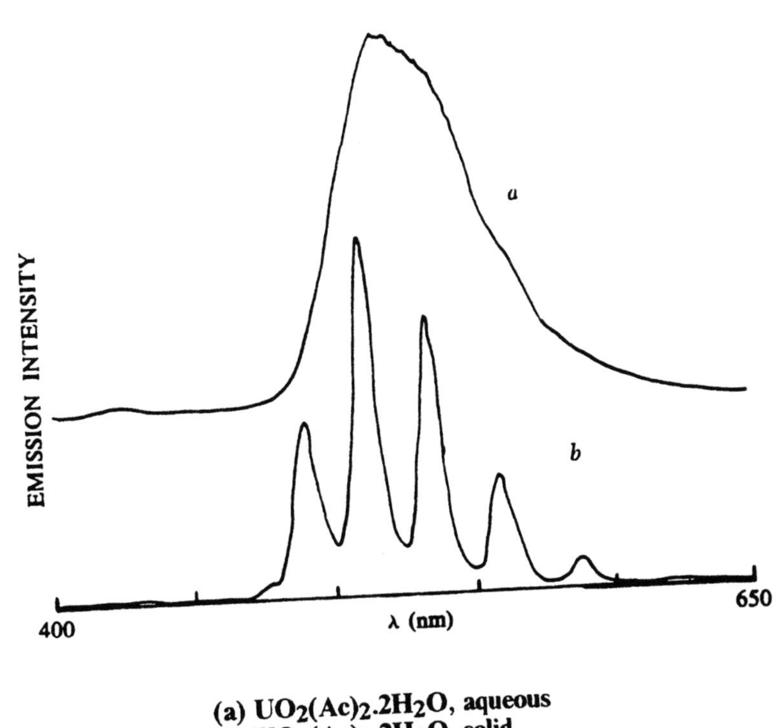

(a) $UO_2(Ac)_2.2H_2O$, aqueous
(b) $UO_2(Ac)_2.2H_2O$, solid

Appendix II

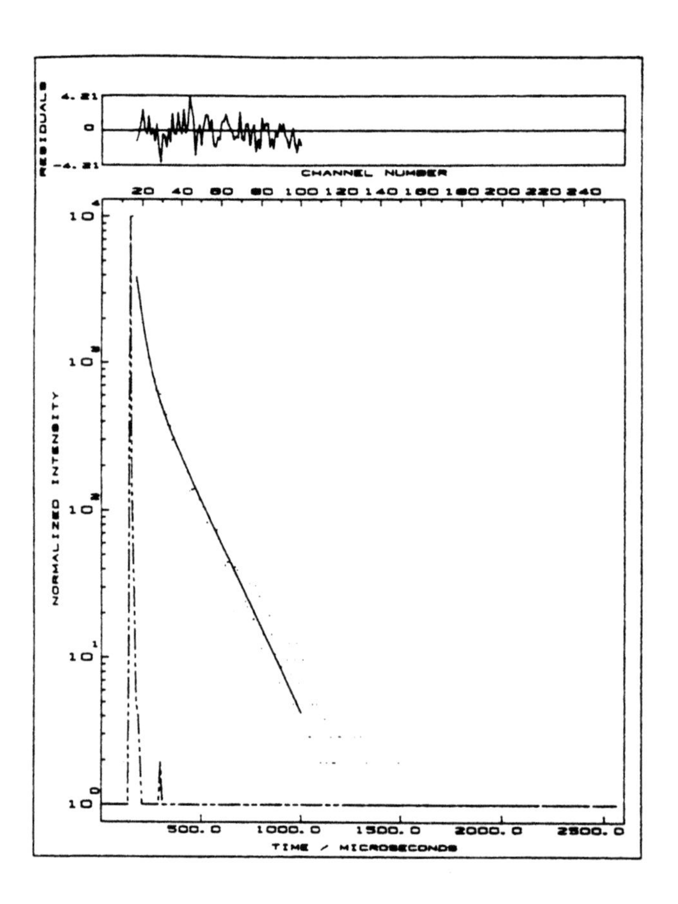

Luminescence Lifetime Decay of Uranyl Solid
Top Graph - Sum of squares of residuals
Bottom Graph - Lifetime decay

X-RAY PHOTOELECTRON SPECTROSCOPY

Equipment needed:	X-ray photoelectron spectrometer pellet press
Chemicals needed:	Coordination compounds (prepared in previous laboratory experiments.)
Time needed:	3 hours
Safety notes:	Make sure a trained operator is available.

Purpose

The object of this experiment is to understand the X-ray photoelectron spectra of transition metal complexes that you have made this semester. In this procedure you will learn the principles of X-ray photoelectron spectroscopy (XPS) and the information that can be obtained from XPS.

Introduction

Surfaces are important as regards the chemical stability and reactivity of solids. In chemistry, we are concerned with the preparation and characterization of compounds. While the bulk properties of a compound such as the structure, geometry, and chemical composition are usually known, the outer layers of materials are seldom well understood.

Surfaces are important in biology, engineering, industrial environments, materials science, metallurgy, and physics. The surfaces of catalysts, semiconductors, polymers, alloys, and microelectronics devices govern the reactivity and stability of these materials. Some recent applications of surface science involve the study of solids that are used in artificial life-sustaining devices.

Principles of X-Ray Photoelectron Spectroscopy

X-ray photoelectron spectroscopy (XPS) is also known as ESCA, electron spectroscopy for chemical analysis. In XPS, X-rays are used to eject photoelectrons from core levels of solids. A diagram of the XPS process is given in Fig. 1.

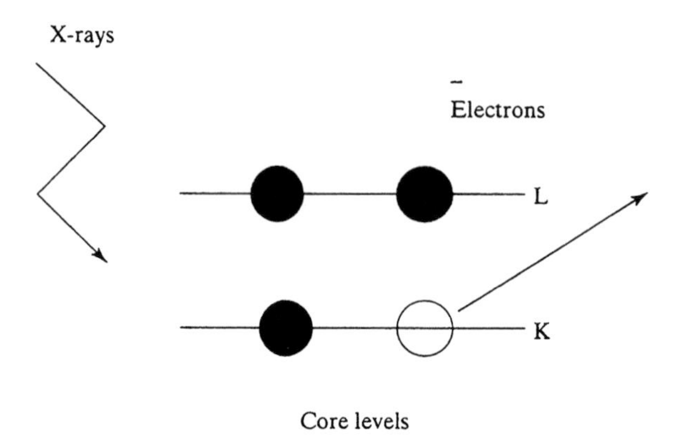

Fig. 1. Ejection of Photoelectrons from Core Levels

The energy of the ejected photoelectron is related to the incident energy ($h\nu$) minus the binding energy of the electron that is ejected from a particular level minus the work function, Φ, of the sample. The following formula applies in this situation:

$$BE = h\nu - KE - \Phi \qquad [1]$$

where BE is the binding energy of the electron in a particular orbital and KE is the kinetic energy of the ejected photoelectron.

The binding energy of the ejected photoelectron depends on the nature of the bonding of the atomic species. The work function of the spectrometer is calibrated by setting the Au 4f 7/2 transition to 83.8 eV and the Cu 2p 3/2 transition to 932.4 eV. This is done with standards of Au and Cu metal.

Energy resolution in the XPS experiment is obtained by varying the widths of the entrance and exit slits of the detector. When electrons are photoejected they have a kinetic energy which is related to the energy of the orbital from which it comes. A hemispherical analyzer is used as a detector as shown in Fig. 2 to measure the kinetic energies of electrons. Reference data for kinetic energies of electrons ejected from various orbitals are available and can be used for the identification of the elements in the sample. Some articles concerning XPS are listed in the *Helpful Background Literature* section at the end of this experiment.

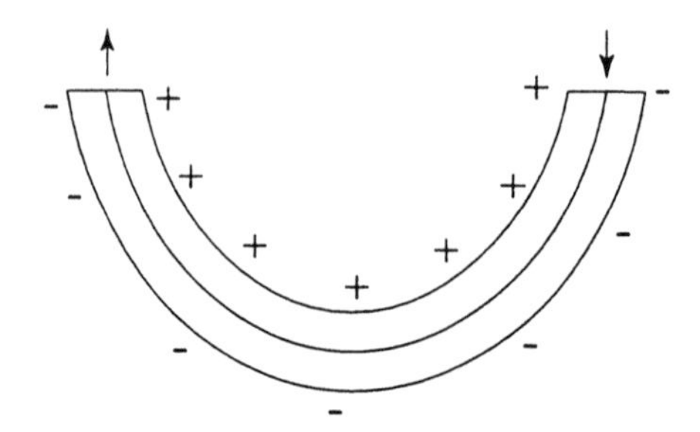

Fig. 2. Hemispherical Analyzer

There are five types of information that can be obtained from surface science experiments. They are listed in Table I.

Table I

Surface Science Data

Qualitative Analysis	-	good
Semiquantitative Analysis		good
Depth Profile Information		poor
Lateral Analysis		poor
Chemical Information		excellent

XPS is useful in determining changes in composition across the surface of a solid (lateral analysis) over a relatively large area of about one cm in diameter because of the inability of focusing the X-ray beam over a smaller area. This inability to focus means that changes in composition occurring over small distances cannot be detected. Depth profile information is difficult to obtain using XPS because the photoelectrons are primarily ejected from surface atoms. However, depth profiling can be done with XPS by bombarding the surface with ions (Ar^+) with subsequent XPS analysis.

XPS can be used to determine what elements are present on the surface of a solid. Because there are different probabilities for the XPS transitions for different elements, it may be easier to detect one element than another.

Quantitative analyses with XPS are semiquantitative at best. Reproducibility is about 10-15% with studies of the same sample in two different laboratories. With improvements in sample handling and data analysis, these studies are becoming more quantitative.

One of the most important contributions of XPS has been the determination of binding energies of electrons of particular orbitals. It turns out that these binding energies are related to the oxidation states of the elements in the sample. In general, the larger the binding energy the higher the oxidation state of an element in a compound.

Experimental Procedure

If possible, XPS experiments of $KCuCl_3$ and $Co(en)_3Cl_3$ should be done. If not, Appendices I and II are data sets collected for these materials.

Since these experiments are done under ultrahigh vacuum conditions, samples need to be loaded into the spectrometer several hours before an analysis. Samples can be pressed into pellets or dispersed in metal foils that are used as supports. The sample is then introduced into the analysis chamber.

During the experiment either Mg K alpha or Al K alpha radiation is used to eject photoelectrons. Initially a wide scan is collected over a large (1100 eV) range of binding energies. This wide scan is used to identfy which elements are present on the surface.

Besides photoelectron transitions, Auger lines are also observed in an XPS experiment. An Auger transition is given in Fig. 3.

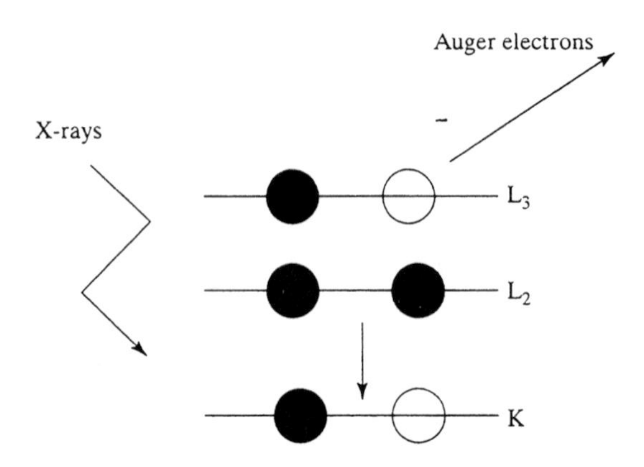

Fig. 3. Auger KLL Transition

In this case the hole (unfilled circle) in the low energy K shell is filled with an electron from the upper L shell. There is an excess energy associated with this process. This excess energy can be given to an electron in a higher shell, L_2, and this electron can be emitted from the solid if it is close enough to the surface. This is a secondary electron or an Auger electron. Energies of Auger electrons are also well known and available.

After wide scans are collected, it is necessary to scan small ranges (5-20 eV) of binding energy at high resolution. With these narrow scans it is possible to assign oxidation states to the different elements in the sample.

A wide scan for a polymer, Kapton, is given below in Fig. 4. Note that the elements C, O, Cu, and Al were detected in the wide scan.

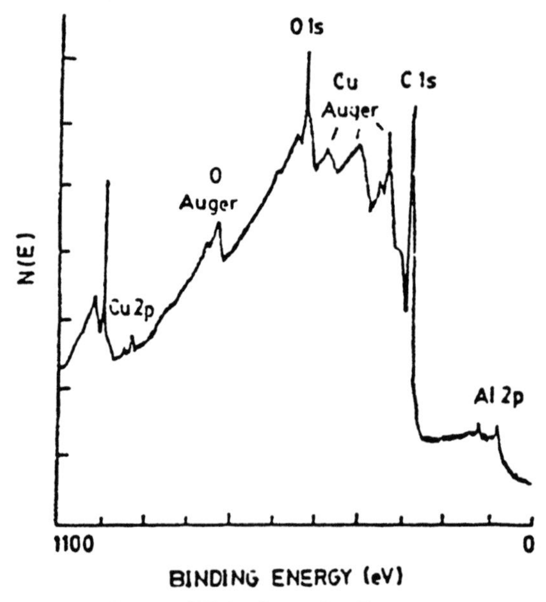

Fig. 4. Wide Scan for Kapton

For semiquantitative and binding energy information, a narrow scan is done over a smaller range of energy. The narrow scan in the carbon 1s region shown below suggests that three types of carbon exist in Kapton. These three types of carbon can be assigned to three chemically different carbon species.

Narrow Scan

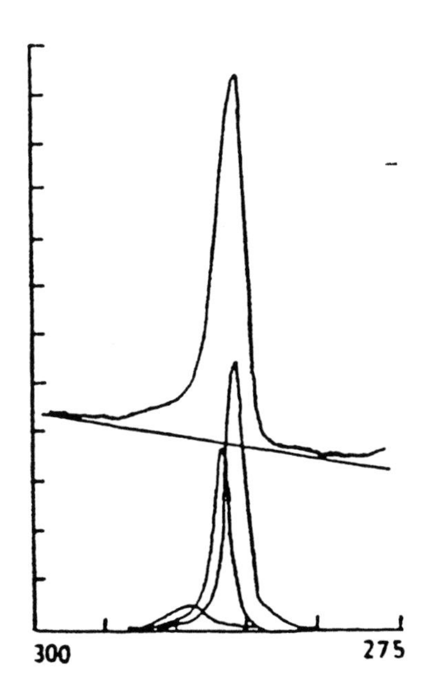

Fig. 5. Types of Carbon in Kapton

Some specific information about the XPS experiment is given below:

Sample Considerations
- Almost any solid
- Grease free

Detection Limits
- $Z > 3$
- 0.1 atomic % (dependent on photoionization cross section)

Volume of Analysis
- 10 layers deep
- 10 – 40 Å depth resolution

Data Collection Time
- 10 – 20 minutes survey

Areas of application
- Semiquantitative analysis
- Depth profiles
- Contaminants, thin films, interfaces, polymers, corrosion

Information Obtained
- Oxidation states of elements, bonding information

Interferences
- For trace work

Questions to be Considered

1. The O 1s and N 1s regions for two different sets of materials of $Co(en)_3Cl_3$ are shown in Appendix I as Sample A and Sample B. Interpret these data.

2. We can determine the relative amount of the individual elements by taking the number of counts (equivalent to the area on the graphs shown in Appendix I) and dividing by the number of scans and by the sensitivity factors of the elements. The sensitivity factor for Co is 5.9, for Cl is 0.58, and for N is 0.3. Using the Co 2p 3/2, Cl 2p, and N 1s transitions, determine the Cl/Co, Cl/N, and N/Co ratios. What do these data tell you about the surface of this complex?

3. Two samples of $KCuCl_3$ were also studied by XPS. One set of data shown in Appendix II is labeled as Sample C and the other set is labelled as Sample D. Which data set is more consistent with pure $KCuCl_3$? Why? For the sample that is less consistent, what do you think went wrong during the synthesis according to the surface data? The transition probabilities for K (1.1) and Cu (4.6) may be useful.

4. The corrected Cu 2p 3/2 transition for a preparation of $KCuCl_3$ is at 933.6 eV. The Cu Auger transition for this sample occurs at 914.15 eV. Using the attached plot of XPS and Auger lines for Cu (Appendix III), what do you conclude about the $KCuCl_3$ sample?

Helpful Background Literature

1. Walls, J. M. Ed. "Methods of Surface Analysis"; Cambridge University Press: Cambridge, 1989.

2. Somorjai, G. A. "Introduction to Surface Chemistry and Catalysis", Wiley Interscience: New York, 1994, pp. 362 – 399.

3. Carlson, T. A. "Photoelectron and Auger Spectroscopy"; Plenum Press: New York, 1975.

4. Reference Data: NIST X-Ray Photoelectron Spectroscopy, http://webbook.nist.gov

5. Somorjai, G. A. "Introduction to Surface Chemistry and Catalysis"; Wiley Interscience: New York, 1994, pp. 362 – 369.

Appendix I

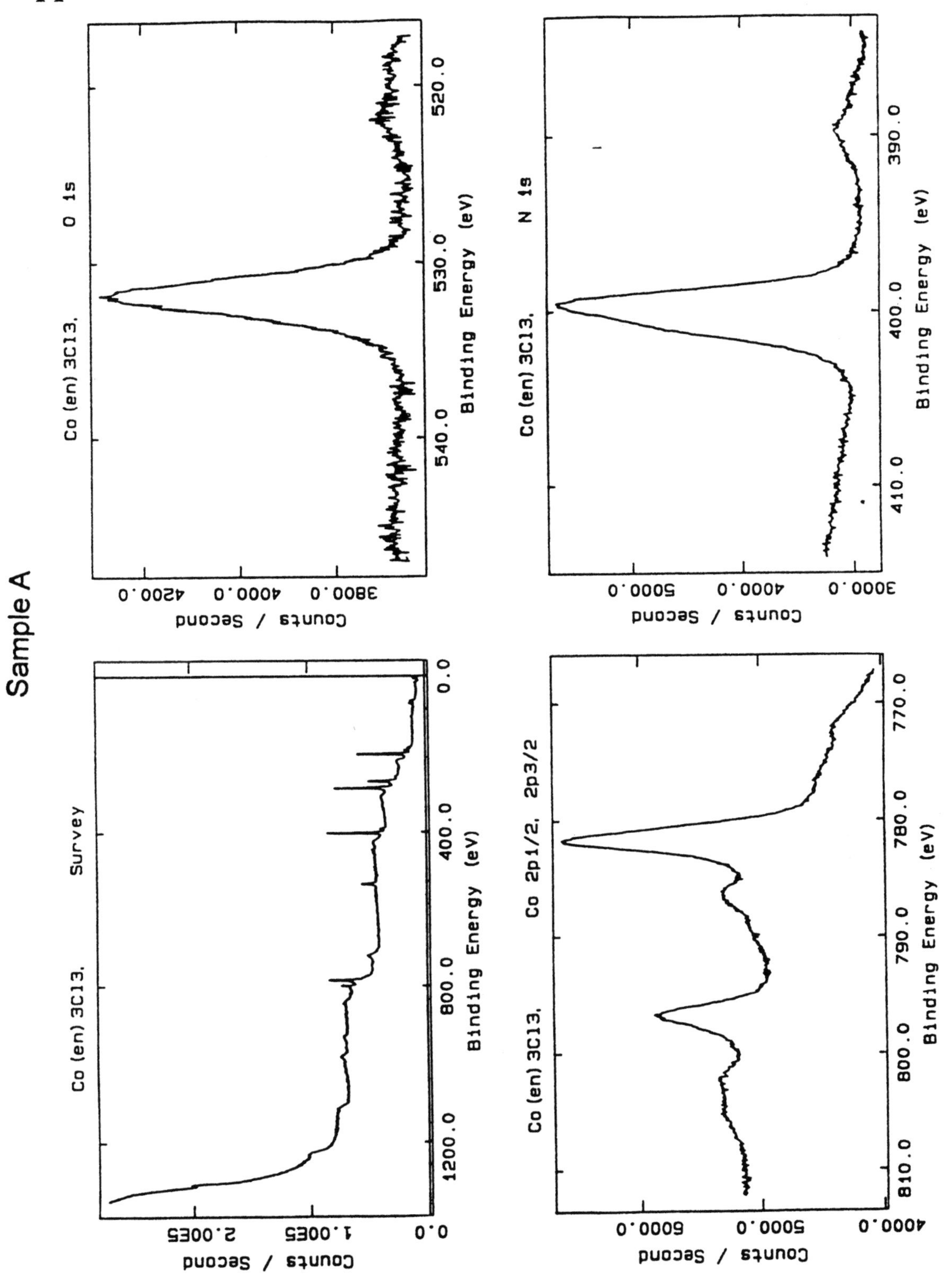
Sample A
Co (en) 3Cl3.
Survey
Counts / Second
Binding Energy (eV)
0.0
1.00E5
2.00E5
1200.0
800.0
400.0
Co (en) 3Cl3.
O 1s
3800.0
4000.0
4200.0
540.0
530.0
520.0
Co (en) 3Cl3.
N 1s
3000.0
4000.0
5000.0
410.0
400.0
390.0
Co (en) 3Cl3.
Co 2p1/2, 2p3/2
4000.0
5000.0
6000.0
810.0
800.0
790.0
780.0
770.0

Appendix I (continued)

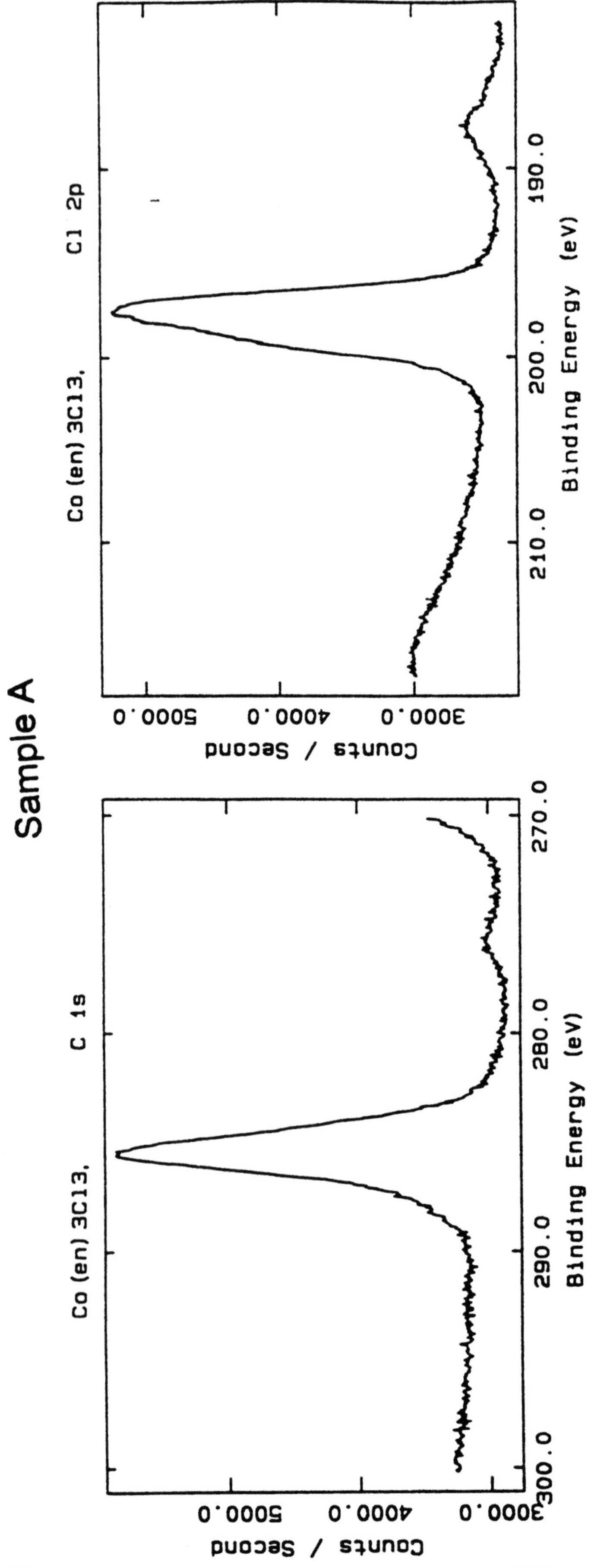

Appendix I (continued)

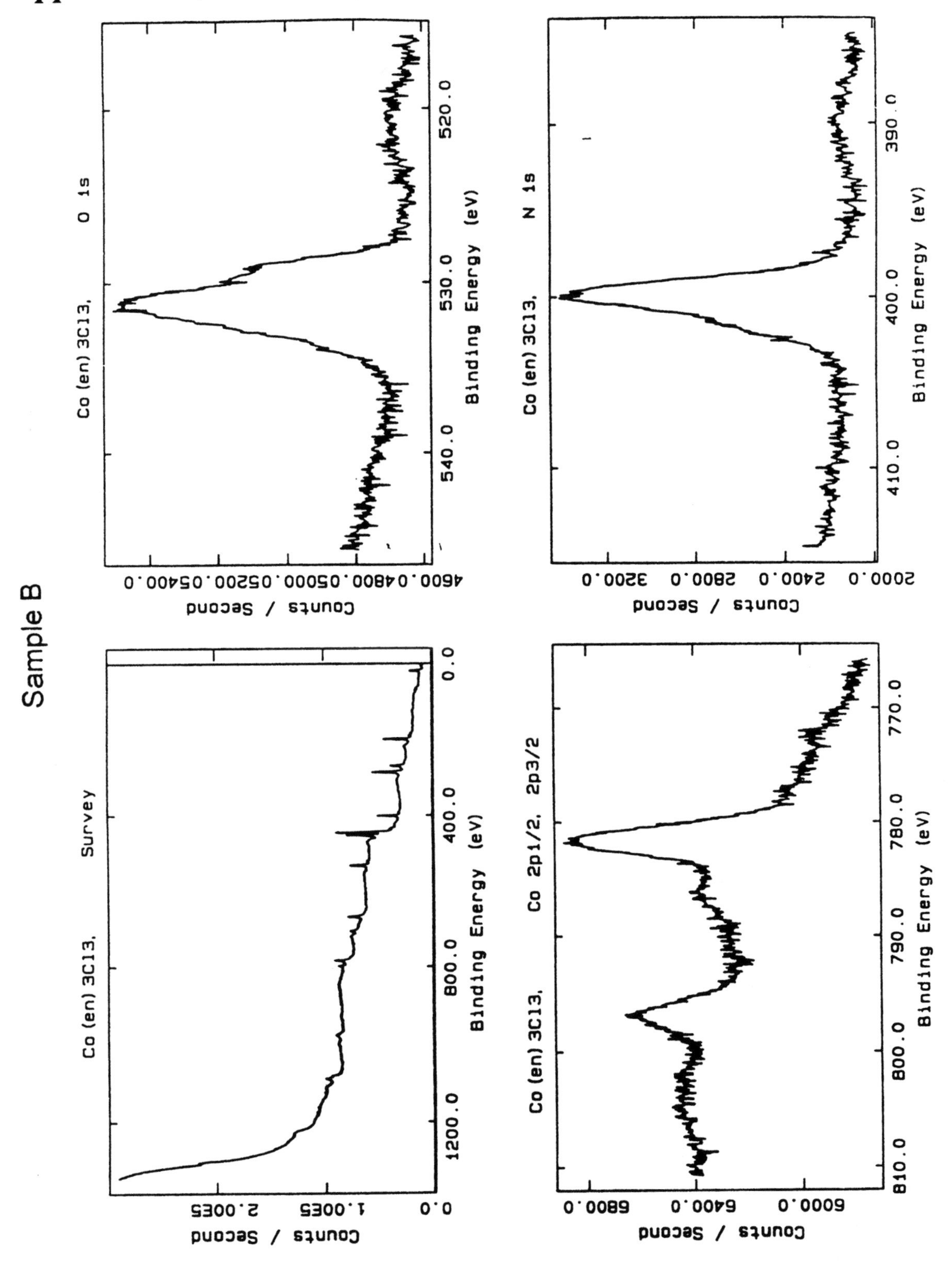

Appendix I (continued)

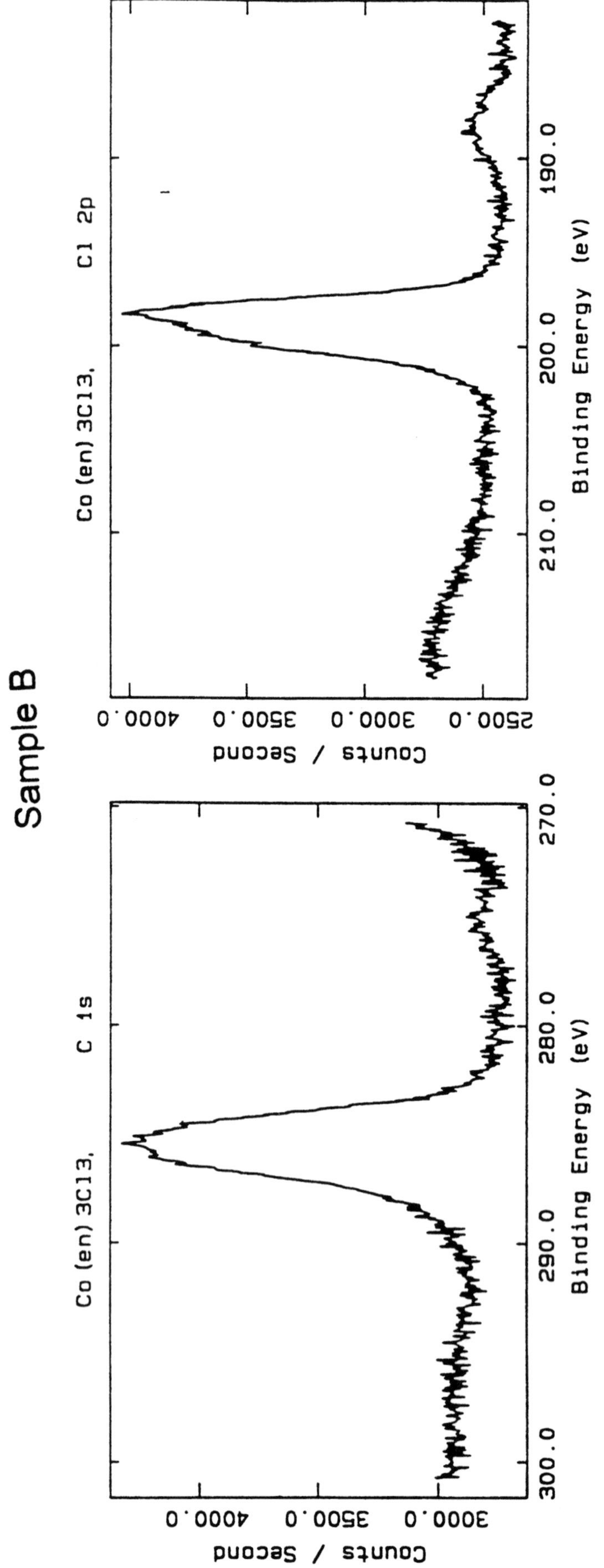

Sample B-

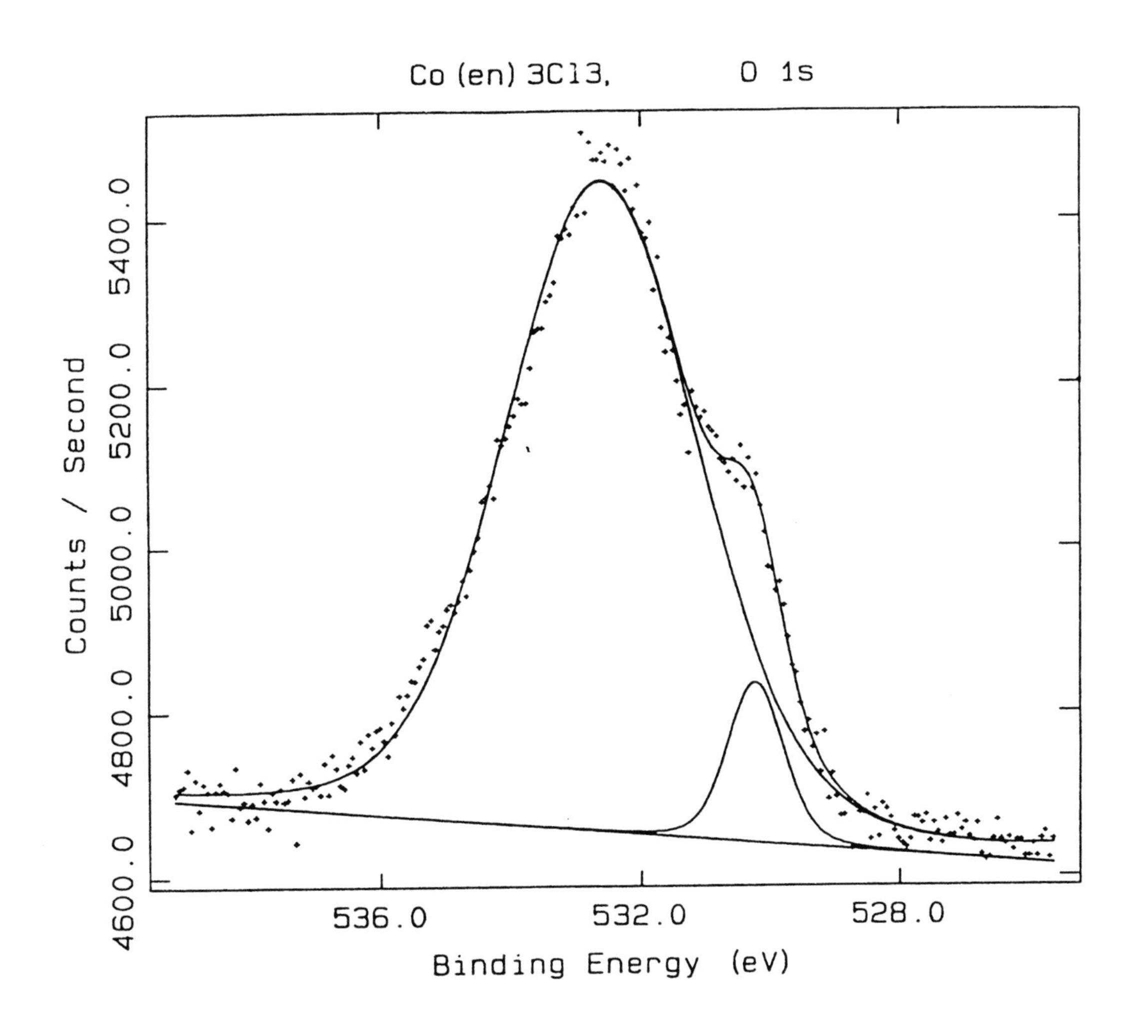
Co (en) 3Cl3, O 1s
Counts / Second
4600.0
4800.0
5000.0
5200.0
5400.0
536.0
532.0
528.0
Binding Energy (eV)

Appendix I (continued)

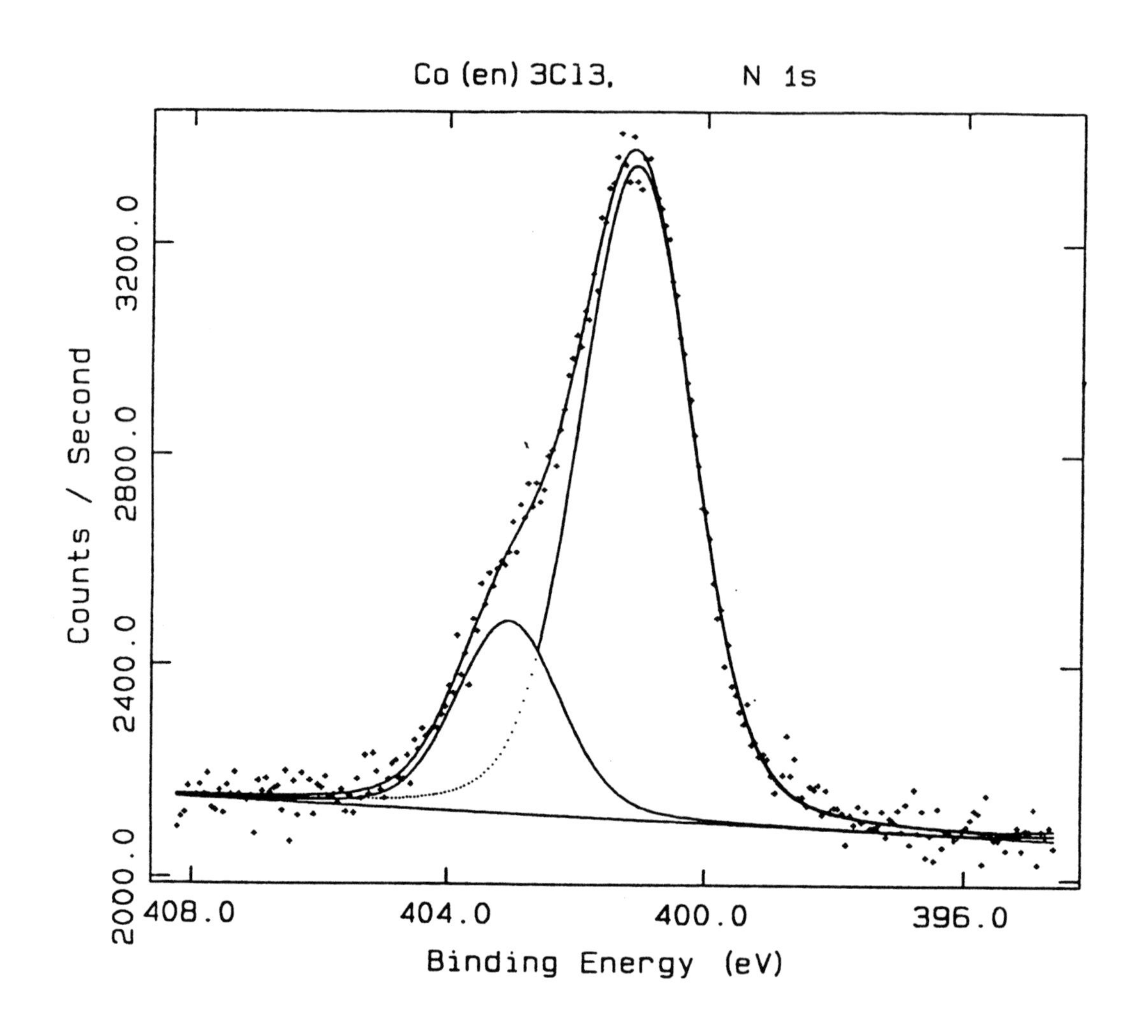

Appendix II

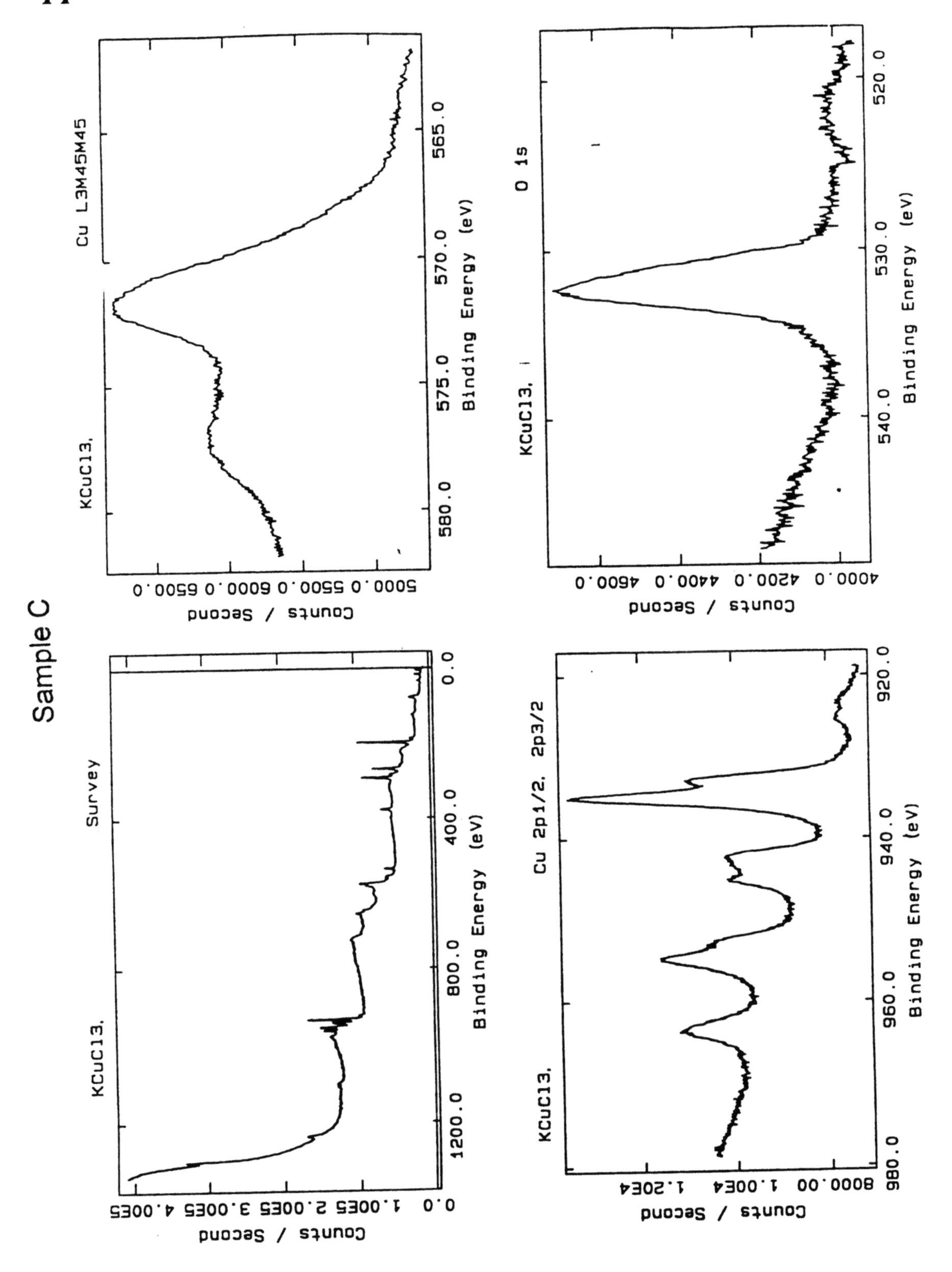

Sample C
KCuCl3.
Survey
Binding Energy (eV)
Counts / Second
KCuCl3.
Cu L3M45M45
Binding Energy (eV)
Counts / Second
KCuCl3.
Cu 2p1/2, 2p3/2
Binding Energy (eV)
Counts / Second
KCuCl3.
O 1s
Binding Energy (eV)
Counts / Second

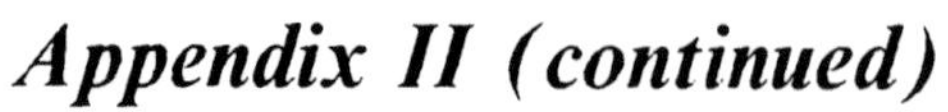

Appendix II (continued)

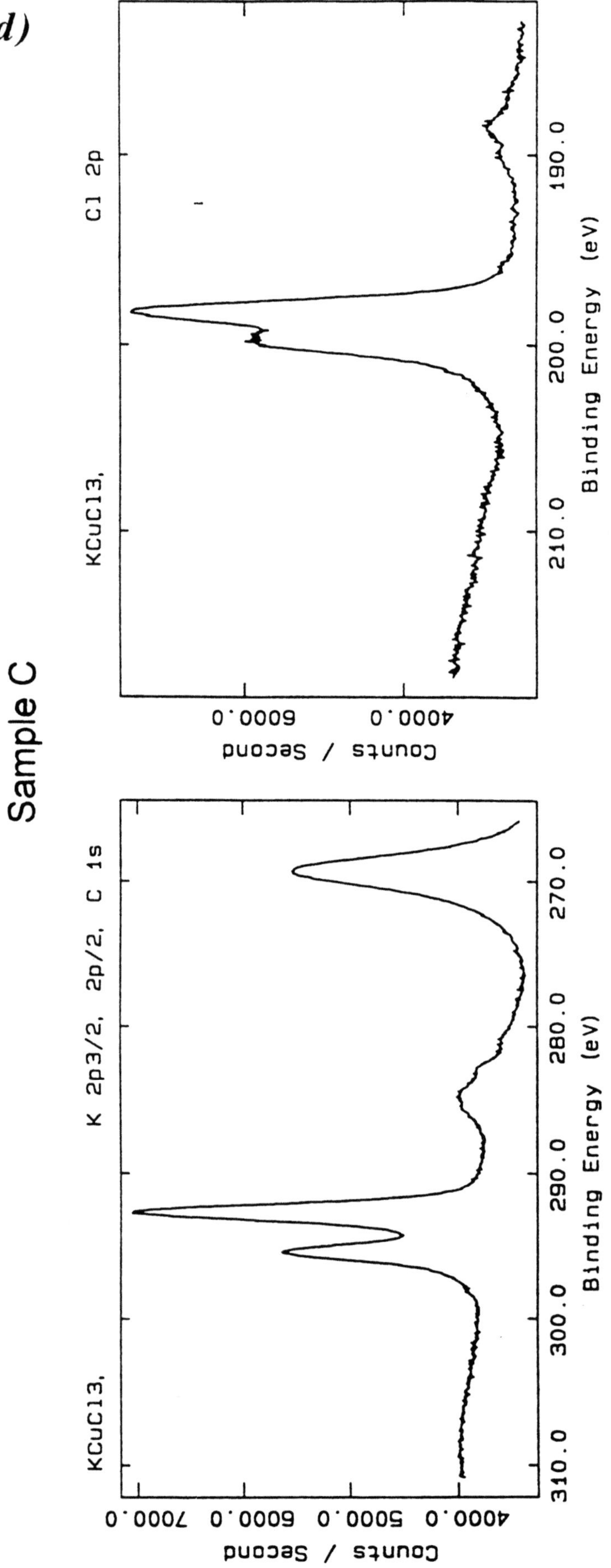

Appendix II (continued)

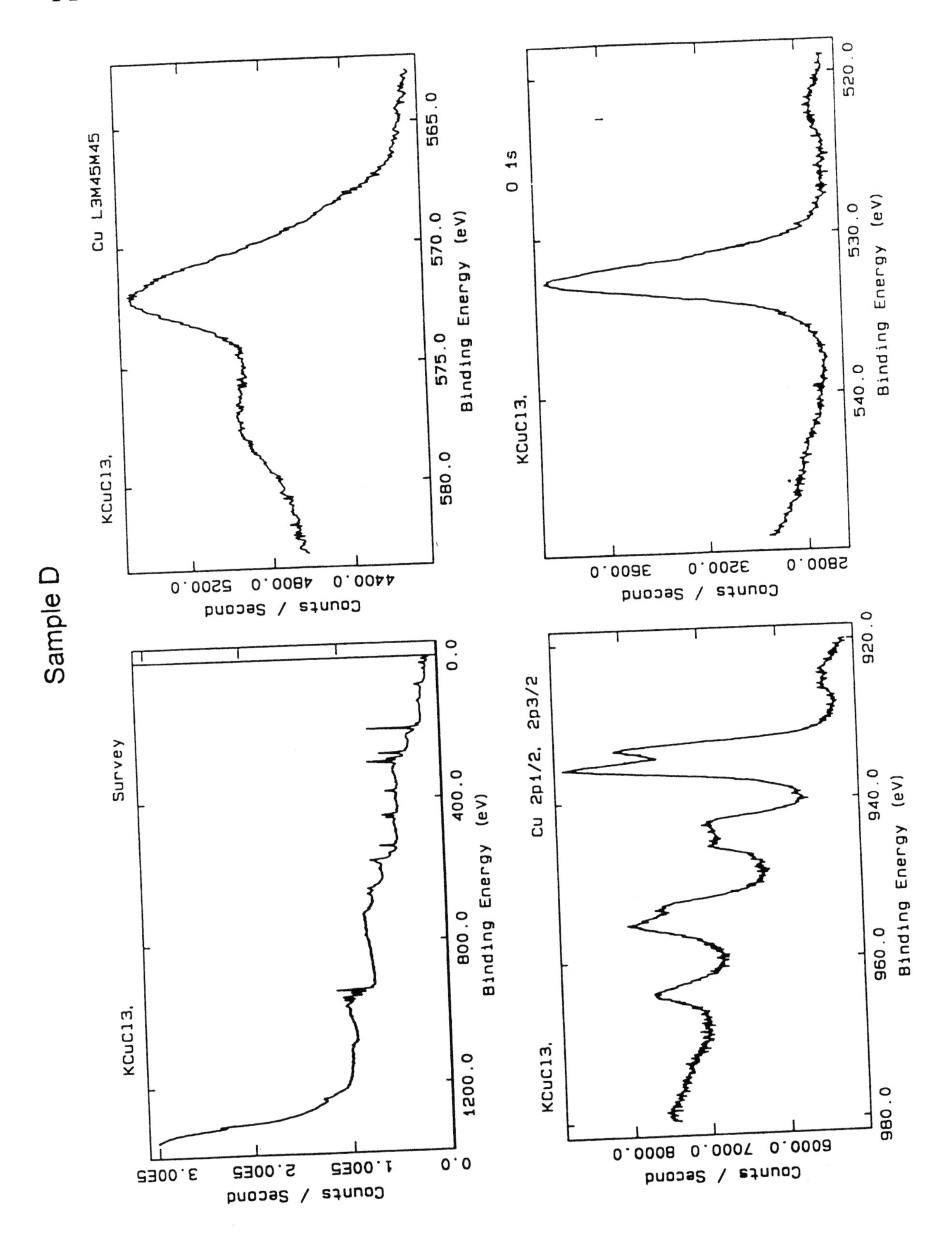

Appendix II (continued)

Sample D

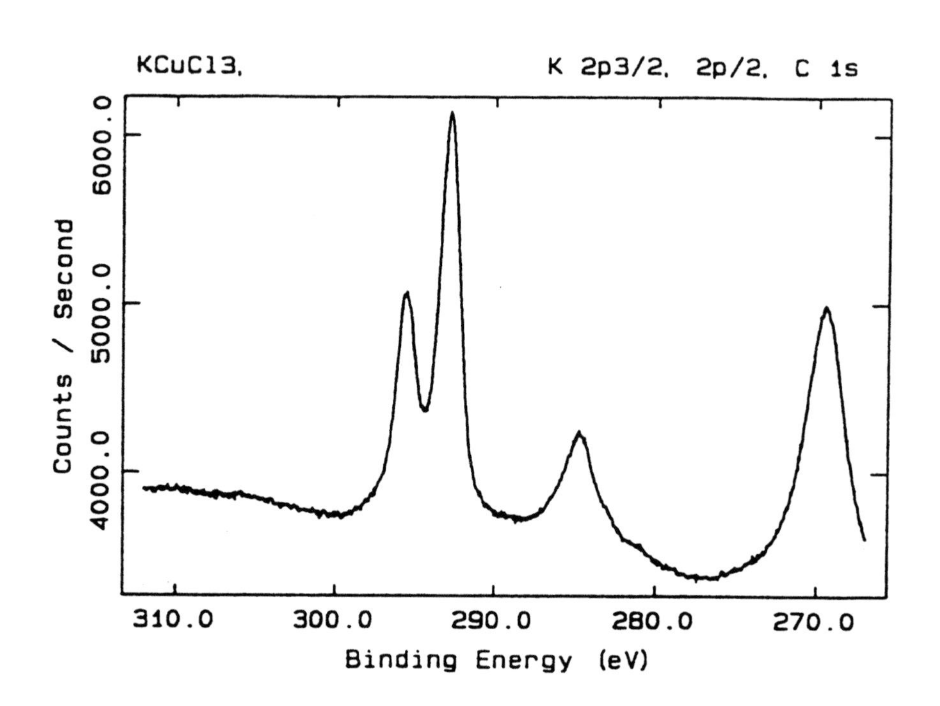

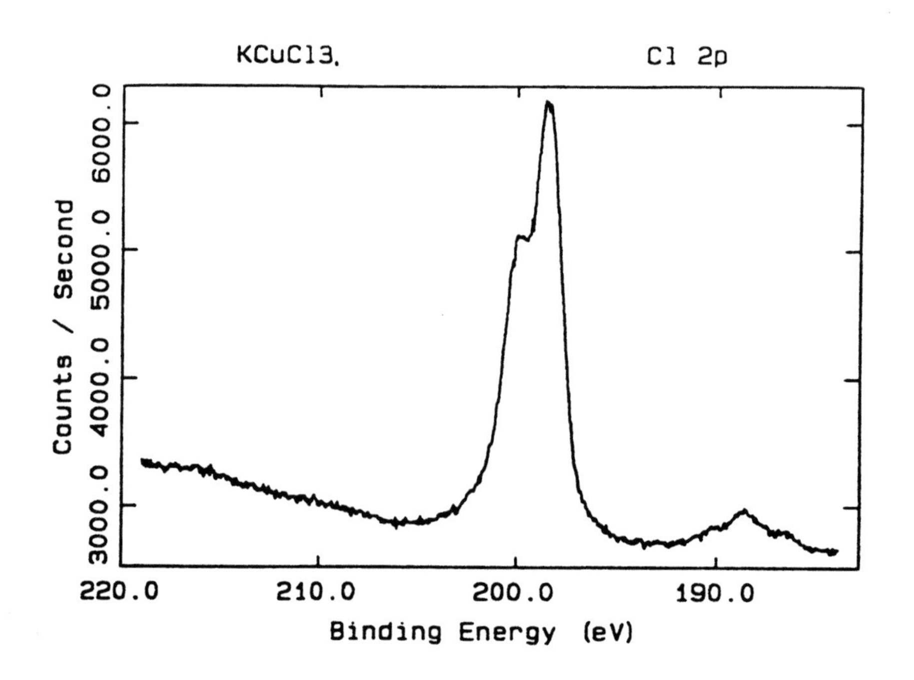

Appendix III

Chemical State Plot for Cu

Copper

Compound	Cu 2p Binding Energy (eV)	Cu LMM Kinetic Energy (eV)
$Cu_2Mo_3O_{10}$	931.6	916.5
Cu_2Se	931.9	917.6
CuAgSe	931.9	917.7
CuSe	932.0	918.4
CuS	932.2	917.9
$CuBr_2$	932.3	916.9
Cu_2S	932.5	917.4
CuCl	932.5	915.0
CuCl	932.5	915.6
Cu_2O	932.5	916.2
Cu_2O	932.5	916.2
Cu_2O	932.5	916.6
Cu_2O	932.5	917.2
Cu	932.6	918.6
Cu	932.6	918.7
$Cu_{64}Zn_{36}$	932.6	918.6
Cu	932.6	918.6
Cu	932.6	918.7
Cu	932.7	918.6
CuCN	933.1	914.5
$CuC(CN)_3$	933.2	914.5
CuO	933.7	918.1
$Cu_3Mo_2O_9$	934.1	916.6
$CuMoO_4$	934.1	916.6
$CuCr_2O_4$	934.6	918.0
$CuSiO_3$	934.9	915.2
$CuCO_3$	935.0	916.3
$Cu(OH)_2$	935.1	916.2
$CuCl_2$	935.2	915.3
$Cu(NO_3)_2$	935.5	915.3
$CuSO_4$	935.5	915.6
CuF_2	936.1	916.0
CuF_2	936.8	914.4
CuF_2	937.0	914.8

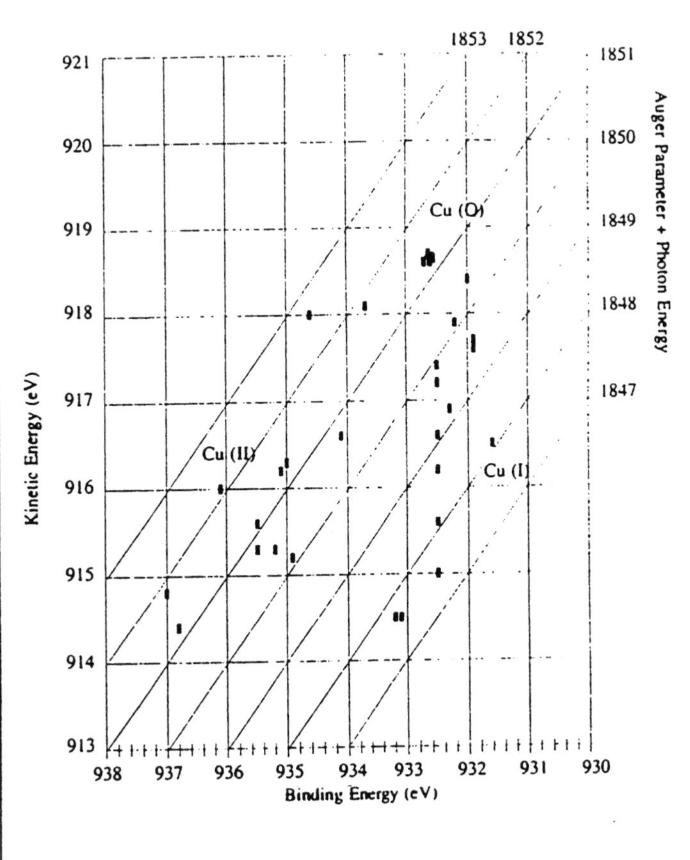

CALIBRATION OF VACUUM LINE COMPONENTS

Equipment Needed:	vacuum line with diffusion pump and McLeod gauge calibrated bulb of approximately 100 mL volume pinhole leak attached to a standard taper joint hood
Chemicals Needed:	2 g dry ice 5 liters liquid nitrogen 200 mL chloroform 200 mL carbon disulfide
Time Needed:	3 hours
Safety Notes:	Safety glasses should be worn around glass vacuum systems because of the possibility of implosions with attendant flying glass. Slush baths should be prepared in the hood so that solvent fumes from possible spills will be carried away. Dry ice should not be handled with bare hands. The -72^oC temperature can cause severe burns. Solid carbon dioxide in a vacuum line should never be allowed to warm to room temperature without being open to a manometer to make sure pressures approaching one atmosphere are not reached.

Purpose

Volatile air sensitive compounds are best handled in a vacuum line. Solutions or solids that are very air sensitive are also best handled in a vacuum line. A glove box is satisfactory if the reactants are not too sensitive. However, a very sensitive compound can act as a "getter' for the small amounts of oxygen or moisture in the glove box and be partially or wholly destroyed. A vacuum line at 10^{-5} torr or better is satisfactory for handling these very sensitive compounds.

Introduction

A vacuum line for quantitative work needs to be calibrated. Because modifications and repairs can change volumes, someone else's calibration should never be accepted without recalibration and confirmation. A McLeod gauge is a useful direct measuring device for vacuum determinations. The principle by which it enables low pressures to be measured is best understood by performing a sample calibration procedure. The purification of carbon dioxide by a low temperature distillation illustrates one use of a vacuum system.

A vacuum line is essential for working with gases and/or air sensitive materials. The chemistry of boron hydrides and silanes could not have been carried out without a vacuum line. The quantitative transfer of gases, the determination of the stoichiometry of reactions involving gaseous reactions, and the manipulation of air sensitive gases such as B_2H_6, NO, or PH_3 are possible only by use of a calibrated vacuum line.

Experimental Procedure

A. Calibration of a McLeod Gauge (A Paper and Pencil Experiment)

A McLeod gauge reads low pressures by compressing the gas molecules in a large bulb (B) at the reduced pressure of the vacuum system into a capillary tubing with a closed end (C1). The pressure of the gas in this capillary tubing can be compared with a reference column of mercury to determine the pressure in millimeters of mercury. In order to correct for the capillary error in the reading of the close ended capillary, a reference capillary (C2) which is closely matched to the close ended capillary is used. In measuring the height difference in these two closely matched capillaries, the errors due to capillary effects are minimized. Since $P_1V_1 = P_2V_2$, the pressure of the gas in the bulb (pressure of the vacuum system) can be calculated if one were to know the volume of the bulb, the volume of the gas in the capillary, and the pressure of the gas in the capillary.

During the construction of the McLeod gauge, the volume of the large bulb is determined. The capillary tubing is selected for uniformity of bore. This is accomplished as follows: A bit of mercury is sucked into a length of capillary tube. By gentle blowing or sucking through an appropriate safety trap, the mercury column is moved down the length of the tube. The length of the mercury column is measured as a function of the position in the capillary. If there is no measurable change in the length of the mercury column, the bore is uniform. If there are measurable changes in the length of the mercury column, the capillary is abandoned and another capillary measured. The length of uniform capillary needed is the combined length of (C1) and (C2).

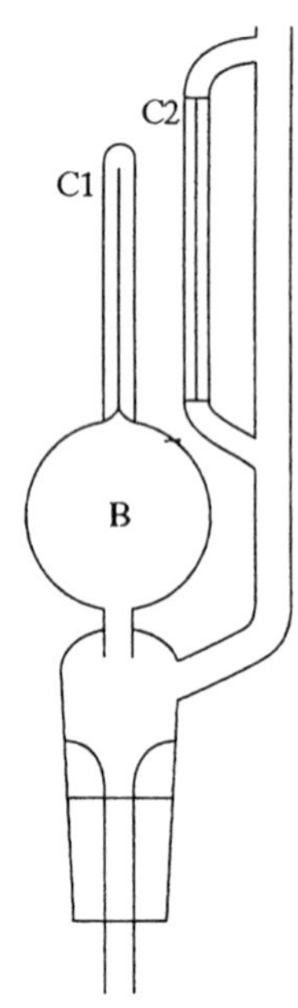

Fig. 1

The capillary selected needs to be calibrated for volume as a function of the length of the bore. This is done by carefully measuring the length of the mercury column and then weighing that same amount of mercury. For purposes of this exercise, assume that a length of mercury was measured to be 20.8 mm long. This amount of mercury was found to weigh 0.3066 g at 24°C.

The glassblower then attached a 15 cm length of capillary to the bulb (B) and closed off the end so that the bore ends in a flat rather than a rounded shape. The bulb with capillary C1 attached was then weighed empty. The bulb was then filled with distilled water to the cut-off tip. When corrected for the small amount of water which entered the closed capillary, the water in the bulb was found to weigh 105.62 g at 24°C.

Calculate the volume of the capillary in terms of mL/mm length and the volume in mL of the bulb (B).

Assume that the level of the mercury in the reference capillary is adjusted so that it is exactly opposite the tip of the close ended bore. The mercury in the closed capillary will be at some height, h, lower than the mercury in the reference capillary because of the pressure of the gas compressed from the bulb (B). The pressure, p, of the gas to be measured was at volume V, where V is the volume of the bulb and the close ended capillary. Since the pressure of the compressed gas in the capillary is given directly by h in mm of mercury and the volume occupied in the capillary is designated by v:

$$pV = hv$$

$$p = \frac{hv}{V}$$

If the capillary is calibrated as mL/mm length (b), then v = hb.

Substituting

$$p = h^2 \frac{b}{V}$$

Since b and V are constants determined for the specific McLeod gauge:

$$\frac{b}{V} = K$$

and

$$p = Kh^2$$

Prepare a calibration scale for the McLeod gauge with the values of b and V which are calculated above.

B. Use of the McLeod Gauge in Vacuum Determination

Use the McLeod gauge to determine the vacuum with the fore pump alone.

Turn on the diffusion pump and determine the vacuum after 10 minutes.

Attach the pin hole leak to the vacuum line and evacuate. Check the vacuum with the McLeod gauge. Use the Tesla Coil to locate the position of the pin hole leak.

C. Purification of Carbon Dioxide

Look up, in the *Handbook of Chemistry and Physics*, the temperatures for the partial vapor pressures of carbon dioxide for about five selected pressures of less than one atmosphere. Copy the data into your laboratory notebook.

Put a piece of dry ice about the size of a pea in the removable trap. Immediately cool the trap with liquid nitrogen and evacuate. Replace the liquid nitrogen trap with a chloroform slush bath (– 63.5°C) and distill through a trap cooled with a carbon disulfide slush bath (– 111°C) to a trap cooled with liquid nitrogen (– 195.8°C). (The carbon disulfide slush bath can be omitted for a smaller vacuum line.)

A slush bath is made by adding liquid nitrogen to the appropriate liquid until a mixture containing both solid and liquid phases is formed. For example, to make the chloroform slush, fill a small wide-mouth Dewar flask about half-full of chloroform. Add liquid nitrogen in small portions, stirring continuously with a spatula. Carry out the operation in a hood since there may be some splattering of the solvent and copious white fumes are formed. A crust of solid chloroform will form on the surface which is broken with the spatula. Any amount of solid phase in the liquid will ensure a temperature of – 63.5°C. To be long lasting, enough liquid nitrogen is added so that the chloroform turns to dry "snow" after which some room temperature chloroform is added to reintroduce some liquid phase. If there is too much solid phase in the slush, it becomes difficult to push the trap into the thick mixture.

The distillation allows gases such as oxygen or nitrogen which do not condense in liquid nitrogen to be separated from carbon dioxide. Impurities such as pump oil or condensed water are held back in the – 63.5°C and – 111°C traps. After the distillation is complete, the pure carbon dioxide in the liquid nitrogen cooled trap is isolated and the impurities in the other traps are condensed back to the removable trap.

Attach a calibrated bulb to the vacuum line and evacuate. A calibration bulb is conveniently made by attaching a vacuum stopcock to a round bottom flask as shown in Fig. 2.

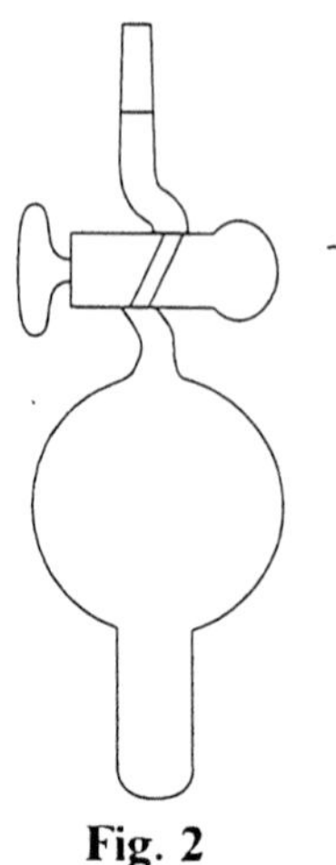

Fig. 2

The sealed off neck at the bottom is a convenient tip to cool with liquid nitrogen should it be desired to condense a gas into the calibration bulb. The calibration is carried out by weighing the bulb empty, then weighing the bulb filled with water.

Close off the line to the pumps so that the carbon dioxide in the liquid nitrogen trap can be warmed and expanded to fill the trap, calibration bulb, and the connecting line. When a pressure of 200 to 400 mm is noted on the manometer, close off the carbon dioxide trap and immediately cool it with liquid nitrogen. Read the manometer carefully to determine the pressure in the calibration bulb. Close the stopcock to the calibration bulb to contain the carbon dioxide at the measured pressure. Condense the residual carbon dioxide in the line to the removable trap. Warm the excess purified carbon dioxide in the liquid nitrogen cooled trap and condense back to the removable trap. Take the removable trap off the line and allow to warm in the hood. When the carbon dioxide has all sublimed away, replace the removable trap on the line and evacuate.

D. Calibration of the Vacuum Line

Thoroughly evacuate the vacuum line. Close off the line to the pumps. Cool one of the traps with liquid nitrogen. Open the stopcock of the calibration bulb and transfer the carbon dioxide from the calibration bulb to the trap. Close off all stopcocks to the trap and allow the carbon dioxide to expand. Measure the pressure on the manometer. Using the ideal gas relationship, calculate the volume of the trap and manometer.

The manometer changes volume with pressure. The volume of the manometer tubing can be calculated as a function of the length of the tube using the internal diameter of the tube. The general practice is to calibrate the manometer to its zero pressure mercury level. At any pressure, the volume of the manometer will be the calibrated volume plus whatever volume increase there is from the zero pressure level to the level of the pressure being read.

If there are other sections of the vacuum line with manometers, the carbon dioxide can be transferred to that section by condensing with liquid nitrogen. The volume of this section can then be determined by reading the pressure caused by the known amount of carbon dioxide.

If there are sections without manometers, the section containing the manometer is opened to the sections not fitted with a manometer. The new volume is determined from the new pressure reading. The volume of the new section is determined by subtracting the volume of the section containing the manometer from the total volume.

Calibrate all the sections of the vacuum line as assigned by the instructor.

Questions to be Considered

1. Obviously, most gases behave ideally at the pressure of the vacuum system. However, the operation of the McLeod gauge depends on Boyle's Law, or ideal gas relationship, holding after compression. Under what circumstances will the McLeod gauge be in serious error?

2. Because of the accuracy with which the height difference can be measured and other such factors, it is generally assumed that a compression ratio of 1×10^{-3} is a practical value for McLeod gauges. If a height difference of 2.5 mm is to be measured in 1 mm bore capillaries, what should be the volume of bulb (B) assuming the above compression ratio?

3. The mercury vapor pressure at 22°C is 1.426×10^{-3} mm of Hg. Explain how this would affect the accuracy of a McLeod gauge.

4. At the very high voltage of the Tesla coil, the gases are ionized. The ionized gas is called a corona and consists of a cascade of electrons and ions. If the tip of the Tesla coil is brought near some grounded object, a gas breakdown can occur. Under these circumstances, explain the phenomena observed when the Tesla coil is brought near a pinhole leak.

5. In the days before thermocouple gauges, ion gauges, and their associated electronics, old-timers would estimate the quality of a vacuum by holding a Tesla coil near the evacuated manifold. A pink glow inside the evacuated tube would mean pressures as high as 10 mm. At 1.5 mm striations become visible. At 0.5 mm the striations are 1 mm apart. At 0.01 mm a green fluorescence is visible on the glass and at pressures less than 1×10^{-3} mm there would be no glow. A "black" vacuum was, therefore considered a good vacuum. Explain this phenomena in terms of atomic spectroscopy.

6. After the carbon dioxide distillation, a grey haze is found at the approximate position of the surface of the liquid nitrogen. This haze remains even after the liquid nitrogen is removed and the carbon dioxide is vaporized. Analysis indicates this haze to be mercury. Explain why this haze is more visible the better the vacuum in the system.

7. In discussing the preparation of the chloroform slush, a statement is made that "Any amount of solid phase in the liquid will ensure a temperature of – 63.5°C." Explain this phenomena using the Phase Rule.

8. Carbon dioxide is used as a calibration gas because it is easily available and readily purified. Explain the disadvantages of carbon monoxide and diethyl ether as calibration gases.

9. A small amount of calibration gas will result in a larger volume error than a large amount of calibration gas. For the practical range of the amount of calibrating gases which can be used in your system, calculate the range of calibration accuracies possible.

Helpful Background Literature:

1. Dushman, S.; Lafferty, J. M. "Scientific Foundations of Vacuum Technique"; John Wiley & Sons, Inc.: New York, 1961; pp. 225 – 233.

2. Shoemaker, D. P.; Garland, C. W.; Steinfeld, J. I.; Nibler, J. W. "Experiments in Physical Chemistry", 4th ed.; McGraw-Hill: New York, 1981; pp. 620 – 628.

3. Daniels, F.; Mathews, J. H.; Williams, J. W. "Experimental Physical Chemistry"; McGraw-Hill: New York, 1941; p. 307.

4. Shriver, D. F. "The Manipulation of Air Sensitive Compounds"; McGraw-Hill: New York, 1969; pp. 27 – 46.

CHARACTERIZATION OF A GASEOUS UNKNOWN

Equipment Needed:

vacuum pump with diffusion pump and McLeod gauge
wide-mouth Dewar flasks
gas IR cell
calibration bulb or other convenient sampling
  bulb for mass spectrometry sample
Stock plunger-type melting point device
molecular weight bulb
cathetometer
low temperature thermometer (– 100°C to
  + 50°C or – 200°C to + 30°C
  or precision thermocouple type K or E
    (make sure the readout device allows
    low temperature readings)
  or vapor pressure thermometers (ethylene,
    carbon dioxide and ammonia)
hand magnet
glass-blowing equipment
sensitive thermometer to read room
  temperature to 0.1°C accuracy

Chemicals Needed:

unknown sample
5 liters liquid nitrogen

Time Needed:

4 hours

Safety Notes:

Safety glasses should be worn around glass vacuum systems because of the possibility of implosions with attendant flying glass. Slush baths should be prepared in the hood so that solvent fumes from possible spills will be carried away. Whenever a gas sample is transferred by condensation, calculate the pressure of that gas sample in the new volume before allowing the condensed material to become gaseous at room temperature. In glass systems, the pressure should always be one atmosphere or less. If gas samples are in break-seal ampoules, always condense the contents with liquid nitrogen before breaking the seal. If the gas should be under pressures greater than one atmosphere, breaking the break-seal while under pressure may lead to the ampoule explosively shattering. The effect is somewhat similar to putting a pinhole in an inflated balloon.

Purpose

In the process of determining the identity of a gaseous unknown, the techniques for handling gas samples will be demonstrated. The amount of the gas unknown will be determined. The gas infrared spectrum will then be determined. This will be followed by the determination of the vapor density molecular weight, vapor pressure – temperature characteristic, melting point, and mass spectra. A positive identification of the unknown gas will then be made from the data accumulated.

Introduction

After carrying out the calibration of a vacuum line, the student should have developed some degree of comfort working with a vacuum system. Although great care must be exercised in the operation of a vacuum system, there is also a reasonable degree of freedom in the way a particular operation can be carried out. The student should understand the principles of handling gases and then plan specific operations to achieve the desired ends. There are very few people who do research with vacuum lines who have not made mistakes. One of the most upsetting is to realize too late that the stopcock to the pumps has been left open and the gaseous material made by a careful and tedious procedure has been lost. Detailed planning can minimize these errors. Part A describes the transfer of the unknown to the vacuum line and the determination of the amount of gas. Following part A, there is an illustrative example of a notebook writeup for a step-by-step procedure to be followed. Write a similar procedure for each of the operations (including a rewrite of this first operation) making sure that your procedures are adapted to the vacuum line you are assigned to use. Prior to commencing the experiments described in parts A through F, the step-by-step procedure in the students′ laboratory notebook should be checked by the instructor.

Experimental Procedure

A. Transfer of Unknown and Determination of Amount

The unknown gas sample will either be given to you in a flask equipped with a vacuum stopcock or in a break-seal ampoule. If the sample is in a break-seal ampoule, place a glass-covered iron rod, hereafter called the breaker, carefully on the break-seal. This is best done by holding the break-seal tube horizontally and pushing the breaker down to the break-seal with a pencil or a similar rod. The ampoule is then carefully brought to a vertical position and clamped on a ring stand. A joint suitable for attaching to the vacuum line is sealed to the ampoule. (The instructor or lab assistant will help you with this operation.) Put the ampoule on the vacuum line and evacuate. Freeze the unknown with liquid nitrogen before breaking the break-seal by raising the iron breaker with a hand magnet and releasing the breaker so that it will drop on the break-seal. Close off the stopcock to the pump and transfer the gas quantitatively from the ampoule to the calibrated trap on the vacuum line. Close off the calibrated trap. Make sure that adjacent calibrated volumes are evacuated and closed off. Allow the unknown to warm slowly from liquid nitrogen temperature to room temperature. If the pressure starts to approach one atmosphere, open a stopcock to the adjacent calibrated volume. Continue this process as needed until the pressure ceases to rise. Calculate the number of millimoles of the unknown gas from the pressure, volume, and temperature determinations. Assume that the gas at equilibrium pressure is at room temperature. This can be read from a sensitive thermometer hanging on the vacuum line near the calibrated trap.

Fig. 1. Sample Detailed Writeup for the Notebook

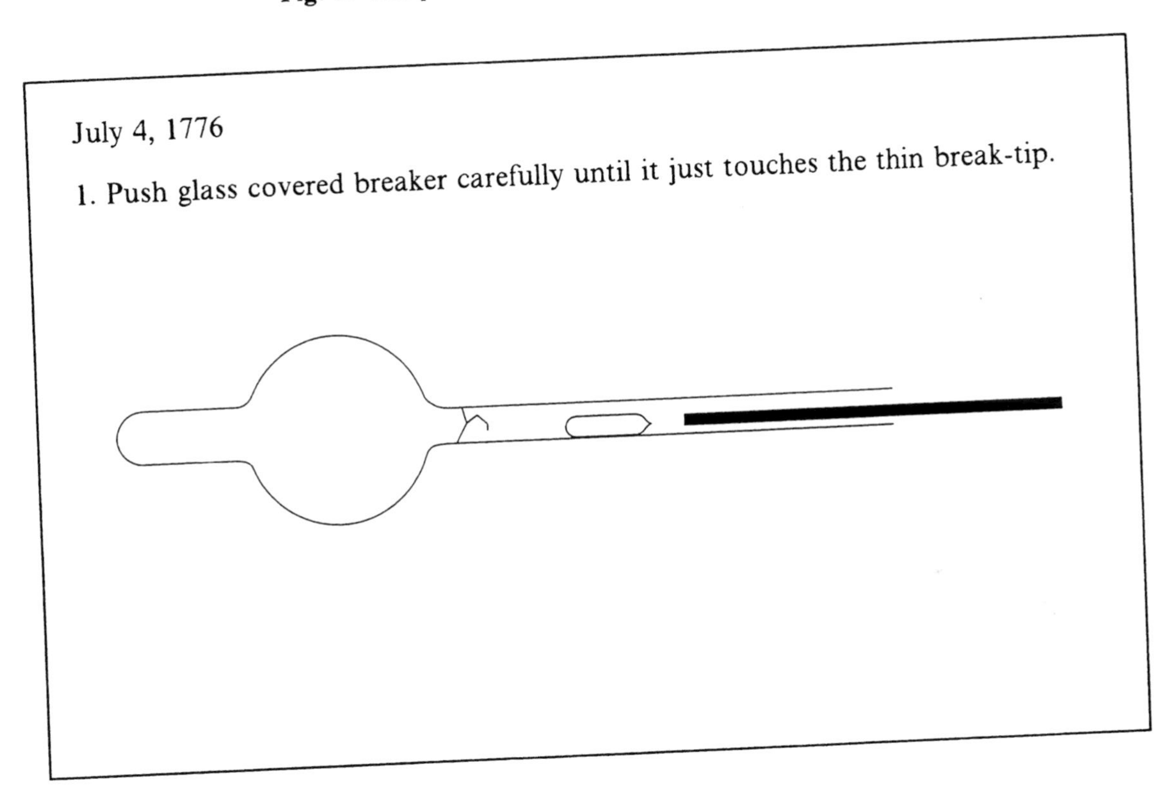

2. Attach a standard taper 12/30 joint to the ampoule.

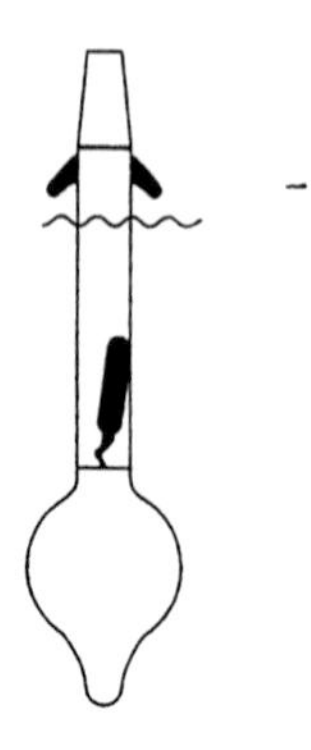

3. Put a light coat of vacuum grease on the standard taper joint and attach to vacuum line at joint A shown below. Be sure to hold the ampoule on the line with rubber bands.

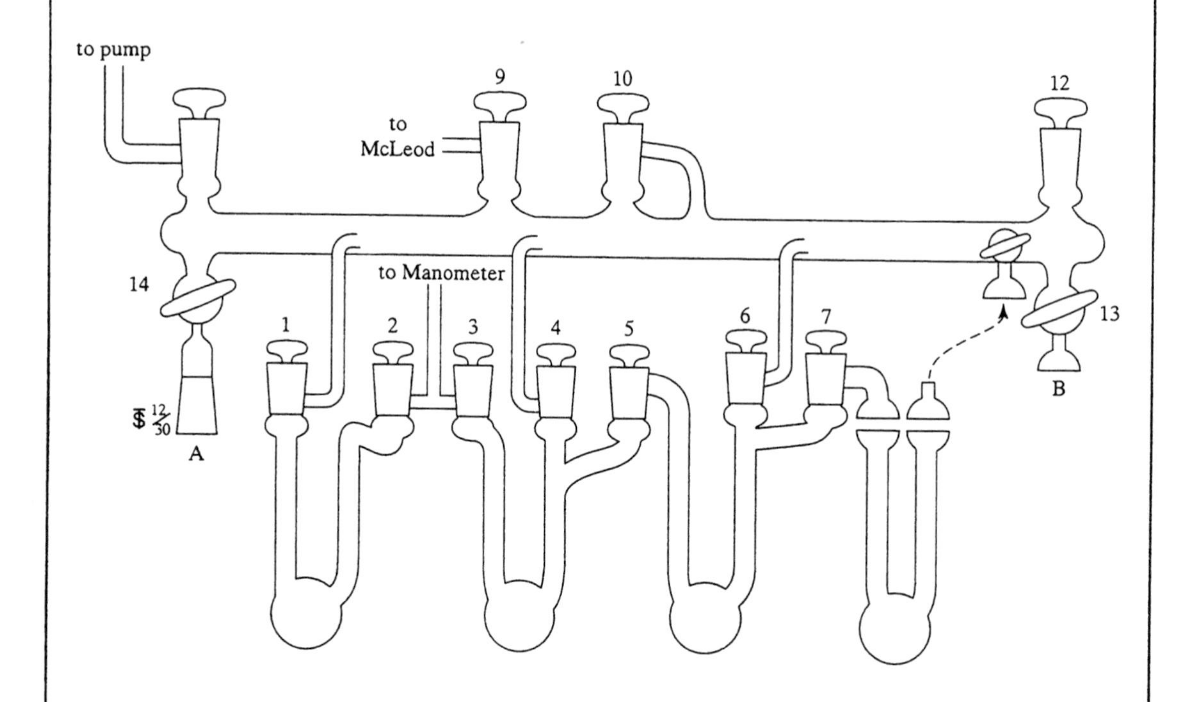

4. Open stopcock #14 and evacuate the region of the ampoule containing the breaker. Check the vacuum.

5. Cool the ampoule tip with liquid nitrogen.

6. After making sure that a good vacuum exists, close stopcock #14.

7. Raise the breaker with a hand magnet and allow it to drop on the break- tip. Repeat if necessary until the tip breaks.

8. Close stopcocks #8, #9, and #10. Open stopcocks #14, #1, and #2. Close stopcock #3. The ampoule is now connected to the calibrated trap. There should be no pressure indicated on the manometer. If by accident, there is a pressure, keep the liquid nitrogen on the ampoule and open stopcock #8 leading to the pumps. A pressure indication is a sign of a problem. Try to figure out what went wrong.

9. Making sure that stopcock #8 to the pump is closed, transfer the liquid nitrogen wide-mouth Dewar flask from the ampoule to the calibrated trap located between stopcocks #1 and #2.

10. When the ampoule is warmed to room temperature, there should be no pressure reading on the manometer. Close stopcocks #1 and #3 so that the gas will be confined to the calibrated trap plus manometer. Remove the liquid nitrogen and allow the calibrated trap to warm to room temperature.

11. Watch the manometer carefully. If the rising pressure approaches 600 mm, close stopcocks #4 and #5 and open stopcock #3. Repeat this process (close stopcocks #6 and #7 and open stopcock #5) if the pressure once again rises to the 600 mm region. Once the pressure stops changing, allow the traps to come to room temperature and carefully read the pressure on the manometer.

12. Calculate the volume the gas is occupying. Be sure to account for the volume change in the manometer.

13. Calculate the number of millimoles of gas using the ideal gas relationship, R = 62.36 mm mL/mmole K.

B. Determination of the Gas Infrared Spectrum

A gas infrared cell consists of a tube 8 to 10 cm in length with salt plate windows attached to the two ends. The infrared cell is attached to the vacuum system and filled with 20 mm to 100 mm pressure. This can be done in two ways: Technique 1: If sufficient gas is available, the evacuated IR cell can be opened carefully to the trap containing the unknown gas. The pressure can be adjusted downward in two ways. Adjacent evacuated and calibrated volumes can be opened up to the gas. This will drop the pressure, but the range of selection is limited. Another way of getting a pressure less than the equilibrium pressure is to freeze the gas in a trap using liquid nitrogen. If the sample is then allowed to warm slowly, the stopcock to the IR cell can be turned off when the desired pressure is reached. If an accurate determination of the pressure in the IR cell is desired, a slightly different procedure can be used. The IR cell should be in contact with a manometer and the unknown frozen in a trap not directly connected to the manometer. The stopcock leading away from the trap containing the sample is turned off when the approximate desired pressure is reached. The isolated trap is then immediately re-cooled with liquid nitrogen to prevent high pressures from forming in the sample trap, and the IR cell and the manometer which are now isolated from the sample trap can be measured carefully. After the measurement, the IR cell can be closed and the stopcock to the cold trap opened to condense the residue of the gas in the manometer and interconnecting tubes back to the cold trap.

If the amount of the unknown is not sufficient to give the desired pressure in the IR cell using the method above, the following technique must be used. This second technique is the most versatile technique and is recommended if the IR cell is equipped with an appendage which can be cooled with liquid nitrogen.

Technique 2: All of the unknown is condensed into a portion of the line not directly connected to a manometer. From the volume of the IR cell provided by the manufacturer or the instructor and the pressure desired for the IR measurement, calculate the pressure of gas needed in the trap-plus-manometer. Connect the trap-plus-manometer to the portion of the line containing the sample. Remove the liquid nitrogen and allow the trap to slowly warm. When the desired pressure is indicated on the manometer, close the stopcock to the source of the gas and quickly re-cool the trap containing the original amount of unknown. The pressure in the trap-plus-manometer is then carefully read. Using Boyle's Law or the ideal gas relationship, recalculate what the pressure of this amount of gas will be in the IR cell. If this calculation agrees with that previously made, cool the tip of the IR cell and transfer this measured amount of gas quantitatively from the measuring trap to the IR cell. Close the stopcock to the IR cell and remove to the infrared spectrometer.

The amount of unknown in the infrared cell can be transferred back to the line after the spectrum has been determined. This is especially important if the original amount of unknown was somewhat small. Attach the cell to the line, evacuate the connections and transfer the gas in the cell to a trap cooled with liquid nitrogen.

C. Determination of the Molecular Weight

The molecular weight bulb is preferably made with a small "O" ring joint. The reason for the preference of the "O" ring over that of a standard taper joint is that the "O" ring joint can be used dry (i.e., without stopcock grease). If an "O" ring joint is not available, the calibration bulb with a standard taper 12/30 joint can be used. If the standard taper 12/30 joint is used, the grease must be carefully removed with toluene from the joint before each weighing. The possibilities of weighing errors would seem to be formidable. However, with care, reasonably reproducible weighings can be made.

The molecular weight determination is made by weighing the empty (evacuated) bulb, filling the bulb with a known pressure of unknown gas, and weighing the filled bulb. The increase in weight is the weight of the gas.

$$PV = nRT$$

$$PV = \frac{g}{MW} RT$$

$$MW = \frac{gRT}{PV}$$

The volume is the calibrated volume of the bulb which will be provided. The gas pressure, gas temperature, and gas weight are experimental observations. R = 62.36 mm mL/mmole K. Thus the molecular weight can be calculated.

The molecular weight bulb can be best weighed by hanging it from the hook just below the point where the pan stirrup is hung on the balance. Although a single-pan balance can be used, accurate determinations are only possible using a two-pan balance and a tare bulb of similar size and shape as the molecular weight bulb. The large size of the bulb introduces a buoyancy correction. More serious is the absorption of moisture on the large glass surface. Using a tare of similar size and shape, these variables are cancelled.

The molecular weight bulb is filled in exactly the same way as the infrared cell. For a bulb of approximately 50 mL, a pressure of 150 mm to 400 mm is appropriate.

D. Vapor Pressure vs. Temperature

The vapor pressure vs. temperature curve is determined by freezing the unknown in a trap connected to a manometer. A cold bath of methylcyclohexane (-126.3°C) is used to replace the liquid nitrogen. A cold bath in this case is methylcyclohexane in a wide-mouth Dewar flask cooled so that there is only a trace of solid phase present. The trap at liquid nitrogen temperature will generally freeze a bit more methylcyclohexane. A thermometer (low temperature or vapor pressure) or a thermocouple is placed in the cold bath. Pressures in equilibrium with the temperature of the cold bath are recorded. If the temperature rise is too slow, small amounts of room temperature methylcyclohexane can be added. In general, this reduces the accuracy of the determination since there is a greater probability of a non-equilibrium conditions existing.

The experiment is concluded either when the pressure rises to 500 to 600 mm or the bath approaches room temperature. Plot the data obtained and extrapolate to 760 mm. The corresponding temperature is the boiling point of the unknown.

The Clausius-Clapeyron equation can be written as [1] for gases which are behaving close to ideality. The pressure-temperature data is related to the ΔH of vaporization.

$$\frac{dp}{dT} = \frac{\Delta H_{vap} P}{RT^2} \qquad [1]$$

The integrated form can be obtained by using the identity:

$$\frac{d \ln P}{dT} = \frac{1}{P}\frac{dP}{dT} \qquad [2]$$

Equation [1] becomes:

$$\frac{d \ln P}{dT} = \frac{\Delta H_{vap}}{RT^2} \qquad [3]$$

Assuming ΔH_{vap} to be constant

$$\int d \ln P = \frac{\Delta H_{vap}}{R} \int \frac{dT}{T^2} \qquad [4]$$

$$\ln P = \frac{-\Delta H_{vap}}{RT} + C \qquad [5]$$

$$\log P = \frac{-\Delta H_{vap}}{2.303\, RT} + C \qquad [6]$$

Thus a plot of log P vs 1/T gives a straight line from which slope ΔH_{vap} can be calculated.

E. Melting Point Determination

The melting point of the unknown will be determined using the plunger technique described by A. Stock (3).

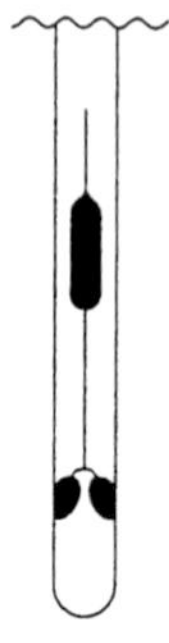

Fig. 2.

The melting point apparatus is a simple tube containing a plunger with a small T at the bottom end and an iron nail encased in glass at the top end as shown in Fig. 2.

A low temperature melting point of a compound which is normally in the gaseous state is accomplished by freezing the compound in such a way that the solid material supports a plunger. If this plunger can be carefully monitored for the first sign of movement, the initiation of the melting point can be determined. This temperature might be thought to correspond to the observation of the first signs of melting observed in a capillary or hot stage in the organic laboratory. Surely by the time the compound is melted, the support for the plunger will have disappeared, and the plunger will fall to the bottom of the tube. The drawing of a plunger low temperature melting point apparatus showing the plunger supported by a ring of solid unknown is shown in Fig. 2. In order that the plunger can be lifted while condensing the unknown supporting ring, an iron nail is encased in glass toward the top end of the plunger. A colored glass tip is useful for detecting the first sign of movement.

First thoroughly evacuate the apparatus. Hold the plunger toward the top of the tube with a hand magnet. Place a liquid nitrogen trap at the bottom of the apparatus so that about one inch of the bottom of the melting point apparatus is bathed in liquid nitrogen. Allow the unknown to enter the melting point apparatus. A solid ring should form in the vicinity of the surface of the liquid nitrogen. Carefully lower the plunger on this solid ring. If the plunger is not supported, there is insufficient unknown condensed. Adjust a cathetometer telescope so that the hairline coincides with the upper tip of the plunger.

Replace the liquid nitrogen with a methylcyclohexane cold bath at -126^oC. Allow the cold bath to warm slowly in the same way as for the determination of the vapor pressure – temperature curve. When the first movement of the plunger is noted through the cathetometer telescope, record the temperature using the low temperature thermometer or thermocouple in the cold bath. Also make note of the temperature at which the plunger falls to the bottom of the melting point apparatus.

One important consideration in carrying out this experiment is that of the vapor pressure of the unknown at its melting point. In general, most compounds at the melting point exhibit less than one atmosphere pressure. Carbon dioxide is a well known exception. If pressures greater than one atmosphere are created in the melting point apparatus, the joint connecting the apparatus to the vacuum line will separate. Having the melting point apparatus open to a manometer provides safety, but enough of the unknown may sublime so that the ring may diminish sufficiently to allow the plunger to drop before the melting point is reached. Use the data gathered for the vapor pressure – temperature curve to make an intelligent decision as to whether to have the stopcock to the melting point apparatus closed or to have the melting point apparatus open to a manometer.

F. Mass Spectrum

Put some or all of the unknown in the calibration bulb equipped with a standard taper 12/30 male joint. It should be noted that the standard inlet joint on most mass spectrometers is a standard taper 12/30 female joint. The pressure is not critical as long as the pressure in the bulb is less than one atmosphere. Submit the properly labeled calibration bulb with the sample to the instructor to have the mass spectrum run. The spectrum will be given to you at the next laboratory period.

Auxiliary Experiment

The rotational infrared lines for diatomic unknowns can be used to calculate the rotational constant B_o and the centrifugal distortion constant D_o. From the value of B_o, r_o, the root mean square internuclear distance in the ground state, can be calculated.

Questions to be Considered

1. If the vacuum is not good, the transfer of a gas sample in a vacuum line is slow. Why?

2. If a trap cooled with liquid nitrogen is left open to the air, a slightly bluish colored liquid condenses in the trap. What is this liquid? Explain why it condenses.

3. Fine structure is observed in gas infrared data not observed in infrared of solids or liquids. Explain.

4. Careful experimenters can determine the molecular weight by the vapor density method to 1 molecular weight unit. Students often turn in results which are in error by

much greater amounts. List the sources of error in the experiment. Which ones do you think most seriously affected your results?

5. If the plot of pressure vs. temperature shows a marked curve and extrapolation is uncertain, how else might the boiling point of the unknown be determined?

6. If the vapor pressure of the unknown at the region of the observed melting point is quite high, the melting point determination may be faulty. Why?

7. From the mass spectrum, pick out the parent peak. Note that the parent peak gives a far more accurate reading of the molecular weight than the vapor density method. However, not all compounds give a parent peak. Fluorocarbons are a notable example.

8. Assign as many of the m/e peaks as possible to the ion species observed in the mass spectrum.

9. In the mass spectrum, are the P + 1 and P + 2 peaks visible? Are they consistent with the expected isotopic abundances of your compound?

Helpful Background Literature

1. Shriver, D. F. "The Manipulation of Air Sensitive Compounds"; McGraw-Hill: New York, 1969.

2. Dodd, R. E.; Robinson, P. L. "Experimental Inorganic Chemistry "; Elsevier: Amsterdam, 1957.

3. Stock, A. "Hydrides of Boron and Silicon"; Cornell University Press: Ithaca, 1957; p. 184.

4. Barrow, G. M. "Introduction to Molecular Spectroscopy"; McGraw-Hill: New York, 1962.

5. McLafferty, F. W. "Interpretation of Mass Spectra"; W. A. Benjamin, Inc.: New York, 1967.

INDEX